Elementary Linear Algebra

TEXTBOOKS in MATHEMATICS

Series Editors: Al Boggess and Ken Rosen

PUBLISHED TITLES

ABSTRACT ALGEBRA: A GENTLE INTRODUCTION
Gary L. Mullen and James A. Sellers

ABSTRACT ALGEBRA: AN INTERACTIVE APPROACH, SECOND EDITION
William Paulsen

ABSTRACT ALGEBRA: AN INQUIRY-BASED APPROACH
Jonathan K. Hodge, Steven Schlicker, and Ted Sundstrom

ADVANCED LINEAR ALGEBRA
Hugo Woerdeman

ADVANCED LINEAR ALGEBRA
Nicholas Loehr

ADVANCED LINEAR ALGEBRA, SECOND EDITION
Bruce Cooperstein

APPLIED ABSTRACT ALGEBRA WITH MAPLE™ AND MATLAB®, THIRD EDITION
Richard Klima, Neil Sigmon, and Ernest Stitzinger

APPLIED DIFFERENTIAL EQUATIONS: THE PRIMARY COURSE
Vladimir Dobrushkin

APPLIED DIFFERENTIAL EQUATIONS WITH BOUNDARY VALUE PROBLEMS
Vladimir Dobrushkin

APPLIED FUNCTIONAL ANALYSIS, THIRD EDITION
J. Tinsley Oden and Leszek Demkowicz

A BRIDGE TO HIGHER MATHEMATICS
Valentin Deaconu and Donald C. Pfaff

COMPUTATIONAL MATHEMATICS: MODELS, METHODS, AND ANALYSIS WITH
MATLAB® AND MPI, SECOND EDITION
Robert E. White

A CONCRETE INTRODUCTION TO REAL ANALYSIS, SECOND EDITION
Robert Carlson

A COURSE IN DIFFERENTIAL EQUATIONS WITH BOUNDARY VALUE PROBLEMS,
SECOND EDITION
Stephen A. Wirkus, Randall J. Swift, and Ryan Szypowski

A COURSE IN ORDINARY DIFFERENTIAL EQUATIONS, SECOND EDITION
Stephen A. Wirkus and Randall J. Swift

PUBLISHED TITLES CONTINUED

PUBLISHED TITLES CONTINUED

Elementary Linear Algebra

James R. Kirkwood and Bessie H. Kirkwood

CRC Press
Taylor & Francis Group
Boca Raton London New York

CRC Press is an imprint of the
Taylor & Francis Group, an **informa** business

A CHAPMAN & HALL BOOK

CRC Press
Taylor & Francis Group
6000 Broken Sound Parkway NW, Suite 300
Boca Raton, FL 33487-2742

First issued in paperback 2022

ISBN 13: 978-1-03-247638-4 (pbk)
ISBN 13: 978-1-4987-7846-6 (hbk)

DOI: 10.4324/9781351253123

Visit the Taylor & Francis Web site at
http://www.taylorandfrancis.com

and the CRC Press Web site at
http://www.crcpress.com

This book is dedicated to Katie and Elizabeth, who have brought indescribable joy, meaning, and

happiness into our lives

Contents

Preface

In beginning to write a linear algebra text, a question that surfaces even before the first keystroke takes place is who is the audience and what do we want to accomplish. The answer to this is more complex than in several other areas of mathematics because of the breadth of potential users. The book was written with the idea that a typical student would be one who has completed two semesters of calculus but who has not taken courses that emphasize abstract mathematics. The goals of the text are to present the topics that are traditionally covered in a first-level linear algebra course so that the computational methods become deeply ingrained and to make the intuition of the theory as transparent as possible.

Many disciplines, including statistics, economics, environmental science, engineering, and computer science, use linear algebra extensively. The sophistication of the applications of linear algebra in these areas can vary greatly. Students intending to study mathematics at the graduate level, and many others, would benefit from having a second course in linear algebra at the undergraduate level.

Some of the computations that we feel are especially important are matrix computations, solving systems of linear equations, representing a linear transformation in standard bases, finding eigenvectors, and diagonalizing matrices. Of less emphasis are topics such as converting the representation of vectors and linear transformations between nonstandard bases and converting a set of vectors to a basis by expanding or contracting the set.

In some cases, the intuition of a proof is more transparent if an example is presented before a theorem is articulated or if the proof of a theorem is given using concrete cases rather than an abstract argument. For example, by using three vectors instead of n vectors.

There are places in Chapters 4 through 7 where there are results that are important because of their applications, but the theory behind the result is time consuming and is more advanced than a typical student in a first exposure would be expected to digest. In such cases, the reader is alerted that the result is given later in the section and omitting the derivation will not compromise the usefulness of the results. Two specific examples of this are the projection matrix and the Gram–Schmidt process.

The exercises were designed to span a range from simple computations to fairly direct abstract exercises.

We expect that most users will want to make extensive use of a computer algebra system for computations. While there are several systems available, MATLAB® is the choice of many, and we have included a tutorial for MATLAB in the appendix. Because of the extensive use of the program R by statisticians, a tutorial for that program is also included.

A Note about Mathematical Proofs

As a text for a first course in linear algebra, this book has a major focus on demonstrating facts and techniques of linear systems that will be invaluable in higher mathematics and fields that use higher mathematics. This entails developing many computational tools.

A first course in linear algebra also serves as a bridge to mathematics courses that are primarily theoretical in nature and, as such, necessitates understanding and, in some cases, developing mathematical proofs. This is a learning process that can be frustrating.

We offer some suggestions that may ease the transition.

First, the most important requirement for abstract mathematics is knowing the important definitions well. These are analogous to the formulas of computational mathematics. A casual perusal will not be sufficient.

You will often encounter the question "Is it true that ...?" Here you must show that a result holds in all cases allowed by the hypotheses or find one example where the claim does not hold. Such an example is called a counter example. If you are going to prove that a result always holds, it is helpful (perhaps necessary) to believe that it holds. One way to establish an opinion for its validity is to work through some examples. This may highlight the essential properties that are crucial to a proof. If the result is not valid, this will often yield a counter example. Simple examples are usually the best. In linear algebra, this usually means working with small dimensional objects, such as 2×2 matrices.

In the text, we will sometimes give the ideas of a proof rather than a complete proof when we feel that "what is really going on" is obscured by the abstraction, for example, if the idea of a proof seems clear by examining 2×2 matrices but would be difficult to follow for $n \times n$ matrices.

In constructing a mathematical proof, there are two processes involved. One is to develop the ideas of the proof, and the second is to present those ideas in appropriate language.

There are different methods of proof that can be broadly divided into direct methods and indirect methods. An example of an indirect method is proof by contradiction. The contrapositive of "if statement A is true, then statement B is true" is "if statement B is false, then statement A is false." Proving the contrapositive of an "if-then" statement is logically equivalent to proving the original statement. The contrapositive of

$$\text{"If } x = 4\text{, then } x \text{ is an integer"}$$

is

$$\text{"If } x \text{ is not an integer, then } x \neq 4\text{."}$$

The converse of an "if-then" statement reverses the hypothesis and the conclusion. The converse of "if statement A is true, then statement B is true" is " if statement B is true, then statement A is true."

Unlike the contrapositive, the converse is not logically equivalent to the original statement. The converse of

$$\text{"If } x = 4\text{, then } x \text{ is an integer"}$$

is

$$\text{"If } x \text{ is an integer, then } x = 4\text{."}$$

Here the original statement is true, but the converse is false. You will sometimes have a problem of the type "show 'A if and only if B.'" In this case, you must prove

$$\text{"If A, then B"}$$

and

"If B, then A"

are both true.

In constructing a proof, it is necessary to completely understand what is given and what is to be shown. Sometimes, what is given can be phrased in different ways, and one way may be more helpful than others. Often in trying to construct a proof, it is worthwhile writing explicitly "this is what I know" and "this is what I want to show" to get started.

In the proofs presented in the text, we make a concerted effort to present the intuition behind the proof and why something is "the reasonable thing to do."

MATLAB® is a registered trademark of The MathWorks, Inc. For product information, please contact:

The MathWorks, Inc.
3 Apple Hill Drive
Natick, MA, 01760-2098 USA
Tel: 508-647-7000
Fax: 508-647-7001
E-mail: info@mathworks.com
Web: www.mathworks.com

1

Matrices

Linear algebra is the branch of mathematics that deals with vector spaces and linear transformations.

For someone just beginning their study of linear algebra, that is probably a meaningless statement. It does, however, convey the idea that there are two concepts in our study that will be of utmost importance, namely, vector spaces and linear transformations. A primary tool in our study of these topics will be matrices. In this chapter, we give the rules that govern matrix algebra.

1.1 Matrix Arithmetic

The term "matrix" was first used in a mathematical sense by James Sylvester in 1850 to denote a rectangular array of numbers from which determinants could be formed. The ideas of using arrays to solve systems of linear equations go as far back as the Chinese from 300 BC to 200 AD.

A matrix is a rectangular array of numbers. Matrices are the most common way of representing vectors and linear transformations and play a central role in nearly every computation in linear algebra. The size of a matrix is described by its dimensions, that is, the number of rows and columns in the matrix, with the number of rows given first. A 3×2 (three by two) matrix has three rows and two columns. Matrices are usually represented by uppercase letters.

The matrix A below is an example of a 3×2 matrix.

$$A = \begin{pmatrix} 0 & -5 \\ 6 & 3 \\ -1 & 3 \end{pmatrix}.$$

We will often want to represent a matrix in an abstract form. The example below is typical.

$$A = \begin{pmatrix} a_{11} & a_{12} & \cdots & a_{1n} \\ a_{21} & a_{22} & \cdots & a_{2n} \\ \vdots & \vdots & \cdots & \vdots \\ a_{m1} & a_{m2} & \cdots & a_{mn} \end{pmatrix}$$

It is customary to denote the numbers that make up the matrix, called the entries of the matrix, as the lowercase version of the letter that names the matrix, with two subscripts. The subscripts denote the position of the entry. The entry a_{ij} occupies the ith row and jth column of the matrix A. The notation $(A)_{ij}$ is also common, depending on the setting.

Two matrices A and B are equal if they have the same dimensions and $a_{ij} = b_{ij}$ for every i and j.

Definition

A square matrix is a matrix that has the same number of rows as columns, that is, an $n \times n$ matrix. If A is a square matrix, the entries $a_{11}, a_{22}, \ldots, a_{nn}$ make up the main diagonal of A. The trace of a square matrix is the sum of the entries on the main diagonal.

A square matrix is a diagonal matrix if the only nonzero entries of A are on the main diagonal. A square matrix is upper (lower) triangular if the only nonzero entries are above (below) or on the main diagonal. A matrix that consists of either a single row or a single column is called a vector. The matrix

$$\begin{pmatrix} 1 \\ 5 \\ 6 \end{pmatrix}$$

is an example of a column vector, and

$$\left(4, -1, 0, 8 \right)$$

is an example of a row vector.

In Figure 1.1, matrix (a) is a diagonal matrix, but the others are not.

$$\begin{pmatrix} 2 & 0 & 0 & 0 \\ 0 & 0 & 0 & 0 \\ 0 & 0 & -1 & 0 \\ 0 & 0 & 0 & 5 \end{pmatrix} \qquad \begin{pmatrix} 0 & 0 & 0 & 0 \\ 0 & 0 & 5 & 0 \\ 0 & 3 & 0 & 0 \\ 2 & 0 & 0 & 0 \end{pmatrix}$$
(a)   (b)

$$\begin{pmatrix} 0 & 1 & 0 & 0 \\ 0 & 3 & 0 & 0 \\ 0 & 0 & 7 & 0 \\ 0 & 0 & 0 & 2 \end{pmatrix} \qquad \begin{pmatrix} 1 & 0 & 0 \\ 0 & 2 & 0 \end{pmatrix}$$
(c)   (d)

FIGURE 1.1
Matrix (a) is diagonal, the others are not.

1.1.1 Matrix Arithmetic

We now define three arithmetic operations on matrices: matrix addition, scalar multiplication, and matrix multiplication.

1.1.1.1 Matrix Addition

In order to add two matrices, the matrices must have the same dimensions, and then one simply adds the corresponding entries. For example

$$\begin{pmatrix} 7 & -1 & 0 \\ -2 & 6 & 6 \end{pmatrix} + \begin{pmatrix} -2 & -3 & 4 \\ 3 & 5 & 0 \end{pmatrix} = \begin{pmatrix} 7-2 & -1-3 & 0+4 \\ -2+3 & 6+5 & 6+0 \end{pmatrix} = \begin{pmatrix} 5 & -4 & 4 \\ 1 & 11 & 6 \end{pmatrix}$$

but

$$\begin{pmatrix} 7 & -1 & 0 \\ -2 & 6 & 6 \end{pmatrix} + \begin{pmatrix} 1 & 0 \\ 2 & -1 \\ 0 & 5 \end{pmatrix}$$

is not defined.

To be more formal—and to begin to get used to the abstract notation—we could express this idea as

$$\left(A+B\right)_{ij} = A_{ij} + B_{ij}.$$

1.1.1.2 Scalar Multiplication

A scalar is a number. Scalar multiplication means multiplying a matrix by a number and is accomplished by multiplying every entry in the matrix by the scalar. For example

$$3 \begin{pmatrix} 1 & -4 \\ 2 & 0 \end{pmatrix} = \begin{pmatrix} 3 & -12 \\ 6 & 0 \end{pmatrix}.$$

1.1.1.3 Matrix Multiplication

Matrix multiplication is not as intuitive as the two prior operations. A fact that we will demonstrate later is that every linear transformation can be expressed as multiplication by a matrix. The definition of matrix multiplication is based in part on the idea that composition of two linear transformations should be expressed as the product of the matrices that represent the linear transformations.

We describe matrix multiplication by stages.

Case 1

Multiplying a matrix consisting of one row by a matrix consisting of one column, where the row matrix is on the left and the column matrix is on the right. The process is

$$\left(a_1, a_2, \ldots, a_n\right)\begin{pmatrix} b_1 \\ b_2 \\ \vdots \\ b_n \end{pmatrix} = a_1 b_1 + a_2 b_2 + \cdots + a_n b_n.$$

This can be accomplished only if there are the same number of entries in the row matrix as in the column matrix. Notice that the product of a $(1 \times n)$ matrix and a $(n \times 1)$ matrix is a number that, in this case, we will describe as a (1×1) matrix.

Case 2

Multiplying a matrix consisting of one row by a matrix consisting of more than one column, where the row matrix is on the left and the column matrix is on the right. Two examples are

$$\left(a_1, \ldots, a_n\right)\begin{pmatrix} b_1 & c_1 \\ \vdots & \vdots \\ b_n & c_n \end{pmatrix} = \left(a_1 b_1 + \cdots + a_n b_n, \quad a_1 c_1 + \cdots + a_n c_n\right)$$

$$\left(a_1, \ldots, a_n\right)\begin{pmatrix} b_1 & c_1 & d_1 \\ \vdots & \vdots & \vdots \\ b_n & c_n & d_n \end{pmatrix} = \left(a_1 b_1 + \cdots + a_n b_n, \quad a_1 c_1 + \cdots + a_n c_n, \quad a_1 d_1 + \cdots + a_n d_n\right).$$

Thus, the product of a $(1 \times n)$ matrix and a $(n \times k)$ matrix is a $(1 \times k)$ matrix.
   It may help to visualize this as applying Case 1 multiple times.

Case 3

Multiplying a matrix consisting of more than one row by a matrix consisting of one column, where the row matrix is on the left and the column matrix is on the right. Two examples of this are

$$(2 \times n) \times (n \times 1)\begin{pmatrix} a_1, \ldots, a_n \\ b_1, \ldots, b_n \end{pmatrix}\begin{pmatrix} c_1 \\ \vdots \\ c_n \end{pmatrix} = \begin{pmatrix} a_1 c_1 + \cdots + a_n c_n \\ b_1 c_1 + \cdots + b_n c_n \end{pmatrix} 2 \times 1$$

$$(3 \times n) \times (n \times 1)\begin{pmatrix} a_1, \ldots, a_n \\ b_1, \ldots, b_n \\ c_1, \ldots, c_n \end{pmatrix}\begin{pmatrix} d_1 \\ \vdots \\ d_n \end{pmatrix} = \begin{pmatrix} a_1 d_1 + \cdots + a_n d_n \\ b_1 d_1 + \cdots + b_n d_n \\ c_1 d_1 + \cdots + c_n d_n \end{pmatrix} 3 \times 1.$$

Case 4

Multiplying any two "compatible" matrices. In order to be compatible for multiplication, the matrix on the left must have the same number of entries in a row as there are entries in a column of the matrix on the right. One example is

$$(2\times3)\times(3\times2)\begin{pmatrix} a_1 & a_2 & a_3 \\ b_1 & b_2 & b_3 \end{pmatrix}\begin{pmatrix} c_1 & c_2 \\ d_1 & d_2 \\ e_1 & e_2 \end{pmatrix} = \begin{pmatrix} a_1c_1 + a_2d_1 + a_3e_1 & a_1c_2 + a_2d_2 + a_3e_2 \\ b_1c_1 + b_2d_1 + b_3e_1 & b_1c_2 + b_2d_2 + b_3e_2 \end{pmatrix},$$

which is a (2×2) matrix.

We describe a formula for computing the product of two compatible matrices.

Suppose that A is an $m\times k$ matrix and B is a $k\times n$ matrix, say

$$A = \begin{pmatrix} a_{11} & a_{12} & \cdots & a_{1k} \\ a_{21} & a_{22} & \cdots & a_{2k} \\ \vdots & \vdots & \vdots & \vdots \\ a_{m1} & a_{m2} & \cdots & a_{mk} \end{pmatrix} \quad B = \begin{pmatrix} b_{11} & b_{12} & \cdots & b_{1n} \\ b_{21} & b_{22} & \cdots & b_{2n} \\ \vdots & \vdots & \vdots & \vdots \\ b_{k1} & b_{k2} & \cdots & b_{kn} \end{pmatrix}.$$

The product matrix AB is $m\times n$ and the formula for the i,j entry of the matrix AB is

$$(AB)_{ij} = \sum_{l=1}^{k} a_{il}b_{lj}.$$

One can think of this as multiplying the ith row of A by the jth column of B.

1.1.1.3.1 A Summary of Matrix Multiplication

If A is an $m \times k$ matrix and B is an $s \times n$ matrix, then AB is defined if and only if $k = s$. If $k = s$, then AB is an $m \times n$ matrix whose i, j entry is

$$(AB)_{ij} = \sum_{s=1}^{k} a_{is}b_{sj}.$$

Another way to visualize matrix multiplication that is useful is to consider the product AB in the following way: Let $\hat{b}_i$ be the vector that is the ith column of the matrix B, so that if

$$B = \begin{pmatrix} b_{11} & b_{12} & b_{13} \\ b_{21} & b_{22} & b_{23} \\ b_{31} & b_{32} & b_{33} \end{pmatrix}$$

then

$$\hat{b}_1 = \begin{pmatrix} b_{11} \\ b_{21} \\ b_{31} \end{pmatrix}, \quad \hat{b}_2 = \begin{pmatrix} b_{12} \\ b_{22} \\ b_{32} \end{pmatrix}, \quad \hat{b}_3 = \begin{pmatrix} b_{13} \\ b_{23} \\ b_{33} \end{pmatrix}.$$

We can then think of the matrix B as a "vector of vectors"

$$B = \begin{bmatrix} \hat{b}_1 & \hat{b}_2 & \hat{b}_3 \end{bmatrix}$$

and we have

$$AB = \begin{bmatrix} A\hat{b}_1 & A\hat{b}_2 & A\hat{b}_3 \end{bmatrix}.$$

This will often be used in the text.

   We verify this in the case

$$A = \begin{pmatrix} a_{11} & a_{12} & a_{13} \\ a_{21} & a_{22} & a_{23} \end{pmatrix}.$$

We have

$$A\hat{b}_1 = \begin{pmatrix} a_{11} & a_{12} & a_{13} \\ a_{21} & a_{22} & a_{23} \end{pmatrix} \begin{pmatrix} b_{11} \\ b_{21} \\ b_{31} \end{pmatrix} = \begin{pmatrix} a_{11}b_{11} + a_{12}b_{21} + a_{13}b_{31} \\ a_{21}b_{11} + a_{22}b_{21} + a_{23}b_{31} \end{pmatrix}$$

$$A\hat{b}_2 = \begin{pmatrix} a_{11} & a_{12} & a_{13} \\ a_{21} & a_{22} & a_{23} \end{pmatrix} \begin{pmatrix} b_{12} \\ b_{22} \\ b_{32} \end{pmatrix} = \begin{pmatrix} a_{11}b_{12} + a_{12}b_{22} + a_{13}b_{32} \\ a_{21}b_{12} + a_{22}b_{22} + a_{23}b_{32} \end{pmatrix}$$

$$A\hat{b}_3 = \begin{pmatrix} a_{11} & a_{12} & a_{13} \\ a_{21} & a_{22} & a_{23} \end{pmatrix} \begin{pmatrix} b_{13} \\ b_{23} \\ b_{33} \end{pmatrix} = \begin{pmatrix} a_{11}b_{13} + a_{12}b_{23} + a_{13}b_{33} \\ a_{21}b_{13} + a_{22}b_{23} + a_{23}b_{33} \end{pmatrix}$$

so

$$\begin{bmatrix} A\hat{b}_1 & A\hat{b}_2 & A\hat{b}_3 \end{bmatrix} = \begin{pmatrix} a_{11}b_{11} + a_{12}b_{21} + a_{13}b_{31} & a_{11}b_{12} + a_{12}b_{22} + a_{13}b_{32} & a_{11}b_{13} + a_{12}b_{23} + a_{13}b_{33} \\ a_{21}b_{11} + a_{22}b_{21} + a_{23}b_{31} & a_{21}b_{12} + a_{22}b_{22} + a_{23}b_{32} & a_{21}b_{13} + a_{22}b_{23} + a_{23}b_{33} \end{pmatrix}.$$

Also,

$$AB = \begin{pmatrix} a_{11} & a_{12} & a_{13} \\ a_{21} & a_{22} & a_{23} \end{pmatrix} \begin{pmatrix} b_{11} & b_{12} & b_{13} \\ b_{21} & b_{22} & b_{23} \\ b_{31} & b_{32} & b_{33} \end{pmatrix}$$

$$= \begin{pmatrix} a_{11}b_{11} + a_{12}b_{21} + a_{13}b_{31} & a_{11}b_{12} + a_{12}b_{22} + a_{13}b_{32} & a_{11}b_{13} + a_{12}b_{23} + a_{13}b_{33} \\ a_{21}b_{11} + a_{22}b_{21} + a_{23}b_{31} & a_{21}b_{12} + a_{22}b_{22} + a_{23}b_{32} & a_{21}b_{13} + a_{22}b_{23} + a_{23}b_{33} \end{pmatrix}.$$

As a final cautionary remark, we note that if A is an $m \times k$ matrix and B is a $k \times n$ matrix, then AB is defined and is an $m \times n$ matrix, but BA is not defined unless $m = n$.

Exercises

1. Suppose

$$A = \begin{pmatrix} 2 & 0 \\ 3 & 1 \end{pmatrix}, \quad B = \begin{pmatrix} -1 & 2 \\ -3 & 2 \\ 1 & 4 \end{pmatrix}, \quad C = \begin{pmatrix} 1 & 5 & 2 \\ -3 & 0 & 6 \end{pmatrix}, \quad D = \begin{pmatrix} 4 & -2 & 2 \\ 0 & 1 & 5 \\ 2 & 0 & -3 \end{pmatrix}, \quad E = \begin{pmatrix} 1 & 7 \\ 4 & -3 \end{pmatrix}.$$

   Where the computations are possible, find
   (a) $3A - 2E$
   (b) BC
   (c) $CB + 4D$
   (d) $C - 4B$
   (e) BA
   (f) EC

2. Compute $A\hat{b}_1$, $A\hat{b}_2$, $\left[A\hat{b}_1, A\hat{b}_2 \right]$, and AB for

$$A = \begin{pmatrix} 2 & -1 \\ 1 & 3 \end{pmatrix}, \quad B = \begin{pmatrix} 1 & 0 & -1 \\ 2 & 1 & 4 \end{pmatrix}.$$

3. Find a 3×3 matrix A for which

   (a) $A \begin{pmatrix} x \\ y \\ z \end{pmatrix} = \begin{pmatrix} 4x - 2y \\ 3z \\ 0 \end{pmatrix}$

   (b) $A \begin{pmatrix} x \\ y \\ z \end{pmatrix} = \begin{pmatrix} z \\ y \\ x \end{pmatrix}$

   (c) $A \begin{pmatrix} x \\ y \\ z \end{pmatrix} = \begin{pmatrix} x + y + z \\ 2x + 2y + 2z \\ -x - y - z \end{pmatrix}$

4. Show that

   (a) $\begin{pmatrix} a_1 & b_1 \\ a_2 & b_2 \end{pmatrix} \begin{pmatrix} x \\ y \end{pmatrix} = x \begin{pmatrix} a_1 \\ a_2 \end{pmatrix} + y \begin{pmatrix} b_1 \\ b_2 \end{pmatrix}$

   (b) $\begin{pmatrix} a_1 & b_1 & c_1 \\ a_2 & b_2 & c_2 \\ a_3 & b_3 & c_3 \end{pmatrix} \begin{pmatrix} x \\ y \\ z \end{pmatrix} = x \begin{pmatrix} a_1 \\ a_2 \\ a_3 \end{pmatrix} + y \begin{pmatrix} b_1 \\ b_2 \\ b_3 \end{pmatrix} + z \begin{pmatrix} c_1 \\ c_2 \\ c_3 \end{pmatrix}$

5. If AB is a 5×7 matrix, how many columns does B have?

6. (a) If A and B are 2×2 matrices, show that

$$(AB)_{11} + (AB)_{22} = (BA)_{11} + (BA)_{22}.$$

   (b) If A and B are $n \times n$ matrices, compare the trace of AB and the trace of BA.

7. If A and B are matrices so that AB is defined, then

$$j\text{th column of } AB = A\left[j\text{th } column \text{ of } B\right]$$

$$i\text{th row of } AB = \left[i\text{th } row \text{ of } A\right]B.$$

Demonstrate this principle by finding the second row and third column of AB for

$$A = \begin{pmatrix} 3 & -1 & 0 \\ 1 & 5 & -2 \\ 4 & 3 & 1 \end{pmatrix} \quad B = \begin{pmatrix} 2 & 5 & -3 \\ -1 & 0 & 4 \\ 4 & 2 & -5 \end{pmatrix}.$$

8. Construct a 5×5 matrix, not all of whose entries are zero, that has the following properties:
   (a) $A_{ij} = 0$ if $i < j$.
   (b) $A_{ij} = 0$ if $i \geq j$.
   (c) $A_{ij} = 0$ if $|i - j| = 1$.

9. Compute

$$(a, b)\begin{pmatrix} e & f \\ g & h \end{pmatrix} \quad \text{and} \quad (c, d)\begin{pmatrix} e & f \\ g & h \end{pmatrix}$$

and compare with

$$\begin{pmatrix} a & b \\ c & d \end{pmatrix}\begin{pmatrix} e & f \\ g & h \end{pmatrix}.$$

Can you make any conjectures based on this example?

10. If AB is a 3×3 matrix, show that BA is defined.

11. If

$$A = \begin{pmatrix} 3 & 8 & 0 \\ 4 & -1 & -2 \end{pmatrix}, \quad B = \begin{pmatrix} 2 & 1 & 1 \\ 3 & -1 & 7 \end{pmatrix}$$

find the matrix C for which

$$3A - 5B + 2C = \begin{pmatrix} 0 & 1 & 2 \\ 2 & -3 & 4 \end{pmatrix}.$$

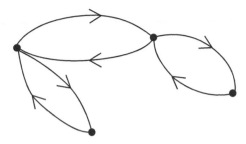

FIGURE 1.2
A simple graph that is not directed.

12. A graph is a set of vertexes and a set of edges between some of the vertices. If the graph is simple, then there is no edge from a vertex to itself. If the graph is not directed, then an edge from vertex i to vertex j is also an edge from j to vertex i. Figure 1.2 shows a simple graph that is not directed.

Associated with each graph is a matrix A defined by

$$A_{ij} = \begin{cases} 1 & \text{if there is an edge from vertex } i \text{ to vertex } j \\ \\ 0 & \text{otherwise} \end{cases}.$$

(a) If there are n vertices, what are the dimensions of the associated matrix A? What else can you conclude about A?

We say there is a path of length k from vertex i to vertex j if it is possible to traverse from vertex i to vertex j by traversing across exactly k edges. (Traversing the same edge more than once is allowed.)

(b) How could the matrix A be used to determine whether there is a path from i to vertex j in exactly 2 steps? In exactly k steps?

(c) How could you tell how many paths there are from i to vertex j in exactly k steps?

　(i) For the graph in Figure 1.2

　(ii) Construct the associated graph.

(d) Use the associated matrix to determine whether it is possible to go from vertex 1 to vertex 4 by traversing across exactly 3 edges using the matrix you constructed.

1.2 The Algebra of Matrices

1.2.1 Properties of Matrix Addition, Scalar Multiplication, and Matrix Multiplication

Matrix addition, scalar multiplication, and matrix multiplication obey the rules of the next theorem.

Theorem 1

If the sizes of the matrices $A, B,$ and C are so that the operations can be performed, and α and β are scalars, then

(a) $A + B = B + A$
(b) $(A + B) + C = A + (B + C)$
(c) $(AB)C = A(BC)$
(d) $A(B + C) = AB + AC$
(e) $(B + C)A = BA + CA$
(f) $\alpha(A + B) = \alpha A + \alpha B$
(g) $(\alpha + \beta)A = \alpha A + \beta A$
(h) $\alpha(\beta A) = (\alpha\beta)A$
(i) $\alpha(AB) = (\alpha A)B = A(\alpha B)$

This theorem is simply saying that these rules of combining matrices obey the usual laws of arithmetic. When we more closely examine rules that govern matrix multiplication, we will see that some of these rules are different from what we might hope.

The proofs of these results are not deep but can be tedious. We demonstrate a proof of part (c).

Suppose that A is a $k \times l$ matrix, B is an $l \times m$ matrix, and C is an $m \times n$ matrix. Then AB is a $k \times m$ matrix, $(AB)C$ is a $k \times n$ matrix, BC is an $l \times n$ matrix, and $A(BC)$ is a $k \times n$ matrix. Thus, $(AB)C$ and $A(BC)$ have the same dimensions. To complete the proof, we must show that the entries of the two matrices are the same. That is, we must show

$$\left[(AB)C\right]_{rs} = \left[A(BC)\right]_{rs} \quad \text{for } r = 1, \ldots, k; \ s = 1, \ldots, n.$$

We have

$$\left[(AB)C\right]_{rs} = \sum_{j=1}^{m}(AB)_{rj}C_{js} = \sum_{j=1}^{m}\left[\sum_{i=1}^{l}A_{ri}B_{ij}\right]C_{js} = \sum_{j=1}^{m}\left[\sum_{i=1}^{l}\left(A_{ri}B_{ij}\right)C_{js}\right] = \sum_{j=1}^{m}\left[\sum_{i=1}^{l}A_{ri}\left(B_{ij}C_{js}\right)\right]$$

$$= \sum_{i=1}^{l}\left[\sum_{j=1}^{m}A_{ri}\left(B_{ij}C_{js}\right)\right] = \sum_{i=1}^{l}A_{ri}\left[\sum_{j=1}^{m}\left(B_{ij}C_{js}\right)\right] = \sum_{i=1}^{l}A_{ri}\left(BC\right)_{is} = \left[A(BC)\right]_{rs}.$$

We have defined the arithmetic operations on matrices. Matrices together with these operations form a mathematical structure. To expand this structure, we need to introduce some additional matrices.

In arithmetic, the numbers 0 and 1 have special significance. The number 0 is important because it is the additive identity, that is,

$$0 + a = a$$

for every number a. The number 1 is important because it is the multiplicative identity, that is,

$$1 \times a = a$$

for every number a.

There are matrices that fulfill these same roles, but we must be sure the dimensions of these matrices are compatible with what we want to accomplish.

A zero matrix is a matrix for which all of the entries are 0. We denote the zero matrix of size $m \times n$ by

$$\tilde{0}_{m \times n}.$$

If the dimensions are obvious, we will simply use $\tilde{0}$. We have the desired property that if A is an $m \times n$ matrix, then

$$A + \tilde{0}_{m \times n} = A.$$

Matrix multiplication presents some challenges that do not occur with matrix addition. The most prominent of these is that matrix multiplication is normally not commutative even if the dimensions allow for multiplication. That is, it is most often the case that for matrices A and B

$$AB \neq BA.$$

We expand on this. If A is an $m \times k$ matrix and B is a $k \times n$ matrix then AB is an $m \times n$ matrix, but BA is not defined unless $m = n$. If $m = n$ then AB is an $m \times m$ matrix, but BA is a $k \times k$ matrix.

Even if A and B are both $m \times m$ matrices, it is most often the case that $AB \neq BA$.

Example

Consider

$$A = \begin{pmatrix} 1 & 0 \\ 0 & 0 \end{pmatrix}, \quad B = \begin{pmatrix} 0 & 0 \\ 1 & 1 \end{pmatrix},$$

We have

$$AB = \begin{pmatrix} 1 & 0 \\ 0 & 0 \end{pmatrix} \begin{pmatrix} 0 & 0 \\ 1 & 1 \end{pmatrix} = \begin{pmatrix} 0 & 0 \\ 0 & 0 \end{pmatrix}$$

and

$$BA = \begin{pmatrix} 0 & 0 \\ 1 & 1 \end{pmatrix} \begin{pmatrix} 1 & 0 \\ 0 & 0 \end{pmatrix} = \begin{pmatrix} 0 & 0 \\ 1 & 0 \end{pmatrix}.$$

Not only do we have $AB \neq BA$, but in the case of AB, we see another point of caution.

If x and y are numbers, and if $xy = 0$, then it must be that $x = 0$ or $y = 0$.

In the case of BA in the earlier example, we have the product of two matrices, neither of which is equal to $\tilde{0}$, but the product is $\tilde{0}$. Among the implications of this is that if

$$AB = AC \quad \text{and} \quad A \neq \tilde{0}$$

we cannot cancel A. That is, we cannot be sure that $B = C$.

We group these results.

Facts

If A and B are matrices for which AB is defined then

     (i) It is not necessarily the case that BA is defined.
    (ii) Even if AB and BA are both defined, it is not necessarily the case that $AB = BA$.
   (iii) It is possible that neither A nor B is the zero matrix, but AB is the zero matrix.
   (iv) It is possible that there are matrices A, B, and C for which $AB = AC$ and $B \neq C$.

1.2.2 The Identity Matrix

The matrix that acts (as closely as possible) like the number 1 as the multiplicative identity is the square matrix that has 1's on the main diagonal and 0's for every other entry. If the identity matrix is an $n \times n$ matrix, we denote it by I_n. For example,

$$I_3 = \begin{pmatrix} 1 & 0 & 0 \\ 0 & 1 & 0 \\ 0 & 0 & 1 \end{pmatrix}.$$

If A is an $n \times n$ matrix, then

$$AI_n = I_n A = A.$$

As an example, it is easy to check that

$$I_3 \begin{pmatrix} a_{11} & a_{12} & a_{13} \\ a_{21} & a_{22} & a_{23} \\ a_{31} & a_{32} & a_{33} \end{pmatrix} = \begin{pmatrix} 1 & 0 & 0 \\ 0 & 1 & 0 \\ 0 & 0 & 1 \end{pmatrix} \begin{pmatrix} a_{11} & a_{12} & a_{13} \\ a_{21} & a_{22} & a_{23} \\ a_{31} & a_{32} & a_{33} \end{pmatrix} = \begin{pmatrix} a_{11} & a_{12} & a_{13} \\ a_{21} & a_{22} & a_{23} \\ a_{31} & a_{32} & a_{33} \end{pmatrix}$$

and

$$\begin{pmatrix} a_{11} & a_{12} & a_{13} \\ a_{21} & a_{22} & a_{23} \\ a_{31} & a_{32} & a_{33} \end{pmatrix} I_3 = \begin{pmatrix} a_{11} & a_{12} & a_{13} \\ a_{21} & a_{22} & a_{23} \\ a_{31} & a_{32} & a_{33} \end{pmatrix} \begin{pmatrix} 1 & 0 & 0 \\ 0 & 1 & 0 \\ 0 & 0 & 1 \end{pmatrix} = \begin{pmatrix} a_{11} & a_{12} & a_{13} \\ a_{21} & a_{22} & a_{23} \\ a_{31} & a_{32} & a_{33} \end{pmatrix}.$$

If A is an $n \times n$ matrix, then $I_n A = A$ and $AI_m = A$.

1.2.3 The Inverse of a Square Matrix

In real numbers, every nonzero number has a multiplicative inverse. For example, the multiplicative inverse of 2 is ½ because

$$2 \times \tfrac{1}{2} = 1,$$

and 1 is the multiplicative identity.

We want to construct the analogous element for matrix multiplication. That is, if A is a nonzero $n \times n$ matrix, we would like to find an $n \times n$ matrix B for which

$$AB = BA = I_n.$$

This is not always possible, as we see in the next example.

Example

Let

$$A = \begin{pmatrix} 1 & 0 \\ 0 & 0 \end{pmatrix}.$$

We seek a 2×2 matrix B for which

$$AB = BA = I_2.$$

Let

$$B = \begin{pmatrix} a & b \\ c & d \end{pmatrix}.$$

Then

$$AB = \begin{pmatrix} 1 & 0 \\ 0 & 0 \end{pmatrix}\begin{pmatrix} a & b \\ c & d \end{pmatrix} = \begin{pmatrix} a & b \\ 0 & 0 \end{pmatrix}$$

and it is impossible to find $a, b, c,$ and d for which

$$AB = \begin{pmatrix} 1 & 0 \\ 0 & 1 \end{pmatrix}.$$

Definition

If A is an $n \times n$ matrix, and B is an $n \times n$ matrix for which

$$AB = I_n = BA$$

then B is said to be the inverse of A.

If the inverse of A exists, then it is unique and it is denoted A^{-1}. A matrix that has an inverse is said to be nonsingular. A square matrix that does not have an inverse is singular.

Theorem 2

If A and B are nonsingular $n \times n$ matrices, then AB is a nonsingular matrix and $(AB)^{-1} = B^{-1}A^{-1}$.

Proof

We have

$$\left(B^{-1}A^{-1}\right)(AB) = \left[\left(B^{-1}A^{-1}\right)A\right]B = \left[B^{-1}\left(A^{-1}A\right)\right]B = \left(B^{-1}I\right)B = B^{-1}B = I$$

and

$$(AB)\left(B^{-1}A^{-1}\right) = A\left[B\left(B^{-1}A^{-1}\right)\right] = A\left[\left(BB^{-1}\right)A^{-1}\right] = A\left(IA^{-1}\right) = AA^{-1} = I.$$

Finding the inverse of a matrix is often a computationally complex task. In this section, we develop an algorithm based on what computers use to determine the inverse of a matrix. In doing so, we introduce several topics that will be important in our later work. These topics include determinants, elementary row operations, and elementary matrices.

Instructions to use the computer program MATLAB®, which performs the computations of linear algebra, are given in the appendix.

1.2.4 Determinants

In Volume 2, we give an in-depth development of determinants. Here, we give an introduction to determinants and basic properties of determinants that are needed for our immediate purposes.

A determinant is a number that is associated with a square matrix.

The most basic way of computing a determinant is expanding by cofactors. We explain this method by way of examples.

For a 2×2 matrix

$$A = \begin{pmatrix} a & b \\ c & d \end{pmatrix}.$$

The determinant of A, denoted $\det(A)$ or $\begin{vmatrix} a & b \\ c & d \end{vmatrix}$, is

$$\det(A) \equiv ad - bc.$$

Suppose we have a 3×3 matrix. We assign a + sign to the upper left-hand position, and alternate signs of the positions as we proceed either to the right or down. We then have

$$\begin{pmatrix} + & - & + \\ - & + & - \\ + & - & + \end{pmatrix}.$$

In expanding by cofactors, one chooses any row or any column and breaks the determinant into pieces as in the next example.

Example

Consider

$$A = \begin{pmatrix} -2 & 3 & 1 \\ 0 & -1 & 2 \\ 4 & 5 & 3 \end{pmatrix}.$$

Suppose we elect to expand along the second column. We compute the determinate by expanding the determinate into three smaller pieces. To get the first piece, we eliminate the entries in the first row and second column to get

$$\begin{pmatrix} 0 & 2 \\ 4 & 3 \end{pmatrix}$$

According to Figure 1.1, the sign assigned to the first row and second column is – and the entry in the first row second column is 3. The first piece in our computation is

$$(-)(3)\begin{vmatrix} 0 & 2 \\ 4 & 3 \end{vmatrix} = -3(0-8) = 24.$$

Since we are expanding along the second column, we move to the A_{22} entry and repeat the procedure above. This gives

$$(+)(-1)\begin{vmatrix} -2 & 1 \\ 4 & 3 \end{vmatrix} = -(-6-4) = 10.$$

Moving to the A_{32} element gives

$$(-)(5)\begin{vmatrix} -2 & 1 \\ 0 & 2 \end{vmatrix} = 20.$$

The determinant of A is the sum of these three numbers, that is,

$$\det(A) = 24 + 10 + 20 = 54.$$

For larger matrices, we follow a procedure analogous to the one above.

It is a somewhat deep fact that one gets the same value regardless of which row or column one chooses for the expansion.

Example

The determinant for upper and lower triangular matrices is easy to compute. Let

$$A = \begin{pmatrix} a_{11} & a_{12} & a_{13} & a_{14} \\ 0 & a_{22} & a_{23} & a_{24} \\ 0 & 0 & a_{33} & a_{34} \\ 0 & 0 & 0 & a_{44} \end{pmatrix}.$$

Then, expanding along the first column

$$\det(A) = a_{11}\begin{vmatrix} a_{22} & a_{23} & a_{24} \\ 0 & a_{33} & a_{34} \\ 0 & 0 & a_{44} \end{vmatrix} - 0\begin{vmatrix} a_{12} & a_{13} & a_{14} \\ 0 & a_{33} & a_{34} \\ 0 & 0 & a_{44} \end{vmatrix} + 0\begin{vmatrix} a_{12} & a_{13} & a_{14} \\ a_{22} & a_{23} & a_{24} \\ 0 & 0 & a_{44} \end{vmatrix}$$

$$- 0\begin{vmatrix} a_{12} & a_{13} & a_{14} \\ a_{22} & a_{23} & a_{24} \\ 0 & a_{33} & a_{34} \end{vmatrix} = a_{11}\begin{vmatrix} a_{22} & a_{23} & a_{24} \\ 0 & a_{33} & a_{34} \\ 0 & 0 & a_{44} \end{vmatrix}$$

$$= a_{11}\left(a_{22}\begin{bmatrix} a_{33} & a_{34} \\ 0 & a_{44} \end{bmatrix} - 0\begin{bmatrix} a_{23} & a_{24} \\ 0 & a_{44} \end{bmatrix} + 0\begin{bmatrix} a_{23} & a_{24} \\ a_{33} & a_{34} \end{bmatrix} \right)$$

$$= a_{11}a_{22}\begin{bmatrix} a_{33} & a_{34} \\ 0 & a_{44} \end{bmatrix} = a_{11}a_{22}a_{33}a_{44}.$$

This is an example of why the next theorem holds.

Theorem 3

The determinant of an upper or lower triangular matrix is the product of the main diagonal entries.

Corollary

The determinant of a diagonal matrix is the product of the main diagonal entries.

Definition

If A is a matrix, the following operations are called elementary row operations on A.

1. Interchanging two rows.
2. Adding a multiple of one row to another.
3. Multiplying a row by a constant.

The next theorem is proven in Chapter 9 of Volume 2.

Theorem 4

Let A be a square matrix and B be the matrix that results when a row operation is applied to A. We have

1. If B is the result of interchanging two rows of A, then $\det(B) = -\det(A)$.
2. If B is the result of adding a multiple of one row of A to another row of A, then $\det(B) = \det(A)$.
3. If B is the result of multiplying a row by a constant k, then $\det(B) = k\det(A)$.

The following corollary is very important.

Corollary

If A is a square matrix, applying elementary row operations will not change the determinant from zero to a nonzero value, and if $k \neq 0$ it will not change the determinant from a nonzero number to zero.

   We will see later that the following is true.

   If A is a square matrix, then it is possible to apply elementary row operations (with $k \neq 0$) to A until the result is either the identity matrix or a matrix with a row of all zeroes.

   In the next chapter, we demonstrate that applying elementary row operations (if $k \neq 0$) does not change whether a matrix is invertible.

   The next example demonstrates how applying elementary row operations can simplify calculating a determinant.

Example

We evaluate

$$
\begin{vmatrix} 2 & 4 & -6 & 0 \\ 1 & 5 & 3 & 6 \\ 3 & 1 & 4 & 5 \\ 2 & 6 & -7 & -3 \end{vmatrix} = 2\begin{vmatrix} 1 & 2 & -3 & 0 \\ 1 & 5 & 3 & 6 \\ 3 & 1 & 4 & 5 \\ 2 & 6 & -7 & -3 \end{vmatrix} = 2\begin{vmatrix} 1 & 2 & -3 & 0 \\ 0 & 3 & 6 & 6 \\ 0 & -5 & 13 & 5 \\ 0 & 2 & -1 & -3 \end{vmatrix}
$$

$$
= 2(3)\begin{vmatrix} 1 & 2 & -3 & 0 \\ 0 & 1 & 2 & 2 \\ 0 & -5 & 13 & 5 \\ 0 & 2 & -1 & -3 \end{vmatrix} = 2(3)\begin{vmatrix} 1 & 2 & -3 & 0 \\ 0 & 0 & 2 & 2 \\ 0 & 0 & 23 & 15 \\ 0 & 0 & -5 & -7 \end{vmatrix} = 2(3)(23)\begin{vmatrix} 1 & 2 & -3 & 0 \\ 0 & 1 & 2 & 2 \\ 0 & 0 & 1 & \frac{15}{23} \\ 0 & 0 & -5 & -7 \end{vmatrix}
$$

$$
= 2(3)(23)\begin{vmatrix} 1 & 2 & -6 & 0 \\ 0 & 1 & 2 & 2 \\ 0 & 0 & 1 & \frac{15}{23} \\ 0 & 0 & 0 & -\frac{86}{23} \end{vmatrix} = 2(3)(23)\left(-\frac{86}{23}\right) = -516.
$$

Theorem 5

Properties of the determinant of $n \times n$ matrices include

1. $\det(I_n) = 1$.
2. $\det(AB) = \det(A)\det(B)$.
3. $\det\left(A^{-1}\right) = \dfrac{1}{\det(A)}$.
4. $\det(cA) = c^n \det(A)$.
5. The determinant of a triangular or diagonal matrix is the product of its diagonal elements.
6. $\det(A) \neq 0$ if and only if A is invertible.
7. If two rows or two columns of a matrix are equal, the determinant is 0.
8. If two rows of a matrix are interchanged, the value of the determinant is multiplied by -1.
9. If a multiple of a row of a matrix is added to another row, the determinant is unchanged.
10. If a row of an $n \times n$ matrix is multiplied by k, the determinant is multiplied by k.

1.2.5 Elementary Matrices

In the next chapter, we study how to use matrices to solve certain types of equations. In the process of solving the equations, we will manipulate the rows of the matrix with the elementary row operations. These manipulations are

1. Interchanging two rows of a matrix
2. Multiplying a row of a matrix by a number
3. Adding a multiple of one row to another row

These operations can be executed by multiplying a matrix on the left by matrices called elementary matrices. In this section, we study elementary matrices.

We use the following notation:

E_{ij} results from interchanging the ith and jth rows of the identity matrix I.

$E_i(c)$ results from multiplying the ith row of the identity matrix I by the scalar c.

$E_{ij}(c)$ results from adding c times the row i to the row j of the identity matrix I.

1.2.6 Matrices That Interchange Two Rows of a Matrix

Suppose that A is a $3 \times n$ matrix. Let E_{ij} denote the matrix that is obtained when the ith and jth rows of the identity matrix I_3 are interchanged. Then $E_{ij}A$ is the matrix that results when the ith and jth rows of A are interchanged.

Example

Let

$$E_{23} = \begin{pmatrix} 1 & 0 & 0 \\ 0 & 0 & 1 \\ 0 & 1 & 0 \end{pmatrix} \quad \text{and} \quad A = \begin{pmatrix} a_{11} & a_{12} & a_{13} & a_{14} \\ a_{21} & a_{22} & a_{23} & a_{24} \\ a_{31} & a_{32} & a_{33} & a_{34} \end{pmatrix}.$$

Then

$$E_{23}A = \begin{pmatrix} 1 & 0 & 0 \\ 0 & 0 & 1 \\ 0 & 1 & 0 \end{pmatrix} \begin{pmatrix} a_{11} & a_{12} & a_{13} & a_{14} \\ a_{21} & a_{22} & a_{23} & a_{24} \\ a_{31} & a_{32} & a_{33} & a_{34} \end{pmatrix} = \begin{pmatrix} a_{11} & a_{12} & a_{13} & a_{14} \\ a_{31} & a_{32} & a_{33} & a_{34} \\ a_{21} & a_{22} & a_{23} & a_{24} \end{pmatrix}.$$

In general, if A is an $m \times n$ matrix and E_{ij} is the matrix that is obtained when the ith and jth rows of the identity matrix I_m are interchanged, then $E_{ij}A$ is the matrix that results when the ith and jth rows of A are interchanged.

We also have $E_{ij}^{-1} = E_{ij}$ since if the same two rows of a matrix are interchanged twice, the matrix is unchanged.

1.2.7 Multiplying a Row of a Matrix by a Constant

If we want to multiply the second row of a $3 \times n$ matrix by 5, we would multiply the matrix by

$$E_2(5) = \begin{pmatrix} 1 & 0 & 0 \\ 0 & 5 & 0 \\ 0 & 0 & 1 \end{pmatrix},$$

which is the matrix that results when the second row of I_3 is multiplied by 5.

We have

$$E_2(5) \begin{pmatrix} a_{11} & a_{12} & a_{13} & a_{14} \\ a_{21} & a_{22} & a_{23} & a_{24} \\ a_{31} & a_{32} & a_{33} & a_{34} \end{pmatrix} = \begin{pmatrix} 1 & 0 & 0 \\ 0 & 5 & 0 \\ 0 & 0 & 1 \end{pmatrix} \begin{pmatrix} a_{11} & a_{12} & a_{13} & a_{14} \\ a_{21} & a_{22} & a_{23} & a_{24} \\ a_{31} & a_{32} & a_{33} & a_{34} \end{pmatrix}$$

$$= \begin{pmatrix} a_{11} & a_{12} & a_{13} & a_{14} \\ 5a_{21} & 5a_{22} & 5a_{23} & 5a_{24} \\ a_{31} & a_{32} & a_{33} & a_{34} \end{pmatrix}.$$

Note that

$$E_2(5)^{-1} = \begin{pmatrix} 1 & 0 & 0 \\ 0 & 1/5 & 0 \\ 0 & 0 & 1 \end{pmatrix} = E_2(1/5).$$

In general, if A is an $m \times n$ matrix and $E_j(k)$ is the matrix that is obtained when the jth row of the identity matrix I_m is multiplied by k, then $E_j(k)A$ is the matrix that results when the jth row of A is multiplied by k.

Also, if $k \neq 0$, then

$$E_j(k)^{-1} = E_j\left(\frac{1}{k}\right).$$

1.2.8 Adding a Multiple of One Row to Another Row

If we want to multiply the second row of a $3 \times n$ matrix by 5 and add it to the third row, we would multiply the matrix by

$$E_{23}(5) = \begin{pmatrix} 1 & 0 & 0 \\ 0 & 1 & 0 \\ 0 & 5 & 1 \end{pmatrix}.$$

If we add k times the ith row of I_m to the jth row of I_m, we obtain the matrix $E_{ij}(k)$. If A is an $m \times n$ matrix, then the product $E_{ij}(k)A$ is the matrix obtained when k times the ith row of A is added to the jth row of A.

For example, we have

$$E_{23}(5)\begin{pmatrix} a_{11} & a_{12} & a_{13} & a_{14} \\ a_{21} & a_{22} & a_{23} & a_{24} \\ a_{31} & a_{32} & a_{33} & a_{34} \end{pmatrix} = \begin{pmatrix} 1 & 0 & 0 \\ 0 & 1 & 0 \\ 0 & 5 & 1 \end{pmatrix}\begin{pmatrix} a_{11} & a_{12} & a_{13} & a_{14} \\ a_{21} & a_{22} & a_{23} & a_{24} \\ a_{31} & a_{32} & a_{33} & a_{34} \end{pmatrix}$$

$$= \begin{pmatrix} a_{11} & a_{12} & a_{13} & a_{14} \\ a_{21} & a_{22} & a_{23} & a_{24} \\ a_{31} + 5a_{21} & a_{32} + 5a_{22} & a_{33} + 5a_{23} & a_{34} + 5a_{24} \end{pmatrix}$$

Also

$$E_{23}(5)^{-1} = \begin{pmatrix} 1 & 0 & 0 \\ 0 & 1 & 0 \\ 0 & -5 & 1 \end{pmatrix}.$$

1.2.9 Computing the Inverse of a Matrix

Normally, we will calculate the inverse of a matrix using computer software. In this subsection, we describe one method of finding the inverse of a matrix by using elementary matrices. These are the ideas involved.

Suppose that A is an $n \times n$ invertible matrix. We concatenate A with I_n to yield the matrix $(I_n \mid A)$.

We multiply the concatenated matrix by elementary matrices to get the matrix $(B \mid I_n)$. The matrix B is the inverse of A.

Example

Let

$$A = \begin{pmatrix} 2 & 4 & 0 \\ 3 & 9 & -6 \\ -4 & 8 & 12 \end{pmatrix}$$

so that

$$(I_3 \mid A) = \left(\begin{array}{ccc|ccc} 1 & 0 & 0 & 2 & 4 & 0 \\ 0 & 1 & 0 & 3 & 9 & -6 \\ 0 & 0 & 1 & -4 & 8 & 12 \end{array}\right).$$

We first multiply the first row by ½. This gives

$$E_1\left(\frac{1}{2}\right)(I_3 \mid A) = \left(\begin{array}{ccc|ccc} 1/2 & 0 & 0 & 1 & 2 & 0 \\ 0 & 1 & 0 & 3 & 9 & -6 \\ 0 & 0 & 1 & -4 & 8 & 12 \end{array}\right).$$

Add -3 times the first row to the second row to get

$$E_{12}(-3)E_1\left(\frac{1}{2}\right)(I_3 \mid A) = \left(\begin{array}{ccc|ccc} 1/2 & 0 & 0 & 1 & 2 & 0 \\ -3/2 & 1 & 0 & 0 & 3 & -6 \\ 0 & 0 & 1 & -4 & 8 & 12 \end{array}\right).$$

Add 4 times the first row to the third row to get

$$E_{13}(4)E_{12}(-3)E_1\left(\frac{1}{2}\right)(I_3 \mid A) = \left(\begin{array}{ccc|ccc} 1/2 & 0 & 0 & 1 & 2 & 0 \\ -3/2 & 1 & 0 & 0 & 3 & -6 \\ 2 & 0 & 1 & 0 & 16 & 12 \end{array}\right).$$

Multiply the second row by $1/3$

$$E_2\left(\frac{1}{3}\right)E_{13}(4)E_{12}(-3)E_1\left(\frac{1}{2}\right)(I_3 \mid A) = \left(\begin{array}{ccc|ccc} 1/2 & 0 & 0 & 1 & 2 & 0 \\ -1/2 & 1/3 & 0 & 0 & 1 & -2 \\ 2 & 0 & 1 & 0 & 16 & 12 \end{array}\right).$$

Add −16 times the second row to the third row to get

$$E_{23}(-16)E_2\left(\frac{1}{3}\right)E_{13}(4)E_{12}(-3)E_1\left(\frac{1}{2}\right)(I_3|\,A) = \begin{pmatrix} 1/2 & 0 & 0 & 1 & 2 & 0 \\ -1/2 & 1/3 & 0 & 0 & 1 & -2 \\ 10 & -16/3 & 1 & 0 & 0 & 44 \end{pmatrix}.$$

Multiply the third row by $1/44$ to get

$$E_3\left(\frac{1}{44}\right)E_{23}(-16)E_2\left(\frac{1}{3}\right)E_{13}(4)E_{12}(-3)E_1\left(\frac{1}{2}\right)(I_3|\,A) = \begin{pmatrix} 1/2 & 0 & 0 & 1 & 2 & 0 \\ -1/2 & 1/3 & 0 & 0 & 1 & -2 \\ 10/44 & -16/32 & 1/44 & 0 & 0 & 1 \end{pmatrix}.$$

Add −2 times the third row to the second row to get

$$E_{23}(2)E_3\left(\frac{1}{44}\right)E_{23}(-16)E_2\left(\frac{1}{3}\right)E_{13}(4)E_{12}(-3)E_1\left(\frac{1}{2}\right)(I_3|\,A) = \begin{pmatrix} 1/2 & 0 & 0 & 1 & 2 & 0 \\ -1/22 & 1/11 & 1/22 & 0 & 1 & 0 \\ 10/44 & -16/32 & 1/44 & 0 & 0 & 1 \end{pmatrix}.$$

Add −2 times the second row to the first row to get

$$E_{21}(-2)E_{23}(2)E_3\left(\frac{1}{44}\right)E_{23}(-16)E_2\left(\frac{1}{3}\right)E_{13}(4)E_{12}(-3)E_1\left(\frac{1}{2}\right)(I_3|\,A)$$

$$= \begin{pmatrix} 13/22 & -2/11 & -1/11 & 1 & 0 & 0 \\ -1/22 & 1/11 & 1/22 & 0 & 1 & 0 \\ 10/44 & -16/132 & 1/44 & 0 & 0 & 1 \end{pmatrix}.$$

Thus

$$A^{-1} = \begin{pmatrix} 13/22 & -2/11 & -1/11 \\ -1/22 & 1/11 & 1/22 \\ 10/44 & -16/132 & 1/44 \end{pmatrix}.$$

One can check that

$$\begin{pmatrix} 13/22 & -2/11 & -1/11 \\ -1/22 & 1/11 & 1/22 \\ 5/22 & -4/33 & 1/44 \end{pmatrix}\begin{pmatrix} 2 & 4 & 0 \\ 3 & 9 & -6 \\ -4 & 8 & 12 \end{pmatrix} = \begin{pmatrix} 2 & 4 & 0 \\ 3 & 9 & -6 \\ -4 & 8 & 12 \end{pmatrix}\begin{pmatrix} 13/22 & -2/11 & -1/11 \\ -1/22 & 1/11 & 1/22 \\ 5/22 & -4/33 & 1/44 \end{pmatrix} = \begin{pmatrix} 1 & 0 & 0 \\ 0 & 1 & 0 \\ 0 & 0 & 1 \end{pmatrix}.$$

If $E_1, \ldots, E_k$ are elementary row operations for which

$$E_k \cdots E_1 A = I$$

then

$$A^{-1} = \left(E_k \cdots E_1 A\right) A^{-1} = \left(E_k \cdots E_1\right)\left(A A^{-1}\right) = \left(E_k \cdots E_1\right) I = \left(E_k \cdots E_1\right).$$

Moreover, since

$$A^{-1} = \left(E_k \cdots E_1\right)$$

then

$$A = \left(E_k \cdots E_1\right)^{-1} = E_1^{-1} \cdots E_k^{-1}.$$

We thus have the following results.

Theorem 6

If A is an $n \times n$ invertible matrix, then A can be expressed as the product of elementary matrices.

Theorem 7

A square matrix has an inverse if and only if the determinant of the matrix is nonzero.

1.2.10 The Transpose of a Matrix

If A is an $m \times n$ matrix, the transpose of the matrix A, denoted A^T, is the $n \times m$ matrix given by

$$\left(A^T\right)_{ij} = A_{ji}.$$

One way to view the transpose of the matrix A is as the matrix that is created when the rows of A become the columns of A^T.

Example

If

$$A = \begin{pmatrix} -1 & 4 & 0 \\ 2 & -3 & 6 \end{pmatrix}$$

then

$$A^T = \begin{pmatrix} -1 & 2 \\ 4 & -3 \\ 0 & 6 \end{pmatrix}.$$

Theorem 8

If the sizes of the matrices A and B are so that the operations can be performed, then

(a) $(A^T)^T = A$.
(b) $(A+B)^T = A^T + B^T$.
(c) $(\alpha A)^T = \alpha(A^T)$ for any scalar α.
(d) $(AB)^T = (B^T)(A^T)$.
(e) If A is a square matrix, then $\det(A) = \det(A^T)$.

Proof of Part (d)

Let A be an $m \times k$ matrix and B be an $k \times n$ matrix. Then AB is an $m \times n$ matrix and $(AB)^T$ is an $n \times m$ matrix and

$$\left((AB)^T \right)_{ji} = (AB)_{ij} = \sum_{l=1}^{k} A_{il} B_{lj}.$$

Furthermore, B^T is an $n \times k$ matrix and A^T is a $k \times m$ matrix so $B^T A^T$ is an $n \times m$ matrix with

$$\left[(B^T)(A^T) \right]_{ji} = \sum_{l=1}^{k} (B^T)_{jl} (A^T)_{li} = \sum_{l=1}^{k} B_{lj} A_{il} = \sum_{l=1}^{k} A_{il} B_{lj} = \left((AB)^T \right)_{ji}.$$

Definition

A matrix A is symmetric if $A^T = A$.
   We will see later that symmetric matrices are particularly important in some applications.

Exercises

1. Show that if A, B, and C are $n \times m$ matrices with $AB = I_n$ and $BC = I_n$ then $A = C = B^{-1}$.
2. Find a nonzero 2×2 matrix A for which $A^2 = 0$.
3. Find all the matrices that commute with

$$\begin{pmatrix} 1 & 0 \\ 1 & 1 \end{pmatrix}.$$

4. Show that matrices of the form

$$\begin{pmatrix} a & 0 \\ 0 & a \end{pmatrix}$$

are the only matrices that commute with all 2×2 matrices.

5. For

$$A = \begin{pmatrix} 1 & -3 \\ 4 & 0 \end{pmatrix}, \quad B = \begin{pmatrix} 2 & 0 \\ -1 & 5 \end{pmatrix}$$

find $A - 3B$, BA^T, AB^T, $(AB^T)^T$.

6. A stochastic matrix is a square matrix whose entries are nonnegative and each row sums to 1. Show that if

$$A = \begin{pmatrix} a_{11} & a_{12} \\ a_{21} & a_{22} \end{pmatrix} \quad \text{and} \quad B = \begin{pmatrix} b_{11} & b_{12} \\ b_{21} & b_{22} \end{pmatrix}$$

are stochastic matrices, then AB is a stochastic matrix. Can you generalize this to larger stochastic matrices?

7. What can you say about the matrix A if A^2 is defined.

8. (a) Suppose that

$$A = \begin{pmatrix} a_{11} & a_{12} & a_{13} \\ 0 & 0 & 0 \\ a_{31} & a_{32} & a_{33} \end{pmatrix} \quad B = \begin{pmatrix} b_{11} & b_{12} & b_{13} \\ b_{21} & b_{22} & b_{23} \\ b_{31} & b_{32} & b_{33} \end{pmatrix}.$$

Compute AB and BA.

(b) Suppose that A and B are square matrices of the same dimension and A has a row of all zeroes. Is there anything you can say about AB or BA?

(c) Show that any matrix with a row of all zeroes or a column of all zeroes cannot be invertible.

9. (a) Suppose that

$$A = \begin{pmatrix} a_{11} & a_{12} & a_{13} \\ a_{11} & a_{11} & a_{11} \\ a_{31} & a_{32} & a_{33} \end{pmatrix} \quad B = \begin{pmatrix} b_{11} & 0 & b_{13} \\ b_{21} & 0 & b_{23} \\ b_{31} & 0 & b_{33} \end{pmatrix}.$$

Compute AB and BA.

(b) Suppose that A and B are square matrices of the same dimension and B has a column of all zeroes. Is there anything you can say about AB or BA?

10. (a) Show that if A and B are matrices with $AB = 0$, and A is invertible, then $B = \tilde{0}$.

    (b) Show that if $A, B,$ and C are matrices with $AB = AC$, and A is invertible, then $B = C$.

    (c) Give an example of nonzero matrices A, B, and C for which $AB = AC$ but $B \neq C$.

11. Let

$$A = \begin{pmatrix} 1 & 2 \\ 3 & 4 \end{pmatrix} \quad B = \begin{pmatrix} 5 & 6 \\ 7 & 8 \end{pmatrix}.$$

    (a) Compute $(A + B)^2$ and $A^2 + 2AB + B^2$.

    (b) Give a necessary and sufficient condition that will ensure

$$\left(A + B\right)^2 = A^2 + 2AB + B^2.$$

12. Let

$$A = \begin{pmatrix} 0 & a & b & c \\ 0 & 0 & d & e \\ 0 & 0 & 0 & f \\ 0 & 0 & 0 & 0 \end{pmatrix}.$$

    Find A^2, A^3, and A^4. If A is an upper triangular $n \times n$ matrix with all zeros on the main diagonal, describe the structure of $A^2, A^3, \ldots, A^n$.

13. If A is an $n \times n$ matrix, C is an $m \times m$ matrix, and B is a matrix so that $A \times B \times C$ is defined, what are the dimensions of B?

14. Let

$$A = \begin{pmatrix} a & b \\ c & d \end{pmatrix}$$

    and suppose that A commutes with the matrix

$$C = \begin{pmatrix} 0 & -1 \\ 1 & 0 \end{pmatrix}.$$

    (a) What can you say about the relationships of the elements of A?

    (b) If A and B each commute with C, show that A commutes with B.

15. Let

$$A = \begin{pmatrix} a_{11} & 0 & 0 \\ a_{21} & a_{22} & 0 \\ a_{31} & a_{32} & a_{33} \end{pmatrix} \quad B = \begin{pmatrix} b_{11} & 0 & 0 \\ b_{21} & b_{22} & 0 \\ b_{31} & b_{32} & b_{33} \end{pmatrix}.$$

(a) Find AB.

(b) Why is it true that $(AB)_{23} = 0$?

(c) Show that the product of two lower triangular matrices is lower triangular.

16. Find a matrix B so that $3A - 2B = C$ if

$$A = \begin{pmatrix} 1 & -2 \\ 7 & 3 \\ -5 & 0 \end{pmatrix} \quad \text{and} \quad C = \begin{pmatrix} 6 & 1 \\ 0 & 3 \\ 2 & -4 \end{pmatrix}.$$

17. Show that if A is an $n \times m$ matrix, then $I_n A = A$ and $A I_m = A$.

18. Which of the matrices has an inverse?

$$\begin{pmatrix} 1 & 0 & 7 & 8 \\ 0 & 3 & 5 & 9 \\ 0 & 0 & -4 & 0 \\ 0 & 0 & 0 & 2 \end{pmatrix}, \quad \begin{pmatrix} 1 & 9 & 2 & 1 \\ 2 & 4 & 0 & 2 \\ 8 & -1 & -6 & 8 \\ 9 & 5 & 5 & 9 \end{pmatrix}, \quad \begin{pmatrix} 1 & 1 & 5 & 7 \\ 0 & 4 & 8 & 0 \\ 0 & 0 & 9 & 3 \\ 0 & 0 & 0 & 0 \end{pmatrix}$$

19. (a) Find the determinant of

$$A = \begin{pmatrix} a & b & c \\ a & b & c \\ d & e & f \end{pmatrix}$$

by expanding along the first column.

(b) Find the determinant of

$$A = \begin{pmatrix} a & b & c \\ d & e & f \\ a & b & c \end{pmatrix}$$

by expanding along the first column and using the answer to part (a).

(c) Make a conjecture about the determinant of a square matrix that has two identical rows.

(d) Make a conjecture about the determinant of a square matrix that has two identical columns.

20. Show that if A and B are $n \times n$ matrices, then AB is invertible if and only if A and B are each invertible.

21. Suppose that A and B are 2×2 symmetric matrices. Show that AB is symmetric if and only if $AB = BA$.

22. Show that if A is an $n \times n$ matrix and if

$$A^3 + 4A^2 + 5A - 2I_n = 0_{n \times n}$$

then A^{-1} exists.

23. (a) Let

$$A = \begin{pmatrix} 1 & 1/2 \\ a & b \end{pmatrix}.$$

Find a and b for which $A^2 = 0$.

(b) Let

$$A = \begin{pmatrix} 1 & 1/n \\ a & b \end{pmatrix}.$$

Find a and b for which $A^2 = 0$.

24. (a) Find the values of a for which

$$\begin{pmatrix} a & a \\ a & a \end{pmatrix}^2 = \begin{pmatrix} a & a \\ a & a \end{pmatrix}.$$

(b) Find the values of a for which

$$\begin{pmatrix} a & a & a \\ a & a & a \\ a & a & a \end{pmatrix}^2 = \begin{pmatrix} a & a & a \\ a & a & a \\ a & a & a \end{pmatrix}.$$

(c) Make a conjecture about the values of a for which the $n \times n$ matrix A, all of whose entries are A satisfies $A^2 = A$.

Can you prove your conjecture is true?

25. If A and B are $n \times n$ matrices, the commutator of A and B, denoted $[A, B]$, is $AB - BA$.

(a) Show that if A and B are 2×2 matrices then the trace of $[A, B]$ is 0.

(b) Show that if A is a 2×2 matrix whose trace is 0 then $A^2 = cI_2$ for some number c.

26. Which matrices below are elementary matrices?

(a) $\begin{pmatrix} 1 & 0 & 2 \\ 0 & 1 & 0 \\ 0 & 0 & 1 \end{pmatrix}$

(b) $\begin{pmatrix} 1 & 0 & -\sqrt{3} \\ 0 & 2 & 0 \\ 0 & 0 & 1 \end{pmatrix}$

(c) $\begin{pmatrix} 2 & 0 & 2 \\ 0 & 1 & 0 \\ 0 & 0 & 1 \end{pmatrix}$

(d) $\begin{pmatrix} -2 & 1 \\ 0 & 1 \end{pmatrix}$

27. Multiply each matrix below by the appropriate elementary matrix to return it to the identity matrix.

(a) $\begin{pmatrix} 1 & 0 & 0 \\ 0 & 3 & 0 \\ 0 & 0 & 1 \end{pmatrix}$

(b) $\begin{pmatrix} 1 & -2 & 0 \\ 0 & 1 & 0 \\ 0 & 0 & 1 \end{pmatrix}$

(c) $\begin{pmatrix} 0 & 1 & 0 \\ 1 & 0 & 0 \\ 0 & 0 & 1 \end{pmatrix}$

28. Give the elementary matrix for which multiplying by that matrix gives the result shown.

(a) $E\begin{pmatrix} 1 & -2 & 0 \\ 3 & -1 & 4 \\ 1 & 2 & 1 \end{pmatrix} = \begin{pmatrix} 1 & 2 & 1 \\ 3 & -1 & 4 \\ 1 & -2 & 0 \end{pmatrix}$

(b) $E\begin{pmatrix} 1 & -2 & 0 \\ 3 & -1 & 4 \\ 1 & 2 & 1 \end{pmatrix} = \begin{pmatrix} 1 & -2 & 0 \\ 0 & -7 & 1 \\ 1 & 2 & 1 \end{pmatrix}$

(c) $E\begin{pmatrix} 1 & -2 & 0 \\ 3 & -1 & 4 \\ 1 & 2 & 1 \end{pmatrix} = \begin{pmatrix} 3 & -6 & 0 \\ 3 & -1 & 4 \\ 1 & 2 & 1 \end{pmatrix}$

29. (a) Find

$$\begin{pmatrix} 1 & -2 & 0 \\ 3 & -1 & 4 \\ 1 & 2 & 1 \end{pmatrix}\begin{pmatrix} 1 & 0 & 0 \\ 0 & 0 & 1 \\ 0 & 1 & 0 \end{pmatrix}.$$

(b) Describe the result when a matrix is multiplied on the right by an elementary matrix.

30. Find the inverse of the matrices below by multiplying the matrix by elementary matrices until the identity is achieved.

(a) $\begin{pmatrix} 2 & 4 \\ 1 & 3 \end{pmatrix}$

(b) $\begin{pmatrix} 0 & -2 \\ 2 & 5 \end{pmatrix}$

(c) $\begin{pmatrix} 1 & 4 \\ 5 & 5 \end{pmatrix}$

(d) $\begin{pmatrix} 1 & -1 & 4 \\ 0 & 3 & 6 \\ -2 & 1 & 0 \end{pmatrix}$

(e) $\begin{pmatrix} 1 & 0 & 4 \\ -2 & 1 & 3 \\ 1 & 2 & 5 \end{pmatrix}$

(f) $\begin{pmatrix} 0 & 2 & 1 \\ 3 & 3 & 0 \\ 1 & 0 & 4 \end{pmatrix}$

(g) $\begin{pmatrix} a & 1 & 0 \\ 0 & b & 1 \\ 0 & 0 & c \end{pmatrix}$ where $abc \neq 0$

31. If A is a matrix and c is a scalar with $cA = 0$, show that $A = 0$ or $c = 0$.

1.3 The *LU* Decomposition of a Square Matrix (Optional)

The material in this section will not be used in later sections and is optional.

In some applications, it may reduce the complexity of the computations to express a square matrix as the product of a lower triangular matrix L and an upper triangular matrix U. This is called the *LU* decomposition of the matrix. In this section, we demonstrate how to accomplish this factorization.

Suppose that A is a $n \times n$ matrix. Our strategy is to apply elementary row operations to A until the result is an upper triangular matrix. The elementary row operations are accomplished by multiplication by elementary matrices, each of which is lower triangular. Suppose these matrices are $E_1, \ldots, E_k$ where the subscript denotes the order in which they are applied. We then have

$$E_k \cdots E_1 A = U$$

so

$$A = E_1^{-1} \cdots E_k^{-1} U.$$

We have shown in the exercises that the product of lower triangular matrices is a lower triangular matrix. Each E_i is lower triangular, so each E_i^{-1} is lower triangular, and thus

$$E_1^{-1} \cdots E_k^{-1} = L$$

is lower triangular.

Example

Let

$$A = \begin{pmatrix} 1 & 2 & -2 \\ 2 & -4 & 6 \\ 2 & 16 & 8 \end{pmatrix}.$$

1. Add (-2) times the first row of A to the second row of A. This is done by multiplying A on the left by

$$E_1 = \begin{pmatrix} 1 & 0 & 0 \\ -2 & 1 & 0 \\ 0 & 0 & 1 \end{pmatrix}.$$

Now

$$E_1 A = \begin{pmatrix} 1 & 2 & -2 \\ 0 & -8 & 10 \\ 2 & 16 & 8 \end{pmatrix}.$$

2. For the matrix found in step 1, Add (-2) times the first row of $E_1 A$ to the third row of $E_1 A$. This is done by multiplying $E_1 A$ on the left by

$$E_2 = \begin{pmatrix} 1 & 0 & 0 \\ 0 & 1 & 0 \\ -2 & 0 & 1 \end{pmatrix}.$$

Now

$$E_2 E_1 A = \begin{pmatrix} 1 & 2 & -2 \\ 0 & -8 & 10 \\ 0 & 12 & 12 \end{pmatrix}$$

3. For the matrix found in step 2, Add (1.5) times the second row of E_2E_1A to the third row of $E_2 E_1A$. This is done by multiplying E_2E_1A on the left by

$$E_3 = \begin{pmatrix} 1 & 0 & 0 \\ 0 & 1 & 0 \\ 0 & 1.5 & 1 \end{pmatrix}.$$

Now

$$E_3E_2E_1A = \begin{pmatrix} 1 & 2 & -2 \\ 0 & -8 & 10 \\ 0 & 0 & 27 \end{pmatrix}$$

which is an upper trianglular matrix.
    We thus have

$$A = E_1^{-1}E_2^{-1}E_3^{-1} \begin{pmatrix} 1 & 2 & -2 \\ 0 & -8 & 10 \\ 0 & 0 & 27 \end{pmatrix}.$$

4. Let

$$L = E_1^{-1}E_2^{-1}E_3^{-1} = \begin{pmatrix} 1 & 0 & 0 \\ 2 & 1 & 0 \\ 2 & -3/2 & 1 \end{pmatrix}; \quad U = \begin{pmatrix} 1 & 2 & -2 \\ 0 & -8 & 10 \\ 0 & 0 & 27 \end{pmatrix}.$$

Then

$$LU = \begin{pmatrix} 1 & 0 & 0 \\ 2 & 1 & 0 \\ 2 & -3/2 & 1 \end{pmatrix}\begin{pmatrix} 1 & 2 & -2 \\ 0 & -8 & 10 \\ 0 & 0 & 27 \end{pmatrix} = \begin{pmatrix} 1 & 2 & -2 \\ 2 & -4 & 6 \\ 2 & 16 & 8 \end{pmatrix} = A.$$

We note that this method does not always work in the original form of the matrix. An example where it does not work is a matrix whose first row is all zeros. In this case, one must permute the rows to arrive at a form where the method will work. It is common to denote the matrix that permutes the rows as P, and one then finds the LU decomposition of PA. Computer programs will do the permutation automatically (sometimes, even when it is unnecessary).

In the next chapter, we study how to solve certain systems of equations using a process called Gaussian elimination. The LU decomposition can provide a more efficient method of solving systems of linear equations than Gaussian elimination in that fewer steps are involved.

The website https://en.wikipedia.org/wiki/LU_decomposition has many enhancements to this process.

Exercises

In Exercises 1–8, find the LU decomposition of the given matrices

1. $\begin{pmatrix} 1 & 3 & 6 \\ 2 & 9 & -2 \\ 3 & 6 & 5 \end{pmatrix}$

2. $\begin{pmatrix} 2 & 4 & 6 \\ 1 & 3 & 5 \\ -1 & 1 & 4 \end{pmatrix}$

3. $\begin{pmatrix} 3 & 1 & 4 \\ 6 & 1 & 8 \\ 0 & 8 & 5 \end{pmatrix}$

4. $\begin{pmatrix} 1 & 2 & 2 \\ 3 & 4 & 8 \\ 2 & 12 & 16 \end{pmatrix}$

5. $\begin{pmatrix} 1 & 2 & -4 \\ 2 & -5 & 8 \\ 4 & 6 & 9 \end{pmatrix}$

6. $\begin{pmatrix} 1 & 7 & -5 \\ 3 & -2 & 1 \\ -4 & 1 & 0 \end{pmatrix}$

7. $\begin{pmatrix} 2 & 2 & 0 & 1 \\ 4 & 0 & -3 & 5 \\ 1 & 3 & 3 & 6 \\ -3 & 5 & 7 & 0 \end{pmatrix}$

8. $\begin{pmatrix} 4 & 4 & 6 & 3 \\ 3 & 1 & -2 & -5 \\ 1 & 2 & -4 & 0 \\ 2 & -1 & 0 & 7 \end{pmatrix}$

2

Systems of Linear Equations

In this chapter, we discuss finding the solution of a system of linear equations. While this is an important topic in itself, the techniques we learn here will carry over to many other types of problems.

2.1 Basic Definitions

Definition

A linear equation in the variables $x_1, \dots, x_n$ is an expression of the form

$$a_1 x_1 + \cdots + a_n x_n = b,$$

where $a_1, \dots, a_n$ and b are constants.

A solution to the equation

$$a_1 x_1 + \cdots + a_n x_n = b$$

is an ordered n-tuple of numbers $(s_1, \dots, s_n)$ for which

$$a_1 s_1 + \cdots + a_n s_n = b.$$

The solution set to the equation

$$a_1 x_1 + \cdots + a_n x_n = b$$

is the set of all solutions to the equation.

A system of linear equations in the variables $x_1, \dots, x_n$ is a finite set of linear equations of the form

$$a_{11} x_1 + \cdots + a_{1n} x_n = b_1$$
$$a_{21} x_1 + \cdots + a_{2n} x_n = b_2$$
$$\vdots$$
$$a_{m1} x_1 + \cdots + a_{mn} x_n = b_m$$

A solution to the system of equations above is an ordered n-tuple of numbers $(s_1, \dots, s_n)$ that is a solution to each of the equations in the system. The set of all possible solutions is called the solution set of the system.

Exercises

In Problems 1 through 3, determine whether the given values form a solution to the system of equations

1. $2x + y - z = 5$
   $x + y = 2$
   $2y - z = 1$
   $x = 2, \quad y = 0, \quad z = -1.$

2. $x - y = 4$
   $2x + 3y = 5$
   $x = \dfrac{17}{5}, \quad y = -\dfrac{3}{5}.$

3. $x - y + 2z = 0$
   $3x - 2y = 4$
   $y + z = 3$
   $x = 1, \quad y = -1, \quad z = -1.$

4. Which of the following are linear equations? If an equation is not a linear equation, tell why.

   (a) $3x - xy + 2z = 7$

   (b) $3a - 4b = \sqrt{7}$

   (c) $\sin(30°)x - 4y - 7z = 0$

   (d) $w - \sqrt{5}x + \dfrac{4}{z} = 0$

   (e) $4\sin x + y^2 + 3^z = -5$

   (f) $3w - 2y = 6z = -8$

2.2 Solving Systems of Linear Equations (Gaussian Elimination)

The solution set of a system of linear equations has an important property that is not necessarily shared by equations that are not linear, namely, there is either

(i) no solution

(ii) exactly one solution

(iii) infinitely many solutions.

We will see why this is the case later in this chapter.

Definition

A system of linear equations that has either exactly one solution or infinitely many solutions is said to be consistent. A system of linear equations that has no solutions is inconsistent.

We give an example of each possibility.

Example

For the system

$$2x - 3y = 5$$
$$x + 3y = -2,$$

there is exactly one solution, namely, $x = 1, y = -1$.
  The graph of the system of equations

$$2x - 3y = 5$$
$$3x - y = 4$$
$$x + y = 1$$

is shown in Figure 2.1a. Since there is no point where the graphs of all three equations meet, there is no solution to the system of equations. The graph of the system of equations

$$2x - 3y = 5$$
$$x + 3y = -2$$
$$5x + 4y = 1$$

is shown in Figure 2.1b. There is a single solution to this system because there is one point of intersection of all three equations.
  The system of equations

$$x + y = 3$$
$$x + y = 4$$

obviously has no solution. The graph is show in Figure 2.2a. The system of equations

$$x + y = 3$$
$$2x + 2y = 6$$

has infinitely many solutions. The graph is shown in Figure 2.2b.

A special and important class of systems of linear equations that always have at least one solution is the class of homogeneous equations. These are systems where every $b_i = 0$. One solution to such a system will always be when each variable is 0. This means for a system of two or three variables, the graph of each equation in a homogeneous system passes through the origin.

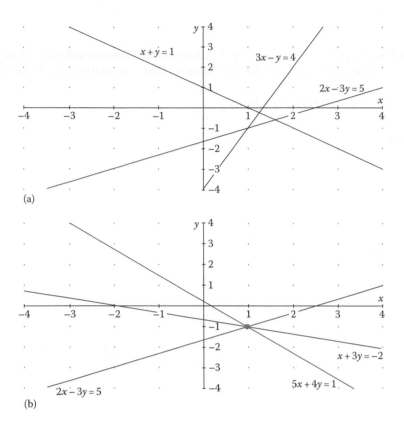

FIGURE 2.1
In (a) the system is inconsistent, in (b) the system has exactly one solution.

2.2.1 Solving Systems of Linear Equations

The most common way of solving a system of linear equations is to convert the given system to an equivalent system where the solution to the equivalent system is obvious. By "an equivalent system," we mean that the two systems have the same variables and the same solution set.

The conversion is accomplished by applying the following operations that do not affect the solution set of a system of equations:

(i) Multiplying an equation by a nonzero number. The equation

$$3x + 4y = 7$$

is equivalent to

$$6x + 8y = 14.$$

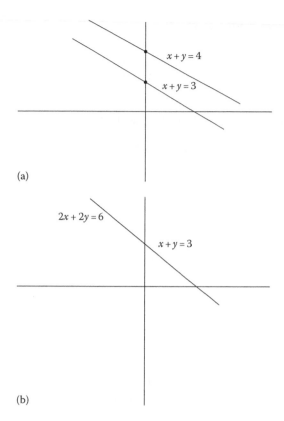

(a)

(b)

FIGURE 2.2
In (a) the system is inconsistent, in (b) the system has infinitely many solutions.

(ii) Interchanging the order of equations. The system of equations

$$2x - 4y + 3z = 5$$
$$9x + 12y + z = -10$$

is equivalent to the system

$$9x + 12y + z = -10$$
$$2x - 4y + 3z = 5.$$

(iii) Replacing one equation by the sum or difference of itself and a multiple of another equation. In the system of equations

$$3x + 2y = 10$$
$$x - y = 4,$$

we replace the second equation by the sum of the first plus twice the second to get

$$3x + 2y = 10$$
$$5x = 18,$$

which is an equivalent system that is easier to solve.

Notice the parallel of these conditions to elementary row operations.

Example

We apply these operations to a system of equations to yield an equivalent system whose solution is obvious. Consider

$$3x + 4y - 2z = 5$$
$$x - 2y + z = 0$$
$$7x + 4y - 3z = 3$$

Interchange the first and second equations to get

$$x - 2y + z = 0$$
$$3x + 4y - 2z = 5$$
$$7x + 4y - 3z = 3$$

Replace the second equation by the second equation minus three times the first equation to get

$$x - 2y + z = 0$$
$$0x + 10y - 5z = 5$$
$$7x + 4y - 3z = 3$$

Replace the third equation by the sum of the third equation and minus seven times the first equation to get

$$x - 2y + z = 0$$
$$0x + 10y - 5z = 5$$
$$0x + 18y - 10z = 3$$

Note that now the variable x appears in only the first equation.
    Multiply the second equation by $1/5$ to get

$$x - 2y + z = 0$$
$$0x + 2y - z = 1$$
$$0x + 18y - 10z = 3$$

Replace the first equation by the sum of the first equation plus the second equation to get

$$x + 0y + 0z = 1$$
$$0x + 2y - z = 1$$
$$0x + 18y - 10z = 3$$

We now know that $x = 1$.
  Replace the third equation by the sum of the third equation minus nine times the second equation to get

$$x + 0y + 0z = 1$$
$$0x + 2y - z = 1$$
$$0x + 0y - 1z = -6$$

We now know that $z = 6$.
  Multiply the third equation by -1 to get

$$x + 0y + 0z = 1$$
$$0x + 2y - z = 1$$
$$0x + 0y + z = 6$$

Replace the second equation by the sum of the second equation and the third equation to get

$$x + 0y + 0z = 1$$
$$0x + 2y + 0z = 7$$
$$0x + 0y + z = 6$$

Multiply the second equation by $1/2$ to get

$$x + 0y + 0z = 1$$
$$0x + y + 0z = 7/2$$
$$0x + 0y + z = 6$$

One typically does not list the terms whose coefficients are 0, so we have

$$x = 1$$
$$y = 7/2$$
$$z = 6$$

The procedure that we have just executed is an example of Gaussian elimination, which is one of the most commonly used methods of solving systems of linear equations. A modified version of Gaussian elimination is what most computer systems and calculators use to solve systems of linear equations.

2.2.2 Using Technology to Accomplish Gaussian Elimination

The operations we did earlier can be done by a computer, which will be our approach in most cases. In order to give the information in a form that is easy to communicate with a computer, it is convenient to use matrices. This also simplifies the notation in solving the system by hand.

Instructions to use the computer program MATLAB®, which performs the computations of linear algebra, are given in the appendix.

Consider the system of equations in the last example. We associate the system of linear equations with the matrix that consists of the coefficients of the variables on the left columns of the matrix and the numbers on the right column for each equation. Thus, we have

$$\begin{matrix} 3x+4y-2z=5 \\ x-2y+z=0 \\ 7x+4y-3z=3 \end{matrix} \leftrightarrow \left(\begin{array}{ccc|c} 3 & 4 & -2 & 5 \\ 1 & -2 & 1 & 0 \\ 7 & 4 & -3 & 3 \end{array}\right). \tag{2.1}$$

The matrix on the right in expression (2.1) is called the augmented matrix for the system of equations. The vertical line is simply to help separate the coefficients of the variables from the numbers on the right side of the equations and will not normally appear.

There is also the matrix of coefficients associated with a system of linear equations that is occasionally used. In expression (2.1), this is

$$\begin{pmatrix} 3 & 4 & -2 \\ 1 & -2 & 1 \\ 7 & 4 & -3 \end{pmatrix}.$$

Representing systems of linear equations by matrices is advantageous for computer solutions of the systems.

Exercises

In Exercises 1 through 3, find the augmented matrix for the systems of equations

1. $3x+5y=9$
   $x-2y=-7$
   $y=5$
2. $x-y+3z=6$
   $2x+4y-3z=2$
   $x+y+z=0$
   $3y-2z=-8$
3. $x+y+z=0$
   $x+y+z=3$
   $x+2y+z=5$

In Exercises 4 through 6, find the systems of equations associated with the augmented matrix. Denote the variables as $x_1, x_2, \ldots$

4. $\begin{pmatrix} 3 & -1 & 2 & 0 \\ 0 & 4 & 1 & 6 \\ -2 & 5 & 1 & -8 \end{pmatrix}$

5. $\begin{pmatrix} 8 & 3 & 0 & 6 \\ -4 & 2 & 1 & 2 \end{pmatrix}$

6. $\begin{pmatrix} 1 & 0 & 4 & 2 & 6 \\ 3 & 2 & 1 & -7 & -3 \\ 0 & 0 & 2 & 0 & 2 \\ -2 & 4 & -1 & 5 & 0 \end{pmatrix}$

2.3 Equivalent Systems of Linear Equations

The manipulations of a system of equations that yield an equivalent system of equations have corresponding manipulations with the associated matrix. These manipulations correspond to multiplying the augmented matrix by a particular elementary matrix. In particular,

(i) Multiplying an equation by a nonzero number corresponds to multiplying the corresponding row of the matrix by the same nonzero number.

(ii) Interchanging the order of equations corresponds to interchanging the corresponding rows of the matrix.

(iii) Replacing one equation by the sum of itself and a multiple of another equation corresponds to replacing the corresponding row by the sum of itself and the multiple of another row.

In Chapter 1, we saw that these operations can be done by multiplying the given matrix on the left by an elementary matrix.

Definition

The operations on a matrix given earlier are called elementary row operations.

Definition

If a matrix B can be obtained from matrix A by applying elementary row operations to matrix A, then A and B are said to be row equivalent.

From what has been noted earlier about elementary row operations and systems of linear equations, we have the following result.

Theorem 1

Two systems of linear equations that are expressed by matrices that are row equivalent have the same solution set.

Example

We redo the previous example, showing how the associated matrices are changed. We have

$$
\begin{matrix}
3x + 4y - 2z = 5 \\
x - 2y + z = 0 \\
7x + 4y - 3z = 3
\end{matrix}
\leftrightarrow
\begin{pmatrix}
3 & 4 & -2 & 5 \\
1 & -2 & 1 & 0 \\
7 & 4 & -3 & 3
\end{pmatrix}
$$

$$
\begin{matrix}
x - 2y + z = 0 \\
3x + 4y - 2z = 5 \\
7x + 4y - 3z = 3
\end{matrix}
\leftrightarrow
\begin{pmatrix}
1 & -2 & 1 & 0 \\
3 & 4 & -2 & 5 \\
7 & 4 & -3 & 3
\end{pmatrix}
$$

$$
\begin{matrix}
x - 2y + z = 0 \\
0x + 10y - 5z = 5 \\
7x + 4y - 3z = 3
\end{matrix}
\leftrightarrow
\begin{pmatrix}
1 & -2 & 1 & 0 \\
0 & 10 & -5 & 5 \\
7 & 4 & -3 & 3
\end{pmatrix}
$$

$$
\begin{matrix}
x - 2y + z = 0 \\
0x + 10y - 5z = 5 \\
0x + 18y - 10z = 3
\end{matrix}
\leftrightarrow
\begin{pmatrix}
1 & -2 & 1 & 0 \\
0 & 10 & -5 & 5 \\
0 & 18 & -10 & 3
\end{pmatrix}
$$

$$
\begin{matrix}
x - 2y + z = 0 \\
0x + 2y - z = 1 \\
0x + 18y - 10z = 3
\end{matrix}
\leftrightarrow
\begin{pmatrix}
1 & -2 & 1 & 0 \\
0 & 2 & -1 & 1 \\
0 & 18 & -10 & 3
\end{pmatrix}
$$

$$
\begin{matrix}
x + 0y + 0z = 1 \\
0x + 2y - z = 1 \\
0x + 18y - 10z = 3
\end{matrix}
\leftrightarrow
\begin{pmatrix}
1 & 0 & 0 & 1 \\
0 & 2 & -1 & 1 \\
0 & 18 & -10 & 3
\end{pmatrix}
$$

$$
\begin{matrix}
x + 0y + 0z = 1 \\
0x + 2y - z = 1 \\
0x + 0y - 1z = -6
\end{matrix}
\leftrightarrow
\begin{pmatrix}
1 & 0 & 0 & 1 \\
0 & 2 & -1 & 1 \\
0 & 0 & -1 & -6
\end{pmatrix}
$$

$$
\begin{matrix}
x + 0y + 0z = 1 \\
0x + 2y - z = 1 \\
0x + 0y + z = 6
\end{matrix}
\leftrightarrow
\begin{pmatrix}
1 & 0 & 0 & 1 \\
0 & 2 & -1 & 1 \\
0 & 0 & 1 & 6
\end{pmatrix}
$$

$$
\begin{matrix}
x + 0y + 0z = 1 \\
0x + 2y + 0z = 7 \\
0x + 0y + z = 6
\end{matrix}
\leftrightarrow
\begin{pmatrix}
1 & 0 & 0 & 1 \\
0 & 2 & 0 & 7 \\
0 & 0 & 1 & 6
\end{pmatrix}
$$

$$
\begin{matrix}
x + 0y + 0z = 1 \\
0x + y + 0z = 7/2 \\
0x + 0y + z = 6
\end{matrix}
\leftrightarrow
\begin{pmatrix}
1 & 0 & 0 & 1 \\
0 & 1 & 0 & 7/2 \\
0 & 0 & 1 & 6
\end{pmatrix}.
$$

For a procedure like this one, we will almost always use computer technology. The typical way that we will solve systems of linear equations is to enter the augmented matrix and command the computer to put the matrix into row reduced form. We must then interpret the output.

Interpreting the output to get a solution is simple when there is exactly one solution as there was in the example. Our next task is to understand the output when there is no solution or there are infinitely many solutions.

2.3.1 Row Reduced Form of a Matrix

The row reduced form of a system of linear equations is the simplest form from which to interpret the solution for a system of linear equations.

A matrix in row reduced form has the following characteristics:

Reading from left to right, the first nonzero entry in each row will be a 1. The first 1 in each row is called a leading 1. The corresponding variable is called a leading variable.

A leading 1 is the only nonzero entry in its column.

A leading 1 in a row is always strictly to the right of a leading 1 of a row above it.

Any rows that are all 0 will appear as the bottom rows of the row reduced form.

The next two matrices are in row reduced form.

$$\begin{pmatrix} 0 & 1 & 3 & 0 & 2 & 0 & 6 \\ 0 & 0 & 0 & 1 & 0 & 5 & 3 \\ 0 & 0 & 0 & 0 & 0 & 0 & 0 \end{pmatrix}$$

If the variables for this matrix are x_1, x_2, x_3, x_4, x_5, and x_6, then the leading variables are x_2 and x_4.

$$\begin{pmatrix} 1 & 0 & 0 & 5 & 0 \\ 0 & 1 & 0 & 4 & 0 \\ 0 & 0 & 1 & 6 & 0 \\ 0 & 0 & 0 & 0 & 0 \end{pmatrix}$$

If the variables for this matrix are x_1, x_2, x_3, and x_4, then the leading variables are x_1, x_2, and x_3. The following matrices are not in row reduced form:

$$\begin{pmatrix} 0 & 1 & 2 & 0 & 3 \\ 1 & 0 & 0 & 0 & 9 \\ 0 & 0 & 0 & 1 & 5 \end{pmatrix}$$

$$\begin{pmatrix} 1 & 1 & 0 & 0 & 3 \\ 0 & 0 & 2 & 0 & 9 \\ 0 & 0 & 0 & 1 & 5 \end{pmatrix}$$

$$\begin{pmatrix} 1 & 0 & 0 & 5 & 0 \\ 0 & 1 & 0 & 1 & 0 \\ 0 & 0 & 0 & 0 & 0 \\ 0 & 0 & 0 & 0 & 1 \end{pmatrix}$$

Exercises

1. Argue that an $n \times n$ matrix in row reduced form either has a row of all 0's or is the identity matrix.

2. Which of the following matrices are in row reduced form?

(a) $\begin{pmatrix} 1 & 0 & 0 & 2 & 3 \\ 0 & 0 & 1 & 0 & 6 \\ 0 & 1 & 0 & 3 & 5 \\ 0 & 0 & 0 & 0 & 0 \end{pmatrix}$

(b) $\begin{pmatrix} 1 & 3 & 0 & 2 & 0 \\ 0 & 0 & 1 & 7 & 0 \\ 0 & 0 & 0 & 0 & 1 \\ 0 & 0 & 0 & 0 & 0 \end{pmatrix}$

(c) $\begin{pmatrix} 0 & 1 & 0 & 4 & 0 \\ 0 & 0 & 0 & 0 & 1 \\ 0 & 0 & 0 & 0 & 0 \\ 0 & 0 & 0 & 0 & 0 \end{pmatrix}$

(d) $\begin{pmatrix} 0 & 0 & 0 & 1 & 0 \\ 0 & 1 & 2 & 0 & 0 \\ 0 & 0 & 0 & 0 & 0 \\ 1 & 0 & 0 & 0 & 0 \end{pmatrix}$

(e) $\begin{pmatrix} 1 & 0 & 0 & 0 & 0 \\ 0 & 1 & 0 & 0 & 0 \\ 0 & 0 & 0 & 1 & 0 \\ 1 & 0 & 0 & 0 & 0 \end{pmatrix}$

2.4 Expressing the Solution of a System of Linear Equations

2.4.1 Systems of Linear Equations That Have No Solutions

The system of equations

$$3x + 4y + 5z = 5$$
$$3x + 4y + 5z = 9$$

has no solution. The augmented matrix for the system is

$$\begin{pmatrix} 3 & 4 & 5 & 5 \\ 3 & 4 & 5 & 9 \end{pmatrix}$$

which, when row reduced, is

$$\begin{pmatrix} 1 & 4/3 & 5/3 & 0 \\ 0 & 0 & 0 & 1 \end{pmatrix}.$$

The row $\begin{pmatrix} 0 & 0 & 0 & 1 \end{pmatrix}$ corresponds to the equation

$$0x + 0y + 0z = 1,$$

which is impossible.

A system of linear equations has no solution if and only if one of the rows in the row reduced matrix of the augmented matrix for the system is $\begin{pmatrix} 0 & 0 & \cdots & 0 & 1 \end{pmatrix}$.

2.4.2 Systems of Linear Equations That Have Exactly One Solution

A system of linear equations will have exactly one solution if and only if each variable is a leading variable and there is no row of the form $\begin{pmatrix} 0 & 0 & \cdots & 0 & 1 \end{pmatrix}$ when the system is in row reduced form.

2.4.3 Systems of Linear Equations That Have Infinitely Many Solutions

Example

Consider the system of equations

$$3x_1 - 2x_3 + x_4 + 6x_5 - x_6 = 3$$
$$x_1 + 5x_2 - 2x_3 - 3x_4 + x_5 + 2x_6 = 10.$$

The augmented matrix and the row reduced matrix for the system are

$$\begin{pmatrix} 3 & 0 & -2 & 1 & 6 & -1 & 3 \\ 1 & 5 & -2 & -3 & 1 & 2 & 10 \end{pmatrix} \rightarrow \begin{pmatrix} 1 & 0 & \dfrac{-2}{3} & \dfrac{1}{3} & 2 & \dfrac{-1}{3} & 1 \\ 0 & 1 & \dfrac{-4}{15} & \dfrac{-2}{3} & \dfrac{-1}{5} & \dfrac{7}{15} & \dfrac{9}{5} \end{pmatrix}$$

There is a solution because there is no row $(0 \quad 0 \quad 0 \quad 0 \quad 0 \quad 0 \quad 1)$.

Because not every variable is a leading variable, there are infinitely many solutions as we will show in more detail below. This, together with the previous cases we have seen, is why a linear system has no solutions, exactly one solution, or infinitely many solutions. Note that when there are more variables than equations, there will be at least one variable that is not a leading variable, and so a consistent system will have infinitely many solutions.

In the case where there are infinitely many solutions, it is sometimes helpful to convert from the row reduced form of the matrix back to the form where the variables are explicitly listed. In this case, we have

$$x_1 - \frac{2}{3}x_3 + \frac{1}{3}x_4 + 2x_5 - \frac{1}{3}x_6 = 1$$
$$x_2 - \frac{4}{15}x_3 - \frac{2}{3}x_4 - \frac{1}{5}x_5 + \frac{7}{15}x_6 = \frac{9}{5}.$$

When there are infinitely many solutions, there is a protocol for expressing the solution in standard form, which we now describe.

The variables that not leading variables are called free variables. In this example, x_1 and x_2 are leading variables and $x_3, x_4, x_5,$ and x_6 are free variables. Any time there are free variables in a consistent system, there will be infinitely many solutions.

The first step to convert to the standard form of the solution is to express each leading variable in terms of the free variables. In this case, we have

$$x_1 = \frac{2}{3}x_3 - \frac{1}{3}x_4 - 2x_5 + \frac{1}{3}x_6 + 1$$
$$x_2 = \frac{4}{15}x_3 + \frac{2}{3}x_4 + \frac{1}{5}x_5 - \frac{7}{15}x_6 + \frac{9}{5}.$$

We generate a solution by choosing any values for the free variables, and those values will determine the values of the leading variables. For example, if we take

$$x_3 = 15, \quad x_4 = 3, \quad x_5 = 5, \quad \text{and} \quad x_6 = 15$$

then

$$x_1 = \frac{2}{3}(15) - \frac{1}{3}(3) - 2(5) + \frac{1}{3}(15) + 1 = 10 - 1 - 10 + 5 + 1 = 5.$$
$$x_2 = \frac{4}{15}(15) + \frac{2}{3}(3) + \frac{1}{5}(5) - \frac{7}{15}(15) + \frac{9}{5} = 4 + 2 + 1 - 7 + \frac{9}{5} = \frac{9}{5}.$$

and one can check that

$$x_1 = 5, \quad x_2 = \frac{9}{5}, \quad x_3 = 15, \quad x_4 = 3, \quad x_5 = 5, \quad \text{and} \quad x_6 = 15$$

is a solution to

$$3x_1 - 2x_3 + x_4 + 6x_5 - x_6 = 3$$
$$x_1 + 5x_2 - 2x_3 - 3x_4 + x_5 + 2x_6 = 10.$$

The preferred format for the infinite solutions case is to assign a parameter to each free variable. In the present example, we let

$$x_3 = r, \quad x_4 = s, \quad x_5 = t, \quad x_6 = u$$

and then express each leading variable in terms of the free variables. Here, we have

$$x_1 = \frac{2}{3}r - \frac{1}{3}s - 2t + \frac{1}{3}u + 1$$
$$x_2 = \frac{4}{15}r + \frac{2}{3}s + \frac{1}{5}t - \frac{7}{15}u + \frac{9}{5}.$$

Example

The augmented matrix for the system of equations

$$3w - 2x + 4y - z = 5$$
$$w + y = 7$$
$$x + y + z = 0$$

is

$$\begin{pmatrix} -3 & -2 & 4 & -1 & 5 \\ 1 & 0 & 1 & 0 & 7 \\ 0 & 1 & 1 & 1 & 1 \end{pmatrix}.$$

The row reduced form of the matrix is

$$\begin{pmatrix} 1 & 0 & 0 & -1/3 & 35/3 \\ 0 & 1 & 0 & 2/3 & 17/3 \\ 0 & 0 & 1 & 1/3 & -14/3 \end{pmatrix}.$$

So w, x, and y are leading variables and z is a free variable. Let $z = t$.
Then

$$w = \frac{1}{3}z + \frac{35}{3} = \frac{1}{3}t + \frac{35}{3}$$

$$x = \frac{-2}{3}z + \frac{17}{3} = \frac{-2}{3}t + \frac{17}{3}$$

$$y = \frac{-1}{3}z - \frac{14}{3} = \frac{-1}{3}t - \frac{14}{3}$$

In the appendix, we give the commands that MATLAB uses to convert a matrix to row reduced form.

2.4.4 Application of Linear Systems to Curve Fitting

Example

We find the equation of the circle that passes through the points $(1,1),(3,1),$ and $(2,4)$. The equation of a circle whose radius r and center is (h,k) is

$$\left(x-h\right)^2 + \left(y-k\right)^2 = r^2$$

or

$$x^2 - 2xh + h^2 + y^2 - 2yk + k^2 = r^2$$

or

$$x^2 + y^2 - 2xh - 2yk + k^2 + h^2 - r^2 = 0,$$

which can be written

$$x\left(-2h\right) + y\left(-2k\right) + \left(k^2 + h^2 - r^2\right) = -x^2 - y^2. \tag{2.2}$$

If we let $a = -2h, b = -2k, c = k^2 + h^2 - r^2$, then Equation 2.2 becomes

$$xa + yb + c = -x^2 - y^2.$$

And each point (x,y) on the circle yields a linear equation in $a, b,$ and c.

We have

$$(1,1) \leftrightarrow a + b + c = -2$$
$$(3,1) \leftrightarrow 3a + b + c = -10$$
$$(2,4) \leftrightarrow 2a + 4b + c = -20.$$

When the augmented matrix

$$\begin{pmatrix} 1 & 1 & 1 & -2 \\ 3 & 1 & 1 & -10 \\ 2 & 4 & 1 & -20 \end{pmatrix}$$

is row reduced, the result is

$$\begin{pmatrix} 1 & 0 & 0 & -4 \\ 0 & 1 & 0 & -14/3 \\ 0 & 0 & 1 & 20/3 \end{pmatrix}$$

so

$$a = -4 \quad \text{and} \quad h = \frac{a}{-2} = 2, \quad b = -\frac{14}{3} \quad \text{and} \quad k = \frac{b}{-2} = \frac{7}{3}, \quad c = \frac{20}{3} \quad \text{and} \quad c = k^2 + h^2 - r^2$$

so

$$r^2 = k^2 + h^2 - c = 4 + \frac{49}{9} - \frac{20}{3} = \frac{36 + 49 - 60}{9} = \frac{25}{9}$$

and

$$r = \frac{5}{3}.$$

Thus, the equation of the circle is

$$(x-2)^2 + \left(y - \frac{7}{3}\right)^2 = \frac{25}{9}.$$

In the next example, we find the equation of a plane that passes through three given points. There will be such a plane if and only if the points are not collinear. If there is a plane, it will have an equation of the form

$$ax + by + cz = d.$$

Our task is, knowing three points on the plane, find a, b, c, and d.

Our method is simple, but there are two cases. Case 1 is where $d \neq 0$ and case 2 is where $d = 0$.

Example

Case 1. $d \neq 0$.

Find the equation of the plane that passes through the points $(8,2,6)$, $(4,4,4)$, and $(2,6,1)$.
The equation of a plane is of the form

$$ax + by + cz = d \quad \text{or} \quad \frac{a}{d}x + \frac{b}{d}y + \frac{c}{d}z = 1.$$

The division by d requires $d \neq 0$.

We will use the second form. We want to find the coefficients of $x, y,$ and z. We substitute the given points, which yield the three equations

$$a'8 + b'2 + c'6 = 1$$
$$a'4 + b'4 + c'4 = 1$$
$$a'2 + b'6 + c'1 = 1,$$

where

$$a' = \frac{a}{d}, \quad b' = \frac{b}{d}, \quad c' = \frac{c}{d}.$$

The augmented matrix for the system of equations is

$$\begin{pmatrix} 8 & 2 & 6 & 1 \\ 4 & 4 & 4 & 1 \\ 2 & 6 & 1 & 1 \end{pmatrix}.$$

When row reduced, this becomes

$$\begin{pmatrix} 1 & 0 & 0 & 1/28 \\ 0 & 1 & 0 & 1/7 \\ 0 & 0 & 1 & 1/14 \end{pmatrix}$$

so

$$a' = \frac{1}{28}, \quad b' = \frac{1}{7}, \quad c' = \frac{1}{14}$$

and the equation of the plane is

$$\frac{1}{28}x + \frac{1}{7}y + \frac{1}{14}z = 1$$

or

$$x + 4y + 2z = 28.$$

If we had thought $d = 0$, our equations would have been

$$a8 + b2 + c6 = 0$$
$$a4 + b4 + c4 = 0$$
$$a2 + b6 + c1 = 0.$$

The augmented matrix would be

$$\begin{pmatrix} 8 & 2 & 6 & 0 \\ 6 & 4 & 4 & 0 \\ 2 & 6 & 1 & 0 \end{pmatrix}$$

which, when row reduced, is

$$\begin{pmatrix} 1 & 0 & 0 & 0 \\ 0 & 1 & 0 & 0 \\ 0 & 0 & 1 & 0 \end{pmatrix}$$

This implies that the equation for the plane is

$$0x + 0y + 0z = 0,$$

which is not a plane after all.

Case 2. $d = 0$.
Find the equation of the plane that passes through the points

$$(1,1,-1), (3,2,0), (1,0,-2/5).$$

In this case, if we did not know $d = 0$ (and when we begin the problem, we typically do not know that) and set up the equations as before, the augmented matrix would be

$$\begin{pmatrix} 1 & 1 & -1 & 1 \\ 3 & -2 & 0 & 1 \\ 1 & 0 & -2/5 & 1 \end{pmatrix}$$

which, when row reduced, is

$$\begin{pmatrix} 1 & 0 & -2/5 & 0 \\ 0 & 1 & -3/5 & 0 \\ 0 & 0 & 0 & 1 \end{pmatrix}.$$

This indicates there is no solution and our assumption that $d \neq 0$ was erroneous.
    We modify the matrix with $d = 0$ to get

$$\begin{pmatrix} 1 & 1 & -1 & 0 \\ 3 & -2 & 0 & 0 \\ 1 & 0 & -2/5 & 0 \end{pmatrix}$$

which, when row reduced, is

$$\begin{pmatrix} 1 & 0 & -2/5 & 0 \\ 0 & 1 & -3/5 & 0 \\ 0 & 0 & 0 & 0 \end{pmatrix}.$$

Thus, c is the free variable

$$a = \frac{2}{5}c, \quad b = \frac{3}{5}c$$

and the equation of the plane is

$$\frac{2}{5}cx + \frac{3}{5}cy + cz = 0$$

or

$$c(2x + 3y + z) = 0.$$

Exercises

In Exercises 1 through 4, we give the row reduced form of the augmented matrix for a system of linear equations. Tell whether the system of equations is consistent, and if it is consistent give the solution in standard form.

1. $\begin{pmatrix} 1 & 2 & 0 & -1 & 3 & 4 \\ 0 & 0 & 1 & 2 & 0 & -2 \end{pmatrix}$

2. $\begin{pmatrix} 1 & 0 & 6 & 0 & 1 \\ 0 & 1 & 2 & 0 & 3 \\ 0 & 0 & 0 & 1 & 0 \\ 0 & 0 & 0 & 0 & 0 \\ 0 & 0 & 0 & 0 & 0 \end{pmatrix}$

3. $\begin{pmatrix} 1 & 0 & 0 \\ 0 & 1 & 0 \\ 0 & 0 & 1 \end{pmatrix}$

4. $\begin{pmatrix} 1 & 4 & 0 & 0 & 6 \\ 0 & 0 & 1 & 0 & 3 \\ 0 & 0 & 0 & 1 & 2 \\ 0 & 0 & 0 & 0 & 0 \end{pmatrix}$

In Exercises 5 through 10, give the augmented matrix for the system, the row reduced form of the matrix, and if a solution exists, give the solution in standard form.

5. $2x - 3y = 6$
   $x + 5y = 9$

6. $x - 2y + 3z = 5$
   $5x - y + 2z = 4$

7. $3x_1 - 5x_2 + x_3 - 2x_4 = 0$
   $6x_1 - 10x_2 + 2x_3 - 4x_4 = -6$

8. $x + 2y - 3z = -1$
   $6x - 2y + 4z = 14$
   $10x + 6y - 8z = 4$

9. $4x + 3y - 2z = 14$
   $-2x + y - 4z = -2$
   $2x + 4y - 6z = 12$

10. $4x - y = 8$
    $2x + 7y = 6$
    $-3x - 5y = 0$

11. (a) Is it possible for a system of three equations and two unknowns to have a unique solution?

    (b) Is it possible for a system of three equations and four unknowns to have a unique solution?

12. A consistent system of linear equations has 4 equations and 6 variables. How many solutions does it have?

13. For what value(s) of h does the system

$$3x - 4y = 7$$
$$2x + hy = 9$$

have exactly one solution?

14. For what value(s) of h does the system

$$x - 2y = 7$$
$$hx + 2y = 9$$

have no solution?

15. Find the line of intersection of the two planes

$$3x - 2y + 5z = 12$$
$$2x + 6y - z = 7$$

16. Find the polynomial $p(x) = ax^2 + bx + c$ for which
    (a) $p(1) = 8, p(2) = -6, p(4) = 16$
    (b) $p(0) = 4, p(-3) = 7, p(2) = 1$
    (c) $p(-2) = 5, p(-4) = 9, p(1) = 0$

17. Find the equation of the plane that passes through the points
    (a) $(3, -9, 2), (-4, 0, 6), (1, 1, 1)$
    (b) $(9, 3, 7), (1, 6, -2), (5, 0, 0)$
    (c) $(4, 9, 6), (2, 7, -8), (4, 4, 12)$
    (d) $(4, 1, 0), (2, 1, 2), (3, 2, 5)$

18. Find a plane that passes through the points

$$(3,-4,5), (5,-3,4), (7,-2,3).$$

Is there a line that passes through these points?

19. Center of mass problems. The center of mass of a group of particles is the unique point that is the average position of all parts of the system weighted according to their masses. This is useful because often systems of many points can be treated as a single point located at the center of mass.

In a system of k particles, the center of mass is computed according to the formula

$$\text{Center of mass} = \frac{1}{m}\left(m_1\hat{v}_1 + \cdots + m_k\hat{v}_k\right),$$

where
m_i is the mass of the ith particle
$\hat{v}_i$ is the location of the ith particle

and

$$m = m_1 + \cdots + m_k.$$

Note that the center of mass is a linear combination of the vectors $\{\hat{v}_1,\ldots,\hat{v}_n\}$

Find the center of mass for the following systems:

(a) Location   Mass
   $(2,-3,7)$ 5 g
   $(5,8,0)$ 8 g
   $(12,-8,-9)$ 10 g

(b) Location   Mass
   $(4,9,1)$ 1 g
   $(7,-6,-10)$ 6 g
   $(-8,0,0)$ 3 g

(c) In parts (a) and (b), find the mass and location of a particle so that if that particle were added to the system the center of mass would be at $(0,0,0)$.

2.5 Expressing Systems of Linear Equations in Other Forms

2.5.1 Representing a System of Linear Equations as a Vector Equation

Matrices are equal when they have the same dimensions and the corresponding entries coincide. In this section, we will use the fact that

$$\begin{pmatrix} a_1 \\ \vdots \\ a_n \end{pmatrix} = \begin{pmatrix} b_1 \\ \vdots \\ b_n \end{pmatrix}$$

if and only if $a_1 = b_1, \ldots, a_n = b_n$.

Consider

$$x \begin{pmatrix} 4 \\ 2 \\ -7 \end{pmatrix} + y \begin{pmatrix} 1 \\ 0 \\ 5 \end{pmatrix} + z \begin{pmatrix} 3 \\ -2 \\ 6 \end{pmatrix} = \begin{pmatrix} 8 \\ 12 \\ -5 \end{pmatrix}. \tag{2.3}$$

This could, for example, be a problem in physics or engineering, where there are forces available in the directions

$$\begin{pmatrix} 4 \\ 2 \\ -7 \end{pmatrix}, \begin{pmatrix} 1 \\ 0 \\ 5 \end{pmatrix}, \begin{pmatrix} 3 \\ -2 \\ 6 \end{pmatrix}$$

and we want to know how much of each force should be applied to achieve the resultant force

$$\begin{pmatrix} 8 \\ 12 \\ -5 \end{pmatrix}.$$

The expression in Equation 2.3 is a vector equation.

If we expand the left side of Equation 2.3, we get

$$x \begin{pmatrix} 4 \\ 2 \\ -7 \end{pmatrix} + y \begin{pmatrix} 1 \\ 0 \\ 5 \end{pmatrix} + z \begin{pmatrix} 3 \\ -2 \\ 6 \end{pmatrix} = \begin{pmatrix} 4x \\ 2x \\ -7x \end{pmatrix} + \begin{pmatrix} 1y \\ 0y \\ 5y \end{pmatrix} + \begin{pmatrix} 3z \\ -2z \\ 6z \end{pmatrix} = \begin{pmatrix} 4x+1y+3z \\ 2x+0y-2z \\ -7x+5y+6z \end{pmatrix},$$

so we must have

$$\begin{pmatrix} 4x+1y+3z \\ 2x+0y-2z \\ -7x+5y+6z \end{pmatrix} = \begin{pmatrix} 8 \\ 12 \\ -5 \end{pmatrix},$$

which is true if and only if each of the equations

$$4x+1y+3z = 8$$
$$2x+0y-2z = 12$$
$$-7x+5y+6z = -5$$

is satisfied. Thus, we have converted a vector equation into a system of linear equations. This is worthwhile knowing because there are instances in which it might be advantageous to formulate a problem in one setting, but easier to solve the problem in another setting.

Example

Express the vector equation

$$x_1 \begin{pmatrix} 2 \\ 0 \\ -1 \\ 1 \end{pmatrix} + x_2 \begin{pmatrix} -6 \\ 4 \\ 3 \\ 7 \end{pmatrix} + x_3 \begin{pmatrix} 4 \\ 9 \\ 0 \\ -4 \end{pmatrix} = \begin{pmatrix} 5 \\ 8 \\ 0 \\ -2 \end{pmatrix}$$

as a system of linear equations.

Following the earlier example, we have

$$2x_1 - 6x_2 + 4x_3 = 5$$
$$4x_2 + 9x_3 = 8$$
$$-x_1 + 3x_2 = 0$$
$$x_1 + 7x_2 - 4x_3 = -2.$$

Example

Express the system of linear equations

$$8w - 3x + 4y + 2z = 1$$
$$w + x + 6y = 9$$

as a vector equation.

Reversing the ideas mentioned earlier, we get

$$w \begin{pmatrix} 8 \\ 1 \end{pmatrix} + x \begin{pmatrix} -3 \\ 1 \end{pmatrix} + y \begin{pmatrix} 4 \\ 6 \end{pmatrix} + z \begin{pmatrix} 2 \\ 0 \end{pmatrix} = \begin{pmatrix} 1 \\ 9 \end{pmatrix}.$$

2.5.2 Equivalence of a System of Linear Equations and a Matrix Equation

The equation

$$\begin{pmatrix} 2 & -1 & 5 \\ 6 & 0 & 3 \end{pmatrix} \begin{pmatrix} x \\ y \\ z \end{pmatrix} = \begin{pmatrix} 3 \\ -4 \end{pmatrix}$$

is an example of a matrix equation.

We want to find x, y, and z. The expression on the left is the product of a 2×3 matrix with a 3×1 matrix which is a 2×1 matrix, so the dimensions are correct, but the answer may not be obvious. If we expand the expression on the left, we get

$$\begin{pmatrix} 2x - y + 5z \\ 6x + 3z \end{pmatrix} = \begin{pmatrix} 3 \\ -4 \end{pmatrix}$$

which gives the system of linear equations

$$2x - y + 5z = 3$$
$$6x + 3z = -4.$$

Similarly, the system of linear equations

$$2w - 3x + y - 5z = 7$$
$$w + 5x - 7y = 0$$

is equivalent to the matrix equation

$$\begin{pmatrix} 2 & -3 & 1 & -5 \\ 1 & 5 & -7 & 0 \end{pmatrix} \begin{pmatrix} w \\ x \\ y \\ z \end{pmatrix} = \begin{pmatrix} 7 \\ 0 \end{pmatrix}.$$

Example

If there are the same number of equations as unknowns, and if the matrix is invertible, then there is another method of solution to a matrix equation. Suppose

$$\begin{pmatrix} 3 & 1 & 9 \\ 0 & -2 & 5 \\ -1 & -3 & 6 \end{pmatrix} \begin{pmatrix} x \\ y \\ z \end{pmatrix} = \begin{pmatrix} -4 \\ 1 \\ 2 \end{pmatrix}.$$

If

$$\begin{pmatrix} 3 & 1 & 9 \\ 0 & -2 & 5 \\ -1 & -3 & 6 \end{pmatrix}$$

is invertible, then

$$\begin{pmatrix} 3 & 1 & 9 \\ 0 & -2 & 5 \\ -1 & -3 & 6 \end{pmatrix}^{-1} \begin{pmatrix} 3 & 1 & 9 \\ 0 & -2 & 5 \\ -1 & -3 & 6 \end{pmatrix} \begin{pmatrix} x \\ y \\ z \end{pmatrix} = \begin{pmatrix} 3 & 1 & 9 \\ 0 & -2 & 5 \\ -1 & -3 & 6 \end{pmatrix}^{-1} \begin{pmatrix} -4 \\ 1 \\ 2 \end{pmatrix}$$

so

$$\begin{pmatrix} x \\ y \\ z \end{pmatrix} = \begin{pmatrix} 3 & 1 & 9 \\ 0 & -2 & 5 \\ -1 & -3 & 6 \end{pmatrix}^{-1} \begin{pmatrix} -4 \\ 1 \\ 2 \end{pmatrix} = \begin{pmatrix} -1/14 \\ -17/14 \\ -2/7 \end{pmatrix}.$$

This demonstrates that the matrix equation $A\hat{x} = \hat{b}$, where A is an $n \times n$ matrix, will have a unique solution if A^{-1} exists.

Thus, we have the equivalency of the three forms: a system of linear equations, a vector equation, and a matrix equation.

Example

Express the vector equation

$$x\begin{pmatrix} 0 \\ 2 \\ -3 \end{pmatrix} + y\begin{pmatrix} 9 \\ 4 \\ 1 \end{pmatrix} + z\begin{pmatrix} -2 \\ 6 \\ -1 \end{pmatrix} = \begin{pmatrix} 3 \\ 8 \\ 5 \end{pmatrix}$$

as a matrix equation and a system of linear equations.

An equivalent matrix equation is

$$\begin{pmatrix} 0 & 9 & -2 \\ 2 & 4 & 6 \\ -3 & 1 & -1 \end{pmatrix}\begin{pmatrix} x \\ y \\ z \end{pmatrix} = \begin{pmatrix} 3 \\ 8 \\ 5 \end{pmatrix}$$

and an equivalent system of linear equations is

$$9x - 2z = 3$$
$$2x + 4y + 6z = 8$$
$$-3x + y - z = 5.$$

We repeat an idea that we have previously seen for emphasis. For A a matrix, the expression $A\hat{x}$ is a linear combination of the columns of A. Similarly, $\hat{y}A$ is a linear combination of the rows of A.

Definition

If $\hat{v}_1, \hat{v}_2, \ldots, \hat{v}_n$ and $\hat{b}$ are vectors, we say that $\hat{b}$ is a linear combination of $\hat{v}_1, \hat{v}_2, \ldots, \hat{v}_n$, if there are scalars $a_1, a_2, \ldots, a_n$ for which

$$a_1\hat{v}_1 + a_2\hat{v}_2 + \cdots a_n\hat{v}_n = \hat{b}.$$

In later chapters, an important question will be whether a given vector can be expressed as a linear combination of a given set of vectors. We now have the tools to answer this type of question.

Example

Determine if $\hat{b}$ is a linear combination of $\hat{v}_1, \hat{v}_2,$ and $\hat{v}_3$ for

$$\hat{v}_1 = \begin{pmatrix} 1 \\ 0 \\ -2 \end{pmatrix}, \quad \hat{v}_2 = \begin{pmatrix} 3 \\ 5 \\ 1 \end{pmatrix}, \quad \hat{v}_3 = \begin{pmatrix} 8 \\ 2 \\ -1 \end{pmatrix}, \quad \hat{b} = \begin{pmatrix} 4 \\ 7 \\ -6 \end{pmatrix}.$$

Another way to phrase this question is does the vector equation

$$a_1\begin{pmatrix} 1 \\ 0 \\ -2 \end{pmatrix} + a_2\begin{pmatrix} 3 \\ 5 \\ 1 \end{pmatrix} + a_3\begin{pmatrix} 8 \\ 2 \\ -1 \end{pmatrix} = \begin{pmatrix} 4 \\ 7 \\ -6 \end{pmatrix} \qquad (2.4)$$

have a solution.

Converting Equation 2.4 to a system of linear equations gives

$$1a_1 + 3a_2 + 8a_3 = 4$$
$$5a_2 + 2a_3 = 7$$
$$-2a_1 + a_2 - a_3 = -6.$$

The augmented matrix for this system of linear equations is

$$\begin{pmatrix} 1 & 3 & 8 & 4 \\ 0 & 5 & 2 & 7 \\ -2 & 1 & -1 & -6 \end{pmatrix}$$

which, when row reduced, is

$$\begin{pmatrix} 1 & 0 & 0 & 233/61 \\ 0 & 1 & 0 & 101/61 \\ 0 & 0 & 1 & -39/61 \end{pmatrix}.$$

Thus, there is a unique solution, namely,

$$a_1 = \frac{233}{61}, \quad a_2 = \frac{101}{61}, \quad a_3 = \frac{-39}{61}$$

Exercises

1. Write the following systems of linear equations as a (i) vector equation and (ii) matrix equation:

   (a) $2x_1 - 5x_2 + x_3 = -2$
   $-4x_1 + x_3 = 7$
   $x_1 + 6x_2 - 4x_3 = 0$

   (b) $x_1 - x_2 + x_3 + x_4 = 0$
   $x_3 = 9$

2. Write the following vector equations as a (i) system of linear equations and (ii) matrix equation:

   (a) $x \begin{pmatrix} 2 \\ 8 \\ 1 \end{pmatrix} + y \begin{pmatrix} -3 \\ 0 \\ 6 \end{pmatrix} + z \begin{pmatrix} 5 \\ -3 \\ -1 \end{pmatrix} = \begin{pmatrix} 4 \\ 9 \\ 0 \end{pmatrix}$

   (b) $x_1 \begin{pmatrix} 0 \\ 3 \\ 0 \\ -5 \end{pmatrix} + x_2 \begin{pmatrix} 1 \\ 4 \\ 2 \\ 7 \end{pmatrix} + x_3 \begin{pmatrix} 0 \\ -2 \\ -3 \\ 6 \end{pmatrix} + x_4 \begin{pmatrix} 3 \\ 1 \\ 1 \\ -2 \end{pmatrix} + x_5 \begin{pmatrix} 1 \\ 8 \\ 4 \\ 2 \end{pmatrix} = \begin{pmatrix} 3 \\ -1 \\ -6 \\ -5 \end{pmatrix}$

3. Write the following matrix equations as a (i) system of linear equations and (ii) vector equation:

(a) $\begin{pmatrix} 0 & -3 & 5 & 2 \\ 2 & 4 & -2 & 3 \\ 1 & 1 & 2 & 6 \end{pmatrix} \begin{pmatrix} w \\ x \\ y \\ z \end{pmatrix} = \begin{pmatrix} -5 \\ 7 \\ 0 \end{pmatrix}$

(b) $\begin{pmatrix} 1 & 5 & 4 \\ -3 & 2 & 1 \\ 6 & 4 & 0 \end{pmatrix} \begin{pmatrix} x \\ y \\ z \end{pmatrix} = \begin{pmatrix} 9 \\ -6 \\ 8 \end{pmatrix}$

In Exercises 4 through 6, solve the vector equations

4. $a \begin{pmatrix} 1 \\ -2 \\ 0 \end{pmatrix} + b \begin{pmatrix} 5 \\ 3 \\ 4 \end{pmatrix} + c \begin{pmatrix} 1 \\ 0 \\ -3 \end{pmatrix} = \begin{pmatrix} 6 \\ 1 \\ -5 \end{pmatrix}$

5. $a \begin{pmatrix} 3 \\ 2 \\ -1 \end{pmatrix} + b \begin{pmatrix} 6 \\ 0 \\ 3 \end{pmatrix} = \begin{pmatrix} 4 \\ 2 \\ 7 \end{pmatrix}$

6. $a \begin{pmatrix} 0 \\ 3 \\ 0 \end{pmatrix} + b \begin{pmatrix} -2 \\ 6 \\ 1 \end{pmatrix} + c \begin{pmatrix} 4 \\ 1 \\ -3 \end{pmatrix} + d \begin{pmatrix} 7 \\ 0 \\ -4 \end{pmatrix} = \begin{pmatrix} 16 \\ -3 \\ -5 \end{pmatrix}$

7. Suppose that A is an $n \times n$ matrix and $\det(A) \neq 0$. Let $\hat{a}_1, \ldots, \hat{a}_n$ denote the columns of A.

   (a) Show that for any $\hat{b}$ in $\mathbb{R}^n$ $A\hat{x} = \hat{b}$ has a solution.

   (b) Show that any vector in $\mathbb{R}^n$ can be written as a linear combination of $\hat{a}_1, \ldots, \hat{a}_n$.

   (c) Show that the only solution to $A\hat{x} = \hat{0}$ is $\hat{x} = \hat{0}$.

   (d) Show that $x_1\hat{a}_1 + \cdots + x_n\hat{a}_n = \hat{0}$ if and only if every $x_i = 0$.

   (e) Show that $A\hat{x} = \hat{b}$ has the same solution set as

   $$x_1\hat{a}_1 + \cdots + x_n\hat{a}_n = \hat{b}$$

8. Determine if $\hat{b}$ is a linear combination of $\hat{v}_1, \hat{v}_2$, and $\hat{v}_3$

   (a) $\hat{v}_1 = \begin{pmatrix} 3 \\ 1 \\ -2 \end{pmatrix}$, $\hat{v}_2 = \begin{pmatrix} 4 \\ -2 \\ 3 \end{pmatrix}$, $\hat{v}_3 = \begin{pmatrix} 10 \\ 0 \\ -1 \end{pmatrix}$, $\hat{b} = \begin{pmatrix} 3 \\ 7 \\ 11 \end{pmatrix}$

   (b) $\hat{v}_1 = \begin{pmatrix} 1 \\ -1 \\ 2 \end{pmatrix}$, $\hat{v}_2 = \begin{pmatrix} 7 \\ 3 \\ 6 \end{pmatrix}$, $\hat{v}_3 = \begin{pmatrix} 1 \\ 0 \\ 5 \end{pmatrix}$, $\hat{b} = \begin{pmatrix} 6 \\ -14 \\ 5 \end{pmatrix}$

9. Is $\hat{b}$ a linear combination of the columns of A if

   (a) $A = \begin{pmatrix} 1 & 2 & 9 \\ 5 & 0 & 7 \\ -3 & 2 & -4 \end{pmatrix}$ $\hat{b} = \begin{pmatrix} 12 \\ 0 \\ -5 \end{pmatrix}$

   (b) $A = \begin{pmatrix} 2 & 0 & 4 \\ 3 & 1 & -6 \\ 8 & 2 & -8 \end{pmatrix}$ $\hat{b} = \begin{pmatrix} 3 \\ 6 \\ -1 \end{pmatrix}$

2.6 Applications

2.6.1 Flow Problems

Linear algebra can be used to model "flow problems." Two examples of flow problems are traffic on a network of streets and flow of water through a water main system.

In such problems, we have a lattice where a quantity can enter the system and exit the system. The quantity entering the system is balanced by the quantity leaving the system.

Consider the traffic flow in Figure 2.3. We hypothesize a direction of flow between the intersections. At each intersection, the input traffic must equal the output traffic. In order to be a valid solution, all variables must be nonnegative.

At intersection $A: x_1 + x_3 = 200$

At intersection $B: x_1 = x_2 + x_4$ or $x_1 - x_2 - x_4 = 0$

At intersection $C: x_2 + x_5 = 100$

At intersection $D: x_5 + x_6 = 100$

At intersection $E: x_4 + x_6 = x_7$ or $x_4 + x_6 - x_7 = 0$

At intersection $F: x_3 + x_7 = 200$

The augmented matrix for this system of equations is

$$
\begin{pmatrix}
1 & 0 & 1 & 0 & 0 & 0 & 0 & 200 \\
1 & -1 & 0 & -1 & 0 & 0 & 0 & 0 \\
0 & 1 & 0 & 0 & 1 & 0 & 0 & 100 \\
0 & 0 & 0 & 0 & 1 & 0 & 0 & 100 \\
0 & 0 & 0 & 1 & 0 & 1 & -1 & 0 \\
0 & 0 & 1 & 0 & 0 & 0 & 1 & 200
\end{pmatrix}
$$

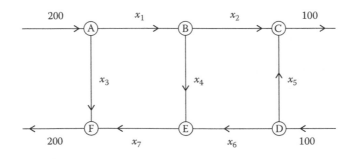

FIGURE 2.3
A traffic flow system with six intersections.

The row reduced form of the matrix is

$$\begin{pmatrix} 1 & 0 & 0 & 0 & 0 & 0 & -1 & 0 \\ 0 & 1 & 0 & 0 & 0 & 0 & -1 & 0 \\ 0 & 0 & 1 & 0 & 0 & 0 & 1 & 200 \\ 0 & 0 & 0 & 1 & 0 & 0 & 0 & 0 \\ 0 & 0 & 0 & 0 & 1 & 0 & 1 & 100 \\ 0 & 0 & 0 & 0 & 0 & 1 & -1 & 0 \end{pmatrix}$$

Thus, x_7 is the free variable and

$$x_1 = x_7$$

$$x_2 = x_7$$

$$x_3 = -x_7 + 200$$

$$x_4 = 0$$

$$x_5 = -x_7 + 100$$

$$x_6 = x_7$$

If we take $x_7 = 50$, then

$$x_1 = 50, \quad x_2 = 50, \quad x_3 = 150, \quad x_4 = 0, \quad x_5 = 50, \quad x_6 = 50.$$

2.6.2 Example: Kirchoff's Laws

Kirchoff's laws are used in the analysis of electrical circuits. These laws give rise to a system of linear equations.

The circuits that we consider will have power source(s) designated V or E and load(s), designated R. The amount of power of a power source is measured in volts, and the amount of load is measured in ohms. When a circuit is activated, current, I, measured in amperes (more often called amps), flows through the circuit and power is dissipated according to $V = IR$. A loop is a path through a circuit that begins and ends at the same point. A junction is a point where at least two circuit paths meet and a branch is a path connecting two junctions.

The junction rule says that the sum of the currents flowing into a node is equal to the sum of currents flowing out of a node. Said another way, the algebraic sum of currents at a node is equal to zero. This says that current is conserved.

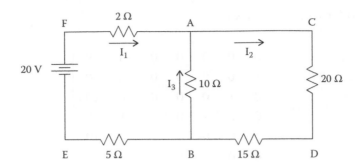

FIGURE 2.4
An electrical system with power sources and resistors.

The closed loop rule says that the algebraic sum of the potential differences around a closed loop is zero.

Example

An electrical circuit is shown in Figure 2.4.

Assign directions for $I_1, I_2,$ and I_3. If these directions are not consistent with the actual current flow, the calculations will give negative values.

Conservation of current: At node A

$$I_1 \rightarrow A \rightarrow I_2$$
$$I_3 \uparrow$$

so

$$I_1 + I_3 = I_2$$

or

$$I_1 - I_2 + I_3 = 0.$$

Node B gives the same equation.

Conservation of energy: There are three loops: *FABEF, ACBDA, and FACDBEF*

For loop *ACDBA* $10I_3 + 20I_2 + 15I_2 = 0$ or $35I_2 + 10I_3 = 0.$
For loop *FABEF* $2I_1 - 10I_3 + 5I_1 = 20$ or $7I_1 - 10I_3 = 20.$

The three equations

$$I_1 - I_2 + I_3 = 0$$
$$35I_2 + 10I_3 = 0$$
$$7I_1 - 10I_3 = 20$$

are sufficient to solve the system.

We create the augmented matrix

$$\begin{pmatrix} 1 & -1 & 1 & 0 \\ 0 & 35 & 10 & 0 \\ 7 & -10 & 0 & 20 \end{pmatrix}$$

which, when row reduced, gives

$$\begin{pmatrix} 1 & 0 & 0 & 180/43 \\ 0 & 1 & 0 & 40/43 \\ 0 & 0 & 1 & -140/43 \end{pmatrix}$$

so

$$I_1 = \frac{180}{43} \approx 4.19, \quad I_2 = \frac{40}{43} \approx .93, \quad I_3 = -\frac{140}{43} \approx -3.26.$$

The fact that I_3 is negative indicates the assumed direction of I_3 in the diagram is incorrect.

2.6.3 Balancing Chemical Equations Using Linear Algebra

Consider the chemical reaction

$$CH_4 + O_2 \rightarrow CO_2 + H_2O.$$

By balancing an equation, we mean assigning integers to each molecule so that there is the same number of atoms of each element on both sides of the equation. As it stands now, for example, there are 2 atoms of oxygen on the left side of the equation and 3 atoms of oxygen on the right side of the equation. We form the reaction

$$wCH_4 + xO_2 \rightarrow yCO_2 + zH_2O.$$

Each element gives rise to an equation as follows:

Oxygen $\quad 2x = 2y + z \quad$ or $\quad 2x - 2y - z = 0.$
Carbon $\quad w = y \quad$ or $\quad w - y = 0.$
Hydrogen $\quad 4w = 2z \quad$ or $\quad 4w - 2z = 0.$

The augmented matrix for the system of equations is

$$\begin{pmatrix} 0 & 2 & -2 & -1 & 0 \\ 1 & 0 & -1 & 0 & 0 \\ 4 & 0 & 0 & -2 & 0 \end{pmatrix}.$$

When row reduced, this gives

$$\begin{pmatrix} 1 & 0 & 0 & -1/2 & 0 \\ 0 & 1 & 0 & -1 & 0 \\ 0 & 0 & 1 & -1/2 & 0 \end{pmatrix}$$

so z is the free variable and

$$w = \frac{1}{2}z, \quad x = z, \quad y = \frac{1}{2}z.$$

Thus, the solution is of the form

$$\begin{pmatrix} w \\ x \\ y \\ z \end{pmatrix} = \begin{pmatrix} \frac{1}{2}z \\ z \\ \frac{1}{2}z \\ z \end{pmatrix} = z \begin{pmatrix} \frac{1}{2} \\ 1 \\ \frac{1}{2} \\ 1 \end{pmatrix}.$$

We are not totally free in our choice of z. Each of the variables must be a positive integer. The simplest way to do this is to set $z = 2$, so that $w = 1$, $x = 2$, and $y = 1$.

Exercises

1. Balance the following chemical equations.
    (a) $Ca + H_3PO_4 \rightarrow Ca_3P_2O_8 + H_2$
    (b) $C_3H_8 + O_2 \rightarrow CO_2 + H_2O$
    (c) $P_2I_4 + P_4 + H_2O \rightarrow PH_4I + H_3PO_4$
    (d) $KI + KClO_3 + HCl \rightarrow I_2 + H_2O + KCl$
    (e) $Al + Fe_2O_3 \rightarrow Al_2O_3 + Fe$
2. Determine the currents in the networks (Figures 2.5 and 2.6).
3. Solve the traffic flow problems depicted in Figure 2.7a through d.

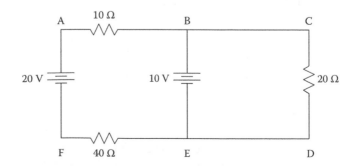

FIGURE 2.5
An electrical system with power sources and resistors.

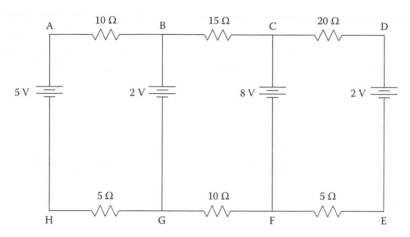

FIGURE 2.6
An electrical system with power sources and resistors.

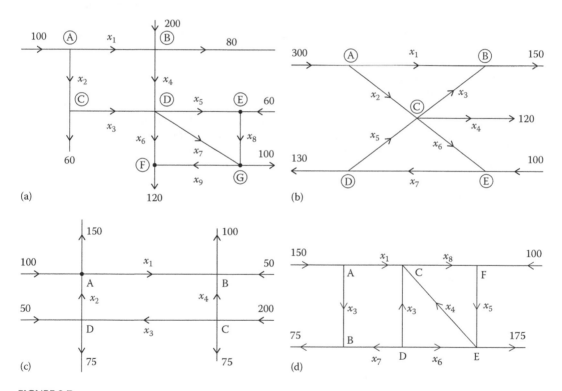

FIGURE 2.7
Traffic flow networks.

2.6.4 Markov Chains

Markov chains is a branch of probability in which matrices and linear algebra play a particularly crucial role. In a Markov chain, we have a collection of states and a process that occupies exactly one of these states at any given time. The model that we consider has a finite number of states, and the process can change states only at a discrete unit of time. Thus, the process is described by the ordered pair (s,t), where s is the state and t is time. There is a collection of transition probabilities that describe how the process changes (possibly remaining in the same state) when time changes. The Markov property is that these probabilities depend only on where the process is at the present time; previous history is irrelevant.

Example

Alice is playing Beth in a game where Alice is somewhat better; in fact, Alice wins 60% of the time. They bet $1 on each play of the game. Suppose that between them they have $5. We will make Alice's fortune the states. The state space would then be $\{0,1,2,3,4,5,\}$.

If Alice's fortune at time t is $2, then at time $t+1$ her fortune will be $3 with probability .6 and will be $1 with probability .4. If her fortune is $0, then it will remain there (she has no money to bet) and if her fortune is $5, then it will remain there (Beth has no money to bet). The states $0 and $5 are called absorbing states, because once the process enters an absorbing state, it never leaves.

From these "transition probabilities," we can construct the transition matrix shown in Figure 2.8.

The number P_{ij} is the probability of going from state i to state j in one step. Since we must go somewhere in one step from state i, the sum of the entries in each row is 1. This is not necessarily true for the columns.

Thinking of an abstract Markov chain rather than the example mentioned earlier, we develop a formula that describes how to go from state i to step j in two steps. This is one place where our definition of matrix multiplication is indispensable. Suppose the process is in state i at time t and we want to compute the probability that the process is in step j at time $t+2$. This occurs if the process goes from state i at time t to some state k at time $t+1$ (which occurs with probability P_{ik}) and then goes to state j at time $t+2$ (which occurs with probability P_{kj}). The probability of both of these transitions occurring is $P_{ik}P_{kj}$. Since this is true for every intermediate step k, the probability of going from state i to step j in two steps is

$$\sum_{k=1}^{n} P_{ik}P_{kj}$$

if there are a total of n states. Note that

$$\sum_{k=1}^{n} P_{ik}P_{kj} = (P^2)_{ij}.$$

$$P = \begin{pmatrix} 1 & 0 & 0 & 0 & 0 & 0 \\ .4 & 0 & .6 & 0 & 0 & 0 \\ 0 & .4 & 0 & .6 & 0 & 0 \\ 0 & 0 & .4 & 0 & .6 & 0 \\ 0 & 0 & 0 & .4 & 0 & .6 \\ 0 & 0 & 0 & 0 & 0 & 1 \end{pmatrix}$$

FIGURE 2.8
The transition matrix for the process.

Extending this reasoning, the probability of going from state i to step j in m steps is

$$\left(P^m\right)_{ij}.$$

It is most often the case that our process has a probability of being in different states initially, and we would like to determine the probability that it will be in a particular state some time later.

Example

Suppose there are three weather conditions, rain (R), snow (S), and clear (C), and we believe that daily weather conditions have the following properties:

If it rains today, then tomorrow it will rain with probability .3, snow with probability .2, and be clear with probability .5

If it snows today, then tomorrow it will rain with probability .1, snow with probability .3, and be clear with probability .6

If it is clear today, then tomorrow it will rain with probability .2, snow with probability .5, and be clear with probability .3

If we enumerate our states as

$$1 = \text{rain}, \quad 2 = \text{snow}, \quad 3 = \text{clear}$$

then the transition matrix is

$$\begin{pmatrix} .3 & .2 & .5 \\ .1 & .3 & .6 \\ .2 & .5 & .3 \end{pmatrix}.$$

Suppose that on October 31, there is a 30% chance of rain, 10% chance it will snow, and 60% chance it will be clear. We determine the likelihood of the different weather states on November 1 using

$$\left(.3, .1, .6\right)\begin{pmatrix} .3 & .2 & .5 \\ .1 & .3 & .6 \\ .2 & .5 & .3 \end{pmatrix} = \left(.22, .39, .39\right).$$

This says that given there was a 30% probability of rain, a 10% probability of snow, and a 60% probability of a clear day on October 31, the probability that it will rain on November 1 is 22%, the probability of snow is 39%, and the probability of a clear day is 39%.

To calculate the probabilities of the weather on November 2, given the same data as before for the weather on October 31, we compute

$$\left(.3, .1, .6\right)\begin{pmatrix} .3 & .2 & .5 \\ .1 & .3 & .6 \\ .2 & .5 & .3 \end{pmatrix}^2 = \left(.183, .356, .461\right).$$

An interesting result occurs in this case if we consider the long-term probabilities. If we project letting the process run for 100 days, we get

$$(.3,.1,.6)\begin{pmatrix} .3 & .2 & .5 \\ .1 & .3 & .6 \\ .2 & .5 & .3 \end{pmatrix}^{100} = (.1810,.3714,.4476) \quad (\text{to 4 decimal places}).$$

More importantly, a high power of the matrix converges to a matrix that has identical rows. In this case

$$\begin{pmatrix} .3 & .2 & .5 \\ .1 & .3 & .6 \\ .2 & .5 & .3 \end{pmatrix}^{100} = \begin{pmatrix} .1810 & .3714 & .4476 \\ .1810 & .3714 & .4476 \\ .1810 & .3714 & .4476 \end{pmatrix}.$$

When this happens, the initial probabilities have no effect on the long-term behavior. If x is the probability it rains on October 31, y is the probability it snows on October 31, and z is the probability it is clear on October 31, then

$$(x,y,z)\begin{pmatrix} .1810 & .3714 & .4476 \\ .1810 & .3714 & .4476 \\ .1810 & .3714 & .4476 \end{pmatrix}$$
$$= (.1810x + .1810y + .1810z, \quad .3714x + .3714y + .3714z, \quad .4476x + .4476y + .4476z)$$
$$= (.1810(x + y + z), \quad .3714(x + y + z), \quad .4476(x + y + z))$$
$$= (.1810, \quad .3714, \quad .4476).$$

if $x + y + z = 1$.

The probability vector $(.1810\ .3714\ .4476)$ is the equilibrium state for the process because

$$(.1810\ .3714\ .4476)\begin{pmatrix} .3 & .2 & .5 \\ .1 & .3 & .6 \\ .2 & .5 & .3 \end{pmatrix} = (.1810\ .3714\ .4476).$$

It predicts that in the long run, 18% of the days will have rain, 37% of the days will have snow, and 45% of the days will be clear.

This type of behavior occurs anytime the entries of the transition matrix (or any power of the transition matrix) are all nonzero entries. This is because when this occurs it is possible to go from any one state to any other state (but not necessarily in one step).

The gambling problem of Alice and Beth has quite a different evolution. The game ends whenever Alice's fortune hits either $0 or $5, so we would expect the game to end eventually. In this case, the most pertinent questions include

1. If we know Alice's initial fortune, what is the probability that Alice will wind up the winner of all the money.
2. If we know Alice's initial fortune, how many plays would we expect before the game ends.

The game ends whenever the process enters the state 5 or the state 0. These are called absorbing states. Absorbing states are identified by having a 1 on the main diagonal.

For purposes of calculation, it is convenient to rearrange the states so that absorbing states are listed first. The modified transition matrix in this example is

$$
P^* = \begin{array}{c}
 \\
0 \\
5 \\
1 \\
2 \\
3 \\
4
\end{array}
\begin{array}{cccccc}
0 & 5 & 1 & 2 & 3 & 4 \\
\begin{pmatrix}
1 & 0 & 0 & 0 & 0 & 0 \\
0 & 1 & 0 & 0 & 0 & 0 \\
.4 & 0 & 0 & .6 & 0 & 0 \\
0 & 0 & .4 & 0 & .6 & 0 \\
0 & 0 & 0 & .4 & 0 & .6 \\
0 & .6 & 0 & 0 & .4 & 0
\end{pmatrix}
\end{array}.
$$

We partition P^* as

$$
P^* = \begin{pmatrix} I & O \\ R & Q \end{pmatrix},
$$

where

$$
I = \begin{pmatrix} 1 & 0 \\ 0 & 1 \end{pmatrix} \quad O = \begin{pmatrix} 0 & 0 & 0 & 0 \\ 0 & 0 & 0 & 0 \end{pmatrix}
$$

$$
R = \begin{pmatrix} .4 & 0 \\ 0 & 0 \\ 0 & 0 \\ 0 & .6 \end{pmatrix}, \quad Q = \begin{pmatrix} 0 & .6 & 0 & 0 \\ .4 & 0 & .6 & 0 \\ 0 & .4 & 0 & .6 \\ 0 & 0 & .4 & 0 \end{pmatrix}.
$$

We have

$$
(I_4 - Q)^{-1} = \frac{1}{211} \begin{pmatrix} 325 & 285 & 225 & 135 \\ 190 & 475 & 375 & 225 \\ 100 & 250 & 475 & 285 \\ 40 & 100 & 190 & 325 \end{pmatrix} \equiv N.
$$

Theorem 2

For a finite state absorbing Markov chain with k nonabsorbing states, the expected number of transitions that the nonabsorbing state i undergoes before absorption is

$$
N_{i1} + N_{i2} + \cdots + N_{ik},
$$

where $N = (I - Q)^{-1}$.

This can also be expressed as follows.

Let t_i be the expected time until absorption, beginning in the nonabsorbing state i, and let

$$\hat{t} = \begin{pmatrix} t_1 \\ \vdots \\ t_k \end{pmatrix}.$$

Then

$$\hat{t} = N \begin{pmatrix} 1 \\ \vdots \\ 1 \end{pmatrix}.$$

In our example,

$$N \begin{pmatrix} 1 \\ \vdots \\ 1 \end{pmatrix} = \frac{1}{211} \begin{pmatrix} 325 & 285 & 225 & 135 \\ 190 & 475 & 375 & 225 \\ 100 & 250 & 475 & 285 \\ 40 & 100 & 190 & 325 \end{pmatrix} \begin{pmatrix} 1 \\ 1 \\ 1 \\ 1 \end{pmatrix} = \frac{1}{211} \begin{pmatrix} 970 \\ 1265 \\ 1110 \\ 655 \end{pmatrix} \approx \begin{pmatrix} 4.60 \\ 6.00 \\ 5.26 \\ 3.10 \end{pmatrix}.$$

So if Alice begins with \$1, the expected number of plays until someone is bankrupt is 4.60; if Alice begins with \$2, the expected number of plays until someone is bankrupt is 6.00, etc.

The previous theorem gives the average number of plays until someone goes bankrupt. The next theorem tells who is likely to win.

Theorem 3

The probability that beginning in the nonabsorbing state i the system is absorbed in the absorbing state j is

$$(NR)_{ij}.$$

In our example,

$$NR = \frac{1}{211} \begin{pmatrix} 325 & 285 & 225 & 135 \\ 190 & 475 & 375 & 225 \\ 100 & 250 & 475 & 285 \\ 40 & 100 & 190 & 325 \end{pmatrix} \begin{pmatrix} .4 & 0 \\ 0 & 0 \\ 0 & 0 \\ 0 & .6 \end{pmatrix} = \frac{1}{211} \begin{pmatrix} 130 & 81 \\ 76 & 135 \\ 40 & 171 \\ 16 & 195 \end{pmatrix} \approx \begin{pmatrix} .62 & .38 \\ .36 & .64 \\ .19 & .81 \\ .08 & .92 \end{pmatrix}.$$

So if Alice begins with \$1 and Beth begins with \$4, Alice will be the ultimate winner 38% of the time.

If Alice begins with \$2 and Beth begins with \$3, Alice will be the ultimate winner 64% of the time.

If Alice begins with \$3 and Beth begins with \$2, Alice will be the ultimate winner 81% of the time.

If Alice begins with \$4 and Beth begins with \$1, Alice will be the ultimate winner 92% of the time.

Exercises

1. Carl and David play a game where Carl wins 55% of the time. They bet $1 at each play of the game and will play until one person goes broke. If Carl starts with $2 and David starts with $3,

   (a) Who is more likely to be the long-term winner?

   (b) What is the expected number of plays until the game ends?

2. Repeat Problem 1 with the assumption that Carl wins 51% of the time.

3. Ellen and Fred are engaged to be married. The ceremony will be held in a location where it either rains or is sunny. If it rains one day, then it will be sunny the next day with probability .7; and if it is sunny one day, then it will rain the next day with probability .4.

   (a) If it is raining 1 week before the ceremony, what is the probability it will be sunny for the wedding?

   (b) If it is raining two weeks before the ceremony, what is the probability it will be sunny for the wedding?

   (c) What is the long-range distribution of the weather in this region?

3

Vector Spaces

The fundamental objects of study in linear algebra are vector spaces and linear transformations. In this chapter, we define what a vector space is, derive additional properties of a vector space beyond those explicitly required by the axioms of the definition, and examine several examples. Linear transformations are a particular type of function between vector spaces and are the topic of Chapter 4.

We will see that vector spaces have different special collections of building blocks called bases. We will spend substantial effort examining characteristics of bases.

We will use the language that includes the term "scalar" as we did in Chapter 1. Recall that in linear algebra, the term "scalar" is synonymous with "number."

A central theme of mathematics is to study a familiar structure that has objects and rules for combining the objects, and then from the familiar structure ascertain what principles are truly important. One then abstracts the principles to create a more general structure. This is what we will do in this chapter. In the first section, we study $\mathbb{R}^n$ to gain some intuition for the more abstract structure. In later sections, we will define a vector space to be a mathematical structure that obeys a group of axioms. A vector will be an object in a vector space that will not necessarily have any geometric properties attached to it.

3.1 Vector Spaces in $\mathbb{R}^n$

To develop some intuition for vector spaces, we start by giving some examples of vector spaces. We first consider the Cartesian plane in two dimensions, which is the set of ordered pairs of real numbers. We denote this set as $\mathbb{R}^2$, so

$$\mathbb{R}^2 = \left\{ (x, y) \mid x, y \in \mathbb{R} \right\}.$$

An advantage of this particular vector space is that it is easily visualized. An aid in the visualization is to associate with each point (x, y) in the plane the unique arrow from the origin $(0, 0)$ to the point (x, y). A description of a vector that is often given in beginning physics is a "vector is a quantity that has magnitude and direction," and arrows provide an excellent visual representation of such quantities. If one takes the definition of equality in this setting as meaning two vectors are equal if they have the same magnitude and direction, then we can move the arrow associated with a vector as long as we don't change the length or orientation of the arrow. It is common to denote a vector in $\mathbb{R}^2$ in one of two ways: either an ordered pair of numbers as originally described, or as a single letter that is usually distinguished by having the letter with a "hat" ($\hat{a}$). Associated with each vector space is a field of scalars. For our purposes, the scalar field will be the real numbers or complex numbers. If the vector space is $\mathbb{R}^n$, then the scalar field is the real numbers. Two arithmetic

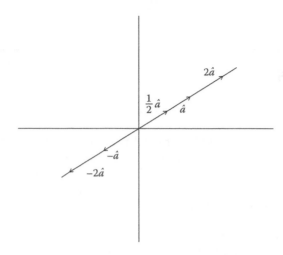

FIGURE 3.1
Scalar multiples of a vector.

operations that one defines on this set are scalar multiplication, that is, multiplication by a real number, and vector addition.

Arithmetically, if $\hat{a} = (a_1,\ a_2), \hat{b} = (b_1,\ b_2)$, and λ is a scalar, then we define

$$\lambda\hat{a} = (\lambda a_1,\ \lambda a_2) \quad \text{and} \quad \hat{a} + \hat{b} = (a_1 + b_1,\ a_2 + b_2).$$

The effect of scalar multiplication by a positive number is to change the length but not the direction of the vector and if the scalar is negative, the effect is to point the arrow in the opposite direction and also change the magnitude by the absolute value of the scalar. See Figure 3.1.

Geometrically, vector addition can be described by placing the tail of the second vector on the head of the first. The sum of the vectors is the arrow from the tail of the first vector to the head of the second. See Figure 3.2.

To create an algebra, we need the zero vector, which is defined as $\hat{0} = (0,\ 0)$. We also need the negative of a vector. If

$$\hat{v} = (v_1,\ v_2)$$

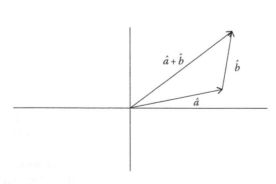

FIGURE 3.2
Diagram of vector addition.

then we define

$$-\hat{v} = (-1)\hat{v} = (-v_1, -v_2).$$

Note that

$$\hat{v} + (-\hat{v}) = \hat{0}.$$

We will normally write

$$\hat{v} + (-\hat{w})$$

as

$$\hat{v} - \hat{w}.$$

Examples

Let

$$\hat{u} = (2, -3), \quad \hat{v} = (0,1), \quad \hat{w} = (-2, -5).$$

Then

$$\hat{u} - \hat{v} = (2, -4)$$

$$3\hat{u} + 4\hat{w} = 3(2, -3) + 4(-2, -5) = (6, -9) + (-8, -20) = (-2, -29)$$

$$4(\hat{u} + 3\hat{v}) = 4\big[(2, -3) + 3(0,1)\big] = 4\big[(2, -3) + (0,3)\big] = 4(2,0) = (8,0).$$

With these definitions, we have the following properties of $\mathbb{R}^2$.

Theorem 1

For

$$\hat{u} = (u_1, u_2), \quad \hat{v} = (v_1, v_2), \quad \hat{w} = (w_1, w_2), \quad \text{and scalars } \alpha, \beta$$

we have

$$(\hat{u} + \hat{v}) \in \mathbb{R}^2, \quad \alpha\hat{u} \in \mathbb{R}^2$$

$$\hat{u} + \hat{v} = \hat{v} + \hat{u}$$

$$\hat{u} + (\hat{v} + \hat{w}) = (\hat{u} + \hat{v}) + \hat{w}$$

$$\hat{u} + \hat{0} = \hat{0} + \hat{u} = \hat{u}$$

$$\hat{u} + \left(-\hat{u}\right) = \hat{0}$$

$$0\hat{u} = \hat{0}$$

$$\alpha\left(\beta\hat{u}\right) = \left(\alpha\beta\right)\hat{u}$$

$$\alpha\left(\hat{u} + \hat{v}\right) = \alpha\hat{u} + \alpha\hat{v}$$

$$\left(\alpha + \beta\right)\hat{u} = \alpha\hat{u} + \beta\hat{u}$$

$$1\hat{u} = \hat{u}$$

We have similar properties in $\mathbb{R}^3$ except using ordered triples instead of ordered pairs. In $\mathbb{R}^n$, we use ordered n–tuples but do not have the capability of drawing vectors as arrows.

For an abstract vector space, we hypothesize a collection of objects (vectors), a scalar field, and two ways of combining the objects that we will call vector addition and scalar multiplication that obey the properties in Theorem 1. This will be the definition of a vector space.

While a vector space will only be required to have the structure mentioned earlier, the vector spaces $\mathbb{R}^n$ have an additional property that enables us to determine the length of a vector and the angle between two vectors. This is the (Euclidean) dot product of vectors.

Definition

If $\hat{u} = \left(u_1, \ldots, u_n\right)$ and $\hat{v} = \left(v_1, \ldots, v_n\right)$ are vectors in $\mathbb{R}^n$, the Euclidean dot product of $\hat{u}$ and $\hat{v}$, denoted $\hat{u} \cdot \hat{v}$, is defined by

$$\hat{u} \cdot \hat{v} = u_1 v_1 + \cdots + u_n v_n.$$

The Euclidean dot product has the following properties.

Theorem 2

For

$$\hat{u} = \left(u_1, \ldots, u_n\right), \quad \hat{v} = \left(v_1, \ldots, v_n\right), \quad \text{and}$$

$$\hat{w} = \left(w_1, \ldots, w_n\right) \text{ and real numbers } \alpha, \beta$$

we have

$$\hat{u} \cdot \hat{v} = \hat{v} \cdot \hat{u}$$

$$\hat{u} \cdot \left(\hat{v} + \hat{w}\right) = \hat{u} \cdot \hat{v} + \hat{u} \cdot \hat{w}$$

$$\alpha\left(\hat{u}\cdot\hat{v}\right)=\left(\alpha\hat{u}\right)\cdot\hat{v}=\hat{u}\cdot\left(\alpha\hat{v}\right)$$

$$\hat{u}\cdot\hat{u}\geq 0 \quad \text{and} \quad \hat{u}\cdot\hat{u}=0, \quad \text{if and only if } \hat{u}=\hat{0}.$$

The dot product provides a way to define the length or norm of a vector in $\mathbb{R}^n$ that is an extension of Pythagoras' theorem.

Definition

If $\hat{u}=\left(u_1,\ldots,u_n\right)$ is a vector in $\mathbb{R}^n$, the Euclidean norm of $\hat{u}$, denoted $\|\hat{u}\|$, is defined by

$$\|\hat{u}\|=\sqrt{\hat{u}\cdot\hat{u}}=\sqrt{u_1^2+\cdots+u_n^2}.$$

The Euclidean distance between $\hat{u}=\left(u_1,\ldots,u_n\right)$ and $\hat{v}=\left(v_1,\ldots,v_n\right)$, denoted $\|\hat{u}-\hat{v}\|$, is defined by

$$\|\hat{u}-\hat{v}\|=\sqrt{\left(u_1-v_1\right)^2+\cdots+\left(u_n-v_n\right)^2}.$$

The length of a vector is often said to be the norm of the vector. A vector of norm 1 is said to be a unit vector.

Theorem 3

The norm has the properties

$$\|\hat{u}\|\geq 0 \quad \text{and}$$

$$\|\hat{u}\|=0 \quad \text{if and only if } \hat{u}=\hat{0}$$

$$\|\alpha\hat{u}\|=|\alpha|\|\hat{u}\|$$

$$\|\hat{u}+\hat{v}\|\leq\|\hat{u}\|+\|\hat{v}\|.$$

We will not prove this result now because we will give proofs in a more general context later.

Theorem 4

In $\mathbb{R}^2$ and $\mathbb{R}^3$, the angle between two nonzero vectors $\hat{a}$ and $\hat{b}$ is determined by

$$\cos\theta=\frac{\hat{a}\cdot\hat{b}}{\|\hat{a}\|\|\hat{b}\|}.$$

Proof

Consider the diagram in Figure 3.3.

By the law of cosines, we have

$$\left\|\hat{a}-\hat{b}\right\|^2 = \left\|\hat{a}\right\|^2 + \left\|\hat{b}\right\|^2 - 2\left\|\hat{a}\right\|\left\|\hat{b}\right\|\cos\theta.$$

Now

$$\left\|\hat{a}-\hat{b}\right\|^2 = \left(\hat{a}-\hat{b}\right)\cdot\left(\hat{a}-\hat{b}\right) = \hat{a}\cdot\hat{a} - \hat{a}\cdot\hat{b} - \hat{b}\cdot\hat{a} - \hat{b}\cdot\hat{b} = \left\|\hat{a}\right\|^2 + \left\|\hat{b}\right\|^2 - 2\hat{a}\cdot\hat{b}$$

so

$$\left\|\hat{a}\right\|^2 + \left\|\hat{b}\right\|^2 - 2\hat{a}\cdot\hat{b} = \left\|\hat{a}\right\|^2 + \left\|\hat{b}\right\|^2 - 2\left\|\hat{a}\right\|\left\|\hat{b}\right\|\cos\theta$$

and

$$\hat{a}\cdot\hat{b} = \left\|\hat{a}\right\|\left\|\hat{b}\right\|\cos\theta.$$

We will see that a particularly important relationship that two nonzero vectors can have with one another is to be orthogonal. In $\mathbb{R}^n$, this means they are perpendicular, and this occurs exactly when $\hat{u}\cdot\hat{v} = 0$.

In abstract settings, there are some vector spaces that have a structure called an inner product, which is similar to the Euclidean dot product. The inner product of vectors $\hat{u}$ and $\hat{v}$ is a number that is denoted $\langle\hat{u}, \hat{v}\rangle$, and the square of the norm of a vector $\hat{u}$ is

$$\left\|\hat{u}\right\|^2 = \langle\hat{u}, \hat{u}\rangle.$$

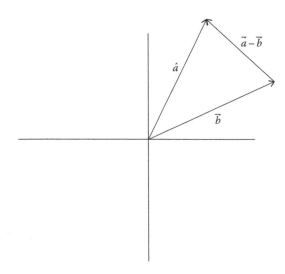

FIGURE 3.3
Diagram showing the difference of vectors.

One can still form

$$\frac{\langle \hat{u}, \hat{v} \rangle}{\|\hat{u}\|\|\hat{v}\|}$$

as long as $\|\hat{u}\|\|\hat{v}\| \neq 0$, but one must be cautious about a geometrical interpretation.

Example

Let $\hat{u} = (3, -1, 6), \hat{v} = (0, 2, 2)$. Then

$$\hat{u} \cdot \hat{v} = (3)(0) + (-1)(2) + (6)(2) = 10$$

$$\|\hat{u}\| = \sqrt{3^2 + (-1)^2 + 6^2} = \sqrt{46}$$

$$\|\hat{v}\| = \sqrt{0^2 + 2^2 + 2^2} = \sqrt{8}$$

$$\cos\theta = \frac{\hat{u} \cdot \hat{v}}{\|\hat{u}\|\|\hat{v}\|} = \frac{10}{\sqrt{8}\sqrt{46}} \approx 0.5213, \quad \theta \approx 58.6^\circ.$$

$$\hat{u} + \hat{v} = (3, -1, 6) + (0, 2, 2) = (3, 1, 8)$$

$$\|\hat{u} + \hat{v}\| = \sqrt{3^2 + 1^2 + 8^2} = \sqrt{74}$$

It is often convenient to have a vector of length 1 that has the same direction as a given vector. If $\hat{v}$ is a nonzero vector, then

$$\frac{1}{\|\hat{v}\|}\hat{v} \quad \text{or} \quad \frac{\hat{v}}{\|\hat{v}\|}$$

is a vector of length 1 that has the same direction as $\hat{v}$, as we show in the exercises.

Example

Let $\hat{v} = (2, -1, -2)$. Then

$$\|\hat{v}\| = \sqrt{2^2 + (-1)^2 + (-2)^2} = 3$$

so

$$\frac{\hat{v}}{\|\hat{v}\|} = \frac{1}{3}(2, -1, -2) = (2/3, -1/3, -2/3).$$

Exercises

1. Let $\hat{u} = (2,3)$, $\hat{v} = (6,-2)$.
   Find $\hat{u} \cdot \hat{v}$, $\|\hat{u}\|$, $\|\hat{v}\|$, $\|\hat{u} + \hat{v}\|$, and the angle between $\hat{u}$ and $\hat{v}$.

2. Let $\hat{u} = (1, -4, 2)$, $\hat{v} = (3, 6, 5)$.
   Find $\hat{u} \cdot \hat{v}$, $\|\hat{u}\|$, $\|\hat{v}\|$, $\|\hat{u} + \hat{v}\|$, and the angle between $\hat{u}$ and $\hat{v}$.

3. $\hat{u} = (1, 0, -4)$. Find

   (a) $\|\hat{u}\|$

   (b) $\dfrac{\hat{u}}{\|\hat{u}\|}$

   (c) $\left\| \dfrac{\hat{u}}{\|\hat{u}\|} \right\|$

   (d) If $\hat{v}$ is any nonzero vector, show that

   $$\left\| \frac{\hat{v}}{\|\hat{v}\|} \right\| = 1$$

4. Find a unit vector in the same direction as the given vectors.

   (a) $(6, -2)$
   (b) $(4, 6)$
   (c) $(3, 1, -3)$
   (d) $(4, 2, 0)$

5. The points $(2, 1, -4)$ and $(-1, -3, \sqrt{11})$ are points on a sphere of radius $\sqrt{21}$. How far is it between the two points traveling on the surface of the sphere?

6. Find y so that $(2, -3)$ is orthogonal to $(4, y)$ with the Euclidean dot product.

7. Find two vectors of length 1 that are orthogonal to the nonzero vector (a, b).

8. Solve for x and y in the equation

$$x(2, 3) + y(1, 4) = (6, -8).$$

9. Why is it not possible to solve for x and y in the equation

$$x(2, 3) + y(4, 6) = (6, 6)?$$

10. Find unit two vectors that are perpendicular to both of the vectors

$$(2, 1, -4) \text{ and } (0, 2, 4).$$

11. Show that if $\hat{u}$ is orthogonal to $\hat{v}$ then $\|\hat{u} + \hat{v}\|^2 = \|\hat{u}\|^2 + \|\hat{v}\|^2$.

12. Show that $\|\hat{u} + \hat{v}\|^2 + \|\hat{u} - \hat{v}\|^2 = 2\|\hat{u}\|^2 + 2\|\hat{v}\|^2$.

13. In subsequent sections, we will define an inner product on the vector space V of continuous functions by

$$\langle f(x), g(x) \rangle = \int_{-1}^{1} f(x)g(x)\,dx$$

(a) Show that this satisfies the properties of the Euclidean dot product.

(b) Show that

$$f(x) = 1 \quad \text{and} \quad g(x) = x^3$$

are orthogonal with respect to this inner product.

3.2 Axioms and Examples of Vector Spaces

In the remainder of the chapter, we will study vector spaces in a more abstract sense.

A vector space V over a scalar field $\mathcal{F}$ is a mathematical structure for which two operations can be performed.

The scalar field will be either the real numbers or complex numbers.

The two operations are vector addition and scalar multiplication.

We hypothesize that if $\hat{u}, \hat{v} \in V$, and $\alpha \in \mathcal{F}$ then

$$\hat{u} + \hat{v} \in V \quad \text{and} \quad \alpha\hat{u} \in V.$$

The first condition says that V is closed under vector addition and the second condition says that V is closed under scalar multiplication.

These operations must satisfy the rules of arithmetic that are given as follows.

Definition

A vector space V over a field $\mathcal{F}$ (the scalars) is a nonempty set of objects (the vectors) together with the two operations, addition (+) and scalar multiplication ($\cdot$)

$$+ : V \times V \to V$$

$$\cdot : \mathcal{F} \times V \to V$$

that satisfy the following axioms for all vectors $\hat{u}, \hat{v}, \hat{w} \in V$, and scalars $a, b \in \mathcal{F}$:

1. $\hat{u} + \hat{v} = \hat{v} + \hat{u}$.
2. $(\hat{u} + \hat{v}) + \hat{w} = \hat{u} + (\hat{v} + \hat{w})$.
3. There is a vector $\hat{0} \in V$ for which $\hat{0} + \hat{u} = \hat{u}$ for all $\hat{u} \in V$.
4. For each vector $\hat{u} \in V$, there is a vector $-\hat{u} \in V$ for which

$$\hat{u} + (-\hat{u}) = \hat{0}.$$

NOTE: We normally denote scalar multiplication by juxtaposition, that is

$$a \cdot b = ab \quad \text{and} \quad a \cdot \hat{u} = a\hat{u}.$$

5. $a(\hat{u} + \hat{v}) = a\hat{u} + a\hat{v}$.

6. $(a+b)\hat{u} = a\hat{u} + b\hat{u}$.

7. $a(b\hat{u}) = (ab)\hat{u}$.

8. For each vector $\hat{u}$, $1\hat{u} = \hat{u}$.

The vector $\hat{0}$ is called the additive identity on V, and the vector $(-\hat{u})$ is called the additive inverse of $\hat{u}$.

Notice that we never say what vectors are, only what they do. This means we can use a single theory to deal with very different objects.

The rules are simply what one would expect of a mathematical structure.

As was stated earlier, the scalar field for the vector spaces we use will be assumed to be either the real numbers $\mathbb{R}$ or the complex numbers $\mathbb{C}$. In some scenarios, there will be differences in results depending on which field is the scalar field. In those cases, the scalar field that is being used will be emphasized.

Examples

Some examples of vector spaces are

$$\mathbb{R}^2 = \left\{ \begin{pmatrix} a \\ b \end{pmatrix} \middle| a, b \in \mathbb{R} \right\} \quad \text{or} \quad \{(a,b) | a, b \in \mathbb{R}\}$$

$$\mathbb{R}^3 = \left\{ \begin{pmatrix} a \\ b \\ c \end{pmatrix} \middle| a, b, c \in \mathbb{R} \right\} \quad \text{or} \quad \{(a,b,c) | a, b, c \in \mathbb{R}\}$$

$$\mathbb{R}^n = \left\{ \begin{pmatrix} x_1 \\ \vdots \\ x_n \end{pmatrix} \middle| x_1,\!,\!,\!,\!,x_n \in \mathbb{R} \right\} \quad \text{or} \quad \{(x_1,\ldots,x_n) | x_1,\ldots,x_n \in \mathbb{R}\}$$

$$\mathbb{C}^n = \left\{ \begin{pmatrix} x_1 \\ \vdots \\ x_n \end{pmatrix} \middle| x_1,\!,\!,\!,\!,x_n \in \mathbb{C} \right\} \quad \text{or} \quad \{(x_1,\ldots,x_n) | x_1,\ldots,x_n \in \mathbb{C}\}$$

Note that the vectors in these vector spaces are particular types of matrices.

The operations on these vector spaces are the matrix addition and scalar multiplication that were defined in Chapter 1.

Less familiar examples of structures that are vector spaces include

(i) The set of $m \times n$ matrices with entries from $\mathbb{R}$, denoted $\mathcal{M}_{m \times n}(\mathbb{R})$ with the usual scalar multiplication and matrix addition

(ii) $F[a,b] =$ real valued functions on $[a,b]$ with the usual addition of functions and multiplication of functions by numbers

(iii) $C[a,b] =$ functions that are continuous for $a \leq x \leq b$ with the usual addition of functions and multiplication of functions by numbers

(iv) Polynomials of degree less than or equal to n with coefficients in $\mathbb{R}$, denoted $\mathcal{P}_n(\mathbb{R})$ with the usual addition of polynomials and multiplication of polynomials by numbers

(v) The set of polynomials $\{P(x) \mid P(0) = 0\}$, with the usual addition of polynomials and multiplication of polynomials by numbers

(vi) Solutions to the differential equation

$$y''(t) + a(t)y'(t) + b(t)y(t) = 0,$$

where $a(t)$ and $b(t)$ are continuous functions with the usual addition of functions and multiplication of functions by numbers

3.2.1 Some Examples of Sets That Are Not Vector Spaces

Polynomials of degree n do not form a vector space because the sum of two such polynomials is not necessarily a polynomial of that type. For example

$$2x^3 + 5x - 3 \quad \text{and} \quad -2x^3 + 7x^2$$

are both polynomials of degree three, but their sum, $7x^2 + 5x - 3$, is a polynomial of degree two.

The 2×2 matrices whose determinant is 0 do not constitute a vector space because the sum of two such matrices of that type is not necessarily a matrix of that type. For example

$$\begin{pmatrix} 1 & 0 \\ 0 & 0 \end{pmatrix} + \begin{pmatrix} 0 & 0 \\ 0 & 1 \end{pmatrix} = \begin{pmatrix} 1 & 0 \\ 0 & 1 \end{pmatrix}.$$

The set of polynomials

$$\{P(x) \mid P(0) = 1\}$$

is not a vector space because if

$$P(0) = 1 \quad \text{and} \quad Q(0) = 1, \text{ then } (P + Q)(0) = 1 + 1 = 2.$$

Example

We show $M_{2 \times 2}$ for matrix of the type

$$\begin{pmatrix} a & a \\ 0 & b \end{pmatrix}$$

is a set that is closed under scalar multiplication and addition. Later, we will show that is sufficient to ensure they form a vector space.

This means we are considering matrices A where the A_{11} and A_{12} entries are the same, the A_{24} entry can be any value, and the A_{23} entry is 0. We want to know whether (i) the sum of two matrices with those characteristics also has those characteristics and (ii) a scalar multiple of a matrix with those characteristics also has those characteristics. Let

$$A = \begin{pmatrix} a_{11} & a_{11} \\ 0 & \alpha \end{pmatrix}, \quad B = \begin{pmatrix} b_{11} & b_{11} \\ 0 & \beta \end{pmatrix}.$$

Then

$$A + B = \begin{pmatrix} a_{11} & a_{11} \\ 0 & \alpha \end{pmatrix} + \begin{pmatrix} b_{11} & b_{11} \\ 0 & \beta \end{pmatrix} = \begin{pmatrix} a_{11} + b_{11} & a_{11} + b_{11} \\ 0 & \alpha + \beta \end{pmatrix}$$

which has the required properties, and

$$cA = c \begin{pmatrix} a_{11} & a_{11} \\ 0 & \alpha \end{pmatrix} = \begin{pmatrix} ca_{11} & ca_{11} \\ 0 & c\alpha \end{pmatrix}$$

which also has the required properties.

However, if we consider matrices of the type

$$\begin{pmatrix} a & a \\ 1 & b \end{pmatrix}$$

then the A_{11} and A_{12} entries are the same, the A_{24} entry can be any value, and the A_{23} entry is 1. We then have for

$$A = \begin{pmatrix} a_{11} & a_{11} \\ 1 & \alpha \end{pmatrix}, \quad B = \begin{pmatrix} b_{11} & b_{11} \\ 1 & \beta \end{pmatrix}$$

that

$$A + B = \begin{pmatrix} a_{11} & a_{11} \\ 1 & \alpha \end{pmatrix} + \begin{pmatrix} b_{11} & b_{11} \\ 1 & \beta \end{pmatrix} = \begin{pmatrix} a_{11} + b_{11} & a_{11} + b_{11} \\ 2 & \alpha + \beta \end{pmatrix}$$

which is not a matrix of that type since the $(A + B)_{23}$ entry is 2.

3.2.2 Additional Properties of Vector Spaces

The next two theorems note additional properties of vector spaces.

Theorem 5

(a) The additive identity of a vector space is unique.
(b) The additive inverse of a vector is unique.

Proof

The typical way to show that something is unique is to suppose that there are two objects that satisfy the defining property, and then show the two objects are equal.

(a) Suppose that $\hat{0}_1$ and $\hat{0}_2$ are both additive identities.
   Since $\hat{0}_1$ is an additive identity, we have

$$\hat{0}_1 + \hat{0}_2 = \hat{0}_2$$

and since $\hat{0}_2$ is an additive identity, we have

$$\hat{0}_1 + \hat{0}_2 = \hat{0}_1.$$

Thus,

$$\hat{0}_1 = \hat{0}_2.$$

(b) Suppose that $\hat{u}$ and $\hat{v}$ are additive inverses of $\hat{w}$. Then

$$\hat{v} = \hat{v} + \hat{0} = \hat{v} + (\hat{w} + \hat{u}) = (\hat{v} + \hat{w}) + \hat{u} = \hat{0} + \hat{u} = \hat{u}.$$

Theorem 6

If V is a vector space over $\mathcal{F}$ and $\hat{u} \in V, a \in \mathcal{F}$, then

(a) $0\hat{u} = \hat{0}$
(b) $a\hat{0} = \hat{0}$
(c) $(-1)\hat{u} = -\hat{u}.$

Proof

(a) We have

$$0\hat{u} = (0+0)\hat{u} = 0\hat{u} + 0\hat{u}$$

so

$$\hat{0} = (-0\hat{u}) + 0\hat{u} = (-0\hat{u}) + (0\hat{u} + 0\hat{u}) = ((-0\hat{u}) + 0\hat{u}) + 0\hat{u} = \hat{0} + 0\hat{u} = 0\hat{u}.$$

(b) We have

$$a\hat{0} = a(\hat{0} + \hat{0}) = a\hat{0} + a\hat{0}$$

so

$$\hat{0} = (-a\hat{0}) + a\hat{0} = (-a\hat{0}) + a\hat{0} + a\hat{0} = ((-a\hat{0}) + a\hat{0}) + a\hat{0} = a\hat{0}.$$

(c) We have

$$\hat{0} = 0\hat{u} = [(-1)+1]\hat{u} = (-1)\hat{u} + 1\hat{u} = (-1)\hat{u} + \hat{u}.$$

Thus, $(-1)\hat{u}$ is the additive inverse of $\hat{u}$, but by definition $-\hat{u}$ is the additive inverse of $\hat{u}$, and additive inverses are unique, so

$$(-1)\hat{u} = -\hat{u}.$$

Example

In a vector space V over a field $\mathcal{F}$, if $a\hat{u} = \hat{0}$, then $a = 0$ or $\hat{u} = \hat{0}$.
Suppose that $a \neq 0$ and $a\hat{u} = \hat{0}$, then

$$\hat{u} = \frac{1}{a} a\hat{u} = \frac{1}{a}\hat{0} = \hat{0}.$$

Exercises

In Exercises 1 through 12, determine whether the set together with the operations defines a vector space. If it is not a vector space, give the axioms that are not satisfied.

1. Polynomials of the form

$$p(t) = \alpha t$$

  with the usual addition and scalar multiplication.

2. Vectors in $\mathbb{R}^2$ with vector addition and scalar multiplication defined as

$$\begin{pmatrix} u_1 \\ u_2 \end{pmatrix} + \begin{pmatrix} v_1 \\ v_2 \end{pmatrix} = \begin{pmatrix} u_1^2 + v_1^2 \\ u_2^2 + v_2^2 \end{pmatrix}, \quad \alpha \begin{pmatrix} u_1 \\ u_2 \end{pmatrix} = \begin{pmatrix} \alpha u_1 \\ \alpha u_2 \end{pmatrix}.$$

3. Polynomials of the form

$$p(t) = \alpha + t$$

  with the usual addition and scalar multiplication.

4. $M_{2\times2}$ matrices of the form

$$\begin{pmatrix} a & a+b \\ a+b & b \end{pmatrix}$$

  with the usual addition and scalar multiplication.

5. $M_{2\times2}$ matrices of the form

$$\begin{pmatrix} a & 1 \\ 0 & b \end{pmatrix}$$

  with the usual addition and scalar multiplication.

6. Continuous functions, with the usual addition and scalar multiplication that satisfy

$$f(2) = 2.$$

7. Continuous functions, with the usual addition and scalar multiplication that satisfy $f(2)=0$.

8. Vectors in $\mathbb{R}^2$ with vector addition and scalar multiplication defined as

$$\begin{pmatrix} u_1 \\ u_2 \end{pmatrix} + \begin{pmatrix} v_1 \\ v_2 \end{pmatrix} = \begin{pmatrix} u_1 + v_1 \\ u_2 + v_2 \end{pmatrix}, \quad \alpha \begin{pmatrix} u_1 \\ u_2 \end{pmatrix} = \begin{pmatrix} \alpha u_1 \\ 0 \end{pmatrix}.$$

9. Vectors $\begin{pmatrix} x \\ y \end{pmatrix}$ in $\mathbb{R}^2$ with $y \geq 0$, with the usual addition and scalar multiplication.

10. Ordered pairs of the form $\begin{pmatrix} 1 \\ y \end{pmatrix}$ with

$$\begin{pmatrix} 1 \\ y_1 \end{pmatrix} + \begin{pmatrix} 1 \\ y_2 \end{pmatrix} = \begin{pmatrix} 1 \\ y_1 + y_2 \end{pmatrix}$$

$$\alpha \begin{pmatrix} 1 \\ y \end{pmatrix} = \begin{pmatrix} 1 \\ \alpha y \end{pmatrix}.$$

11. Vectors of the form $\begin{pmatrix} 3t \\ 0 \\ 5t \end{pmatrix}$ with the usual addition and scalar multiplication.

12. Vectors of the form $\begin{pmatrix} 3t \\ 1 \\ 5t \end{pmatrix}$ with usual addition and scalar multiplication.

13. Show that even polynomials are a vector space, but odd polynomials are not.

14. (a) Show that diagonal $n \times n$ matrices are a vector space.

    (b) Show that $n \times n$ matrices with trace equal to zero are a vector space.

15. Show that if $\hat{u} + \hat{v} = \hat{u} + \hat{w}$, then $\hat{v} = \hat{w}$.

3.3 Subspaces of a Vector Space

Definition

Let V be a vector space and W a nonempty subset of V. W is a subspace of V if W is a vector space under the same operations as V.

The central problem of this section is to determine when a nonempty subset W of a vector space V is a subspace. We now show that some of the axioms of a vector space will always be satisfied by the elements of W by virtue of those elements being in V.

If $\hat{w}_1, \hat{w}_2 \in W$ then

$$\hat{w}_1 + \hat{w}_2 = \hat{w}_2 + \hat{w}_1$$

because $\hat{w}_1 + \hat{w}_2 = \hat{w}_2 + \hat{w}_1$ for any elements in V. However, we do not automatically know that $\hat{w}_1 + \hat{w}_2$ is an element of W, and this is required in order for W to be a vector space.

Similarly, if $\hat{w}_1, \hat{w}_2, \hat{w}_3 \in W$ then

$$\left(\hat{w}_1 + \hat{w}_2\right) + \hat{w}_3 = \hat{w}_1 + \left(\hat{w}_2 + \hat{w}_3\right)$$

but we do not automatically know that $\left(\hat{w}_1 + \hat{w}_2\right) + \hat{w}_3$ is an element of W, and this is required in order for W to be a vector space. Thus, in order for W to be a vector space, it must be that W is closed under vector addition.

In a similar vein, if $\hat{w}_1, \hat{w}_2 \in W$, and $a, b \in \mathcal{F}$, then

$$a\left(b\hat{w}_1\right) = \left(ab\right)\hat{w}_1 \quad \text{and} \quad a\left(\hat{w}_1 + \hat{w}_2\right) = a\hat{w}_1 + a\hat{w}_2$$

but there is nothing that ensures any of these are in W. Thus, in order for W to be a vector space, it must be closed under scalar multiplication.

The next theorem shows that closure under vector addition and scalar multiplication is sufficient for a nonempty set of vectors from a vector space to be a subspace.

Theorem 7

A nonempty subset W of the vector space V is a subspace of V, if and only if

  (i) For every $\hat{u}, \hat{v} \in W$, we have $\hat{u} + \hat{v} \in W$ and
  (ii) For every $\hat{u} \in W$ and $a \in \mathcal{F}$, we have $a\hat{u} \in W$.

Proof

We must verify that the axioms of a vector space are satisfied by W.

If $\hat{0}$ is the only vector in W, then the proof is immediate.

All of the axioms except for (3) and (4) are immediate since the elements in the subset follow the rules of vector addition and scalar multiplication as in the vector space.

If $\hat{u} \in W$, then $-1\hat{u} = -\hat{u} \in W$, so axiom (4) is satified.

If $\hat{u} \in W$, then $-\hat{u} \in W$, so $\hat{u} + (-\hat{u}) = \hat{0} \in W$ and axiom (3) is satisfied.

In showing that a set of vectors is nonempty, one can usually observe that the set contains the zero vector.

In Section 3.2, the examples of sets that were not vector spaces failed to be closed under one of the vector operations.

Example

In $\mathbb{R}^3$, the only subspaces are lines through the origin, planes through the origin, $\{\hat{0}\}$ and $\mathbb{R}^3$.

Definition

Let $\{\hat{v}_1, \ldots, \hat{v}_n\}$ be a collection of vectors from the vector space V. A linear combination of $\{\hat{v}_1, \ldots, \hat{v}_n\}$ is a vector of the form

$$\hat{u} = c_1\hat{v}_1 + \cdots + c_n\hat{v}_n \quad \text{where } c_i \in \mathcal{F}.$$

The next theorem highlights a very important class of vector spaces.

Theorem 8

Let $\{\hat{v}_1, \ldots, \hat{v}_n\}$ be a collection of vectors from the vector space V. The set of linear combinations of these vectors is a subspace of V.

Proof

NOTE: Something that is often an impediment to beginners in proving this type of result is choosing representations. In this proof, we will use the previous theorem but we need to represent $\hat{u}$ and $\hat{v}$ in that theorem as they pertain to the present case.

We first observe that

$$\hat{0} = 0\hat{v}_1 + \cdots + 0\hat{v}_n.$$

Let $\hat{u}$ and $\hat{v}$ be linear combinations of $\{\hat{v}_1, \ldots, \hat{v}_n\}$. Then we can write

$$\hat{u} = a_1\hat{v}_1 + \cdots + a_n\hat{v}_n$$
$$\hat{v} = b_1\hat{v}_1 + \cdots + b_n\hat{v}_n,$$

where $a_i, b_i \in \mathcal{F}; i = 1, \ldots, n.$

So

$$\hat{u} + \hat{v} = \left(a_1\hat{v}_1 + \cdots + a_n\hat{v}_n\right) + \left(b_1\hat{v}_1 + \cdots + b_n\hat{v}_n\right) = \left(a_1 + b_1\right)\hat{v}_1 + \cdots + \left(a_n + b_n\right)\hat{v}_n$$

which is a linear combination of $\{\hat{v}_1, \ldots, \hat{v}_n\}$.

Thus, the set is closed under vector addition.

If $c \in \mathcal{F}$, then

$$c\hat{u} = c\left(a_1\hat{v}_1 + \cdots + a_n\hat{v}_n\right) = \left(ca_1\right)\hat{v}_1 + \cdots + \left(ca_n\right)\hat{v}_n$$

which is a linear combination of $\{\hat{v}_1, \ldots, \hat{v}_n\}$.

Thus, the set is closed under scalar multiplication.

Example

Let A be an $m \times n$ matrix and let

$$W = \left\{\hat{x} \in \mathcal{F}^n \mid A\hat{x} = \hat{0}\right\}.$$

The set W is called the null space of A. It will usually be denoted $\mathcal{N}(A)$ and will play an important role in our theory.

We show that W is a subspace of $\mathcal{F}^n$.

We have that W is nonempty since $A\hat{0} = \hat{0}$.

Suppose

$$\hat{u}, \hat{v} \in W, \quad \text{i.e., } A\hat{u} = \hat{0} \text{ and } A\hat{v} = \hat{0}.$$

We must show $\hat{u} + \hat{v} \in W$ and $a\hat{u} \in W$, i.e., we must show

$$A(\hat{u} + \hat{v}) = \hat{0} \quad \text{and} \quad A(a\hat{u}) = \hat{0}.$$

But

$$A(\hat{u}+\hat{v})=A(\hat{u})+A(\hat{v})=\hat{0}+\hat{0}=\hat{0}$$

and

$$A(a\hat{u})=aA(\hat{u})=a\hat{0}=\hat{0}.$$

Another note about proofs. The idea of rephrasing a hypothesis is often helpful. We rephrased the condition $\hat{v} \in W$, as $A\hat{v}=\hat{0}$. It also helps one focus on the idea that part of what needs to be shown is $A(\hat{u}+\hat{v})=\hat{0}$.

Example

(1) We show that vectors of the form

$$\begin{pmatrix} 2a+5b \\ 4a-2b \\ 3b \end{pmatrix}$$

form a subspace of $\mathbb{R}^3$ by showing such vectors are a linear combination of vectors. We have

$$\begin{pmatrix} 2a+5b \\ 4a-2b \\ 3b \end{pmatrix}=\begin{pmatrix} 2a \\ 4a \\ 0 \end{pmatrix}+\begin{pmatrix} 5b \\ -2b \\ 3b \end{pmatrix}=a\begin{pmatrix} 2 \\ 4 \\ 0 \end{pmatrix}+b\begin{pmatrix} 5 \\ -2 \\ 3 \end{pmatrix}.$$

Thus, the set consists of the linear combinations of the vectors

$$\begin{pmatrix} 2 \\ 4 \\ 0 \end{pmatrix} \quad \text{and} \quad \begin{pmatrix} 5 \\ -2 \\ 3 \end{pmatrix}$$

and the set of all linear combinations of a (nonempty) set of vectors form a subspace.

(2) Vectors of the form

$$\begin{pmatrix} 2a \\ 4a \\ 1 \end{pmatrix}$$

do not form a subspace because the sum of two vectors of that type is not a vector of that type. In particular

$$\begin{pmatrix} 2a \\ 4a \\ 1 \end{pmatrix}+\begin{pmatrix} 2b \\ 4b \\ 1 \end{pmatrix}=\begin{pmatrix} 2(a+b) \\ 4(a+b) \\ 2 \end{pmatrix}.$$

Also note that the zero vector is not in the set.

(3) Vectors of the form $(a,b,0)$ form a subspace of $\mathbb{R}^3$, but $\mathbb{R}^2$ is not a subspace of $\mathbb{R}^3$, because elements in $\mathbb{R}^2$ are not in $\mathbb{R}^3$.

Example

We show that

$$\{(x, y, z)|2x - 2y - 3z = 0\}$$

is a subspace of $\mathbb{R}^3$.

The row reduced form of the augmented matrix associated with

$$2x - 2y - 3z = 0$$

is

$$\begin{pmatrix} 1 & -1 & -3/2 & 0 \end{pmatrix}$$

so y and z are free variables and x is a leading variable and we have

$$x = y + \frac{3}{2}z.$$

If $y = r$ and $z = s$, then

$$\begin{pmatrix} x \\ y \\ z \end{pmatrix} = \begin{pmatrix} r + \frac{3}{2}s \\ r \\ s \end{pmatrix} = \begin{pmatrix} r \\ r \\ 0 \end{pmatrix} + \begin{pmatrix} \frac{3}{2}s \\ 0 \\ s \end{pmatrix} = r\begin{pmatrix} 1 \\ 1 \\ 0 \end{pmatrix} + s\begin{pmatrix} \frac{3}{2} \\ 0 \\ 1 \end{pmatrix}.$$

Thus the set is a linear combination of the vectors

$$\begin{pmatrix} 1 \\ 1 \\ 0 \end{pmatrix} \quad \text{and} \quad \begin{pmatrix} \frac{3}{2} \\ 0 \\ 1 \end{pmatrix}$$

and so is a subspace.

Exercises

1. Tell whether the following collections of vectors form a subspace.

   (a) Vectors of the form $\begin{pmatrix} a - b \\ 3a + 2b \\ b \end{pmatrix}$

   (b) Vectors of the form $\begin{pmatrix} 2a - b \\ 0 \\ 3b \end{pmatrix}$

   (c) Vectors of the form $\begin{pmatrix} a - b \\ 3 \\ b \end{pmatrix}$

2. Let S be the set of all 2×2 matrices for which $A = A^T$.

   Show that S is a subspace of the set of all 2×2 matrices

3. Tell whether matrices of the following form constitute a subspace of $M_{2 \times 2}(\mathbb{R})$

   (a) $\begin{pmatrix} a & b \\ 0 & 0 \end{pmatrix}$

   (b) $\begin{pmatrix} a & b \\ -b & a \end{pmatrix}$

   (c) Invertible 2×2 matrices.

4. Let $A \in M_{2 \times 2}(\mathbb{R})$. Tell whether the following sets of matrices form a subspace of $M_{2 \times 2}(\mathbb{R})$.

   (a) $\left\{ B \in M_{2 \times 2}(\mathbb{R}) \mid AB = \hat{0} \right\}$

   (b) $\left\{ B \in M_{2 \times 2}(\mathbb{R}) \mid AB = BA \right\}$

   (c) $\left\{ B \in M_{2 \times 2}(\mathbb{R}) \mid A + B = \hat{0} \right\}$

5. (a) Show that if U and W are subspaces of the vector space V, then $U + W$ is a subspace of V where

$$U + W = \left\{ \hat{u} + \hat{v} \mid \hat{u} \in U, \ \hat{v} \in V \right\}.$$

   (b) Give an example of nonempty subsets U and W of the vector space V for which $U + W$ is not a subspace of V.

6. Show that the set of matrices $A \in M_{2 \times 2}(\mathbb{R})$ for which $A^2 = A$ does not form a subspace.

7. (a) Show that $\{(x, y, z) \mid 2x - y - z = 0\}$ is a subspace of $\mathbb{R}^3$.

   (b) Show that $\{(x, y, z) \mid x - 6y + 3z = 0\}$ is a subspace of $\mathbb{R}^3$.

   (c) Show that $\{(x, y, z) \mid 3x - z = 0\}$ is a subspace of $\mathbb{R}^3$.

   (d) Show that $\{(x, y, z) \mid 2x - y - z = 1\}$ is not a subspace of $\mathbb{R}^3$.

8. Show that the solution sets for the systems of equations form subspaces of $\mathbb{R}^3$.

   (a) $x - 3y + 4z = 0$

   $2x + y = 0s$

   (b) $7x + 5y - 3z = 0$

   $4x + y - 2z = 0$

9. Let U and W be subspaces of the vector space V.

   (a) Show that $U \cap W$ is a subspace of V.

   (b) Given an example where $U \bigcup W$ is not a subspace of V.

   Hint: $V = \mathbb{R}^2$ is a good place to look.

3.4 Spanning Sets, Linearly Independent Sets and Bases

We have shown that the linear combinations of a set of vectors form a subspace of a vector space. In this section, we describe which sets of vectors can build (through linear combinations) the entire vector space in the most efficient way. For a given vector space, there will be many such sets of vectors. Each such set is called a basis of the vector space.

To make the central ideas more intuitive, we present the following metaphor.

Consider a toy construction set that contains parts from which you can make toys—Legos, erector sets, Lincoln logs, and tinker toys are some examples. Such a construction set typically comes with an instruction booklet that tells how to make particular toys from the set. The manufacturer of the construction set has two major considerations. First, the set must contain all of the parts necessary to build each of the toys in the catalog; otherwise, the customers will feel cheated. Second, there should be no surplus parts; otherwise, there is an unnecessary expense and profits will suffer.

We begin with a vector space V. We want to find a collection of vectors $\{\hat{v}_1, \ldots, \hat{v}_n\}$ in V so that

1. Each vector in V can be written as a linear combination of vectors in $\{\hat{v}_1, \ldots, \hat{v}_n\}$
2. The set $\{\hat{v}_1, \ldots, \hat{v}_n\}$ is minimal in the sense that if any one of the vectors is removed from the set, then not all of the vectors in V are linear combinations of the reduced set.

Definition

A vector $\hat{v} \in V$ is said to be in the span of $\{\hat{v}_1, \ldots, \hat{v}_n\}$ if $\hat{v}$ can be written as a linear combination of $\{\hat{v}_1, \ldots, \hat{v}_n\}$. A set of vectors $\{\hat{v}_1, \ldots, \hat{v}_n\}$ is a spanning set for V if every vector in V can be written as a linear combination of $\{\hat{v}_1, \ldots, \hat{v}_n\}$.

Another way to say that vectors $\{\hat{v}_1, \ldots, \hat{v}_n\}$ are a spanning set for V is to say that $\{\hat{v}_1, \ldots, \hat{v}_n\}$ spans V.

Definition

Let V be vector space. A set of vectors $\{\hat{v}_1, \ldots, \hat{v}_n\}$ is linearly independent if the only way

$$a_1\hat{v}_1 + \cdots + a_n\hat{v}_n = \hat{0}$$

is for $a_i = 0$ for every $i = 1, \ldots, n$.

A set of vectors $\{\hat{v}_1, \ldots, \hat{v}_n\}$ is linearly dependent if it is not linearly independent. This means there are scalars $a_1, \ldots, a_n$ not all of which are 0, with

$$a_1\hat{v}_1 + \cdots + a_n\hat{v}_n = \hat{0}.$$

Definition

A set of vectors $\{\hat{v}_1, \ldots, \hat{v}_n\}$ is a basis for V if it is a spanning set for V and is linearly independent.

Relating the definition of a basis to the construction toy metaphor, saying $\{\hat{v}_1, \ldots, \hat{v}_n\}$ is a spanning set is analogous to saying we have enough parts to build all the toys in the catalog and saying the set $\{\hat{v}_1, \ldots, \hat{v}_n\}$ is linearly independent is analogous to saying there are no surplus parts.

Example

The vector space $\mathcal{F}^3$ consists of the vectors

$$\left\{ \begin{pmatrix} a \\ b \\ c \end{pmatrix} \middle| a, b, c \in \mathcal{F} \right\}$$

with the operations of vector addition and scalar multiplication. The most natural basis for this vector space is the set of vectors

$$\left\{ \begin{pmatrix} 1 \\ 0 \\ 0 \end{pmatrix}, \begin{pmatrix} 0 \\ 1 \\ 0 \end{pmatrix}, \begin{pmatrix} 0 \\ 0 \\ 1 \end{pmatrix} \right\}.$$

This is called the usual (or standard) basis for $\mathcal{F}^3$, and we will use the notation

$$\hat{e}_1 = \begin{pmatrix} 1 \\ 0 \\ 0 \end{pmatrix}, \quad \hat{e}_2 = \begin{pmatrix} 0 \\ 1 \\ 0 \end{pmatrix}, \quad \hat{e}_3 = \begin{pmatrix} 0 \\ 0 \\ 1 \end{pmatrix}$$

for the usual basis.

To show that this is a basis, we must show that

1. The set spans $\mathcal{F}^3$.
2. The set is linearly independent.

To show the set spans $\mathcal{F}^3$, let

$$\begin{pmatrix} a \\ b \\ c \end{pmatrix} \in \mathcal{F}^3.$$

Then

$$\begin{pmatrix} a \\ b \\ c \end{pmatrix} = a \begin{pmatrix} 1 \\ 0 \\ 0 \end{pmatrix} + b \begin{pmatrix} 0 \\ 1 \\ 0 \end{pmatrix} + c \begin{pmatrix} 0 \\ 0 \\ 1 \end{pmatrix}.$$

To show the set is linearly independent, suppose that

$$a \begin{pmatrix} 1 \\ 0 \\ 0 \end{pmatrix} + b \begin{pmatrix} 0 \\ 1 \\ 0 \end{pmatrix} + c \begin{pmatrix} 0 \\ 0 \\ 1 \end{pmatrix} = \begin{pmatrix} a \\ b \\ c \end{pmatrix} = \begin{pmatrix} 0 \\ 0 \\ 0 \end{pmatrix}.$$

This is true if and only if $a=0$, $b=0$, and $c=0$.

We will use analogs of this example repeatedly.

Theorem 9

A set of vectors $\{\hat{v}_1, \ldots, \hat{v}_n\}$ is a basis for V if and only if every vector in V can be written as a linear combination of $\{\hat{v}_1, \ldots, \hat{v}_n\}$ in exactly one way.

Proof

Suppose that $\{\hat{v}_1, \ldots, \hat{v}_n\}$ is a basis for V.

Since $\{\hat{v}_1, \ldots, \hat{v}_n\}$ spans V, every vector in V can be written as a linear combination of $\{\hat{v}_1, \ldots, \hat{v}_n\}$.

Suppose that there is a vector $\hat{v} \in V$ that can be written as a linear combination of $\{\hat{v}_1, \ldots, \hat{v}_n\}$ in two ways, say

$$\hat{v} = a_1 \hat{v}_1 + \cdots + a_n \hat{v}_n$$

and

$$\hat{v} = b_1 \hat{v}_1 + \cdots + b_n \hat{v}_n.$$

Then

$$\hat{0} = \hat{v} - \hat{v} = \left(a_1 \hat{v}_1 + \cdots + a_n \hat{v}_n \right) - \left(b_1 \hat{v}_1 + \cdots + b_n \hat{v}_n \right)$$
$$= \left(a_1 - b_1 \right) \hat{v}_1 + \cdots + \left(a_n - b_n \right) \hat{v}_n.$$

So we have written $\hat{0}$ as a linear combination of $\{\hat{v}_1, \ldots, \hat{v}_n\}$. Since $\{\hat{v}_1, \ldots, \hat{v}_n\}$ is a basis, the set is linearly independent, so $(a_i - b_i) = 0$; or $a_i = b_i$, $i = 1, \ldots, n$.

Conversely, suppose that every vector in V can be written as a linear combination of $\{\hat{v}_1, \ldots, \hat{v}_n\}$ in exactly one way.

Because every vector in V can be written as a linear combination of $\{\hat{v}_1, \ldots, \hat{v}_n\}$, the set spans V.

Because

$$\hat{0} = 0\hat{v}_1 + \cdots + 0\hat{v}_n$$

and there is only one linear combination of $\{\hat{v}_1, \ldots, \hat{v}_n\}$ that gives $\hat{0}$, the set $\{\hat{v}_1, \ldots, \hat{v}_n\}$ is linearly independent.

The next theorem says that a linearly dependent set of vectors has some vectors that are, in some sense, superfluous.

Theorem 10

If $\{\hat{v}_1, \hat{v}_2, \ldots, \hat{v}_n\}$ is a linearly dependent set of vectors, then one of the vectors can be written as a linear combination of the others.

The implication of this theorem is that a linearly dependent set of vectors can be reduced so that the reduced set will span the same set as the original set. This is because if

$$\hat{v} = a_1 \hat{v}_1 + a_2 \hat{v}_2 + \cdots + a_n \hat{v}_n$$

and

$$\hat{v}_1 = b_2\hat{v}_2 + \cdots + b_n\hat{v}_n$$

then

$$\hat{v} = a_1\hat{v}_1 + a_2\hat{v}_2 + \cdots + a_n\hat{v}_n = a_1\left(b_2\hat{v}_2 + \cdots + b_n\hat{v}_n\right) + a_2\hat{v}_2 + \cdots + a_n\hat{v}_n = \left(a_1 b_2 + a_2\right)\hat{v}_2 + \cdots + \left(a_1 b_n + a_n\right)\hat{v}_n$$

and the last expression is a linear combination of $\{\hat{v}_2, \ldots, \hat{v}_n\}$.

Proof

Since $\{\hat{v}_1, \hat{v}_2, \ldots, \hat{v}_n\}$ is a linearly dependent set of vectors, it is possible to find scalars $c_1, c_2, \ldots, c_n$, not all of which are zero so that

$$c_1\hat{v}_1 + c_2\hat{v}_2 + \cdots + c_n\hat{v}_n = \hat{0}.$$

Suppose we have arranged the vectors so that $c_1 \neq 0$. Then

$$c_1\hat{v}_1 = -c_2\hat{v}_2 - \cdots - c_n\hat{v}_n$$

and since $c_1 \neq 0$ we may divide by c_1 to get

$$\hat{v}_1 = -\left(\frac{c_2}{c_1}\right)\hat{v}_2 - \cdots - \left(\frac{c_n}{c_1}\right)\hat{v}_n.$$

Thus $\hat{v}_1$ is a linear combination of $\{\hat{v}_2, \ldots, \hat{v}_n\}$.

We will soon be precise about what is meant by the dimension of a vector space. For now, simply realize that it means something about the size of a vector space.

Every vector space has a basis. (This is a relatively deep fact.) A finite dimensional vector space is defined to be a vector space that has a finite spanning set. We will show that such a vector space has a basis. A vector space that is not finite dimensional is said to be infinite dimensional. (Showing that an infinite dimensional vector space has a basis is proven using Zorn's lemma or a logically equivalent axiom, which is beyond the scope of this text.)

From now on, we will assume that the vector spaces we consider are finite dimensional.

In the remainder of this section and the next section, we want to prove the following facts.

1. Every basis of a given vector space has the same number of vectors. We call this number the dimension of the vector space.
2. Every linearly independent set of vectors that is not a basis can be expanded to be a basis. We give an argument for why this is true and an algorithm that accomplishes the expansion.
3. Every spanning set of vectors that is not linearly independent can be reduced to a basis. We give an argument for why this is true and an algorithm that accomplishes the reduction.

Our immediate task is to show that all bases of a given vector space have the same number of vectors. This is accomplished by Theorems 11 and 12.

Theorem 11

Suppose that $\{v_1, \ldots, v_n\}$ is a basis for the vector space V. Then any set of more than n vectors will be linearly dependent.

Proof

We demonstrate the proof in the case where a basis consists of two vectors. Suppose $\{\hat{v}_1, \hat{v}_2\}$ is a basis for the vector space V. Suppose that $\hat{w}_1$, $\hat{w}_2$, and $\hat{w}_3$ are vectors in V. We will show they are linearly dependent, that is, we will show that there are b_1, b_2, b_3 not all zero, with

$$b_1\hat{w}_1 + b_2\hat{w}_2 + b_3\hat{w}_3 = \hat{0}.$$

Since $\{\hat{v}_1, \hat{v}_2\}$ spans V, there are a_{11} and a_{12} with

$$\hat{w}_1 = a_{11}\hat{v}_1 + a_{12}\hat{v}_2$$

a_{21} and a_{22} with

$$\hat{w}_2 = a_{21}\hat{v}_1 + a_{22}\hat{v}_2$$

and a_{31} and a_{32} with

$$\hat{w}_3 = a_{31}\hat{v}_1 + a_{32}\hat{v}_2.$$

So

$$b_1\hat{w}_1 + b_2\hat{w}_2 + b_3\hat{w}_3 = b_1\left(a_{11}\hat{v}_1 + a_{12}\hat{v}_2\right) + b_2\left(a_{21}\hat{v}_1 + a_{22}\hat{v}_2\right) + b_3\left(a_{31}\hat{v}_1 + a_{32}\hat{v}_2\right)$$

$$= \left(b_1a_{11} + b_2a_{21} + b_3a_{31}\right)\hat{v}_1 + \left(b_1a_{12} + b_2a_{22} + b_3a_{33}\right)\hat{v}_2 = \hat{0}.$$

Now the a_{ij} are all fixed, and the question comes down to whether it is possible to find b_1, b_2, and b_3 not all zero, with

$$b_1a_{11} + b_2a_{21} + b_3a_{31} = 0$$

and

$$b_1a_{12} + b_2a_{22} + b_3a_{32} = 0.$$

Since there are two equations with three unknowns, there will be a free variable in the solution, and thus there is a solution with not all of b_1, b_2, b_3 equal to zero.

The general case uses the same ideas, and we leave the write-up as an exercise.

Theorem 12

If $\{\hat{v}_1,\ldots,\hat{v}_n\}$ is a basis for the vector space V, then any set of fewer than n vectors will not span V.

Proof

We again present a simple case that highlights the idea of the proof and leave the proof of the general case as an exercise.

Suppose that $\{\hat{v}_1, \hat{v}_2, \hat{v}_3\}$ is a basis for the vector space V and suppose that $\{\hat{w}_1, \hat{w}_2\}$ spans V. Then

$$\hat{v}_1 = a_{11}\hat{w}_1 + a_{12}\hat{w}_2$$
$$\hat{v}_2 = a_{21}\hat{w}_1 + a_{22}\hat{w}_2$$
$$\hat{v}_3 = a_{31}\hat{w}_1 + a_{32}\hat{w}_2.$$

We will show that $\{\hat{v}_1, \hat{v}_2, \hat{v}_3\}$ is not a linearly independent set.

Now,

$$b_1\hat{v}_1 + b_2\hat{v}_1 + b_3\hat{v}_3 = b_1\left(a_{11}\hat{w}_1 + a_{12}\hat{w}_2\right) + b_2\left(a_{21}\hat{w}_1 + a_{22}\hat{w}_2\right) + b_3\left(a_{31}\hat{w}_1 + a_{32}\hat{w}_2\right)$$

$$= \left(b_1a_{11} + b_2a_{21} + b_3a_{31}\right)\hat{w}_1 + \left(b_1a_{12} + b_2a_{22} + b_3a_{32}\right)\hat{w}_2.$$

Consider the system of equations

$$b_1a_{11} + b_2a_{21} + b_3a_{31} = 0$$
$$b_1a_{12} + b_2a_{22} + b_3a_{32} = 0$$

There are two equations and three unknowns (b_1, b_2, b_3) so it is possible to find a solution such that not every $b_i = 0$. This contradicts the assumption that $\{\hat{v}_1, \hat{v}_2, \hat{v}_3\}$ is a basis.

Corollary

In a finite dimensional vector space, all bases have the same number of vectors.

Corollary

The number of vectors in a basis for $\mathcal{F}^n$ is n.

Proof

The standard basis for $\mathcal{F}^n$ has n vectors.

Definition

The number of vectors in a basis of a vector space is the dimension of the vector space.

Definition

Two vector spaces U and V over the same field are isormorphic if there is a one-to-one and onto function

$$T : U \to V$$

for which

$$T\left(\alpha_1 \hat{u}_1 + \alpha_2 \hat{u}_2\right) = \alpha_1 T\left(\hat{u}_1\right) + \alpha_2 T\left(\hat{u}_2\right)$$

For all $\hat{u}_1, \hat{u}_2 \in U$; $\alpha_1, \alpha_2 \in \mathcal{F}$.

Isomorphic vector spaces are structurally the same; the only real difference is how the elements are named. The next theorem gives a succinct categorization of finite dimensional vector spaces.

Theorem 13

Every n-dimensional *vector* space V over a field $\mathcal{F}$ is isomorphic to $\mathcal{F}^n$.

Proof

Let V be a vector space with basis $\{\hat{v}_1, \ldots, \hat{v}_n\}$ and suppose $\hat{v} \in V$. There is a unique $n -$ tuple of scalars $(\alpha_1, \ldots, \alpha_n) \in \mathcal{F}^n$ for which

$$\hat{v} = \alpha_1 \hat{v}_1 + \cdots + \alpha_n \hat{v}_n.$$

Define

$$T : V \to \mathcal{F}^n$$

By

$$T\left(\hat{v}\right) = T\left(\alpha_1 \hat{v}_1 + \cdots + \alpha_n \hat{v}_n\right) = \left(\alpha_1, \ldots, \alpha_n\right).$$

The function T is one-to-one and onto.

We show $T\left(\gamma \hat{v} + \delta \hat{w}\right) = \gamma T\left(\hat{v}\right) + \delta T\left(\hat{w}\right)$.

Suppose $\hat{w} = \beta_1 \hat{v}_1 + \cdots + \beta_n \hat{v}_n$. Then

$$
\begin{aligned}
T\left(\gamma \hat{v} + \delta \hat{w}\right) &= T\left(\gamma\left(\alpha_1 \hat{v}_1 + \cdots + \alpha_n \hat{v}_n\right) + \delta\left(\beta_1 \hat{v}_1 + \cdots + \beta_n \hat{v}_n\right)\right) \\
&= T\left(\left(\gamma \alpha_1 + \delta \beta_1\right) \hat{v}_1 + \cdots + \left(\gamma \alpha_n + \delta \beta_n\right) \hat{v}_n\right) \\
&= \left(\left(\gamma \alpha_1 + \delta \beta_1\right), \ldots, \left(\gamma \alpha_n + \delta \beta_n\right)\right) \\
&= \left(\gamma\left(\alpha_1, \ldots, \alpha_n\right) + \delta\left(\beta, \ldots, \beta_n\right)\right) \\
&= \gamma T\left(\hat{v}\right) + \delta T\left(\hat{w}\right).
\end{aligned}
$$

The question that should arise is, why bother with vector spaces such as $\mathcal{P}_n$. Part of the answer is that often an intuition that is tied to a particular vector space could be lost by considering only $\mathcal{F}^n$.

We would like to be able to easily identify when a set of vectors is a basis for a subspace. The next theorems provide some ways of doing that.

Theorem 14

A set of vectors that contains the zero vector is not linearly independent, and thus is not a basis.

Proof

The proof is left as an exercise.

The most common vector spaces that we encounter are of the form $\mathcal{F}^n$. A set of vectors that does not have exactly n vectors cannot be a basis for $\mathcal{F}^n$ because we have identified the standard basis of $\mathcal{F}^n$ as having n vectors. If the set does have n vectors, then we must do further analysis.

The next theorem gives an easy way to tell whether a set of n vectors forms a basis of $\mathcal{F}^n$.

Theorem 15

A set of n vectors of $\mathcal{F}^n$ is a basis for $\mathcal{F}^n$ if and only if the determinant of the matrix whose columns (or rows) are the given vectors is not 0.

The essential idea of the proof is given in Exercise 8.

Example

Determine whether the following sets of vectors form a basis for the appropriate vector spaces:

(a) $\left\{ \begin{pmatrix} 1 \\ 3 \\ -2 \end{pmatrix}, \begin{pmatrix} 5 \\ -9 \\ 0 \end{pmatrix}, \begin{pmatrix} 7 \\ -3 \\ -4 \end{pmatrix} \right\}$. There are three vectors and the vector space is $\mathcal{F}^3$ so it is possible that the set is a basis. We form the matrix

$$A = \begin{pmatrix} 1 & 5 & 7 \\ 3 & -9 & -3 \\ -2 & 0 & -4 \end{pmatrix}$$

and find $det(A)=0$, so the set is not a basis.

(b) $\left\{ \begin{pmatrix} 1 \\ -1 \\ 2 \\ 0 \end{pmatrix}, \begin{pmatrix} 0 \\ 3 \\ 3 \\ -5 \end{pmatrix}, \begin{pmatrix} 2 \\ 0 \\ -1 \\ -4 \end{pmatrix}, \begin{pmatrix} 6 \\ 1 \\ 1 \\ 1 \end{pmatrix} \right\}$. There are four vectors and the vector space is $\mathcal{F}^4$ so it is possible that the set is a basis. We form the matrix

$$B = \begin{pmatrix} 1 & 0 & 2 & 6 \\ -1 & 3 & 0 & 1 \\ 2 & 3 & -1 & 1 \\ 0 & -5 & -4 & 1 \end{pmatrix}$$

and find $det(B)=-302$, so the set is a basis.

Exercises

In Exercises 1 through 6, determine whether the sets of vectors are linearly independent.

1. $\left\{ \begin{pmatrix} 1 \\ 3 \end{pmatrix}, \begin{pmatrix} 2 \\ 7 \end{pmatrix} \right\}$ in $\mathbb{R}^2$

2. $\left\{ \begin{pmatrix} 1 \\ 3 \\ 2 \end{pmatrix}, \begin{pmatrix} 4 \\ 6 \\ 9 \end{pmatrix}, \begin{pmatrix} 9 \\ 15 \\ 20 \end{pmatrix} \right\}$ in $\mathbb{R}^3$

3. $\left\{ \begin{pmatrix} 1 \\ 1 \\ 2 \end{pmatrix}, \begin{pmatrix} 2 \\ -1 \\ 0 \end{pmatrix}, \begin{pmatrix} 1 \\ -2 \\ -2 \end{pmatrix} \right\}$ $\mathbb{R}^3$

4. $\{2, \sin^2 x, \cos^2 x\}$ in continuous functions.

5. $\{1+x, 2x+x^2, x^2\}$ in $P_2(x)$

6. $\left\{ \begin{pmatrix} 2 & 0 \\ 3 & 1 \end{pmatrix}, \begin{pmatrix} -1 & 4 \\ 0 & 6 \end{pmatrix}, \begin{pmatrix} 0 & 8 \\ 3 & 13 \end{pmatrix} \right\}$ in $M_{2\times2}(\mathbb{R})$

7. Show that if $\{\hat{v}_1, \hat{v}_2, \ldots, \hat{v}_n\}$ is a linearly independent set of vectors, then any subset of $\{\hat{v}_1, \hat{v}_2, \ldots, \hat{v}_n\}$ is a linearly independent set of vectors.

8. In this exercise, we justify the claim that the determinant of a matrix is zero if and only if one of the rows is a linear combination of the other rows. This will mean that the rows are linearly dependent and thus do not form a basis. We demonstrate this in the case of a 3×3 matrix.

   (a) Show that

   $$det \begin{pmatrix} a_1 & a_2 & a_3 \\ b_1 & b_2 & b_3 \\ 0 & 0 & 0 \end{pmatrix} = 0.$$

   (b) A linear combination of the first two rows of the matrix in part (a) is of the form

   $$\alpha(a_1, a_2, a_3) + \beta(b_1, b_2, b_3) = (\alpha a_1, \alpha a_2, \alpha a_3) + (\beta b_1, \beta b_2, \beta b_3).$$

   Show that

   $$det \begin{pmatrix} a_1 & a_2 & a_3 \\ b_1 & b_2 & b_3 \\ 0 & 0 & 0 \end{pmatrix} = det \begin{pmatrix} a_1 & a_2 & a_3 \\ b_1 & b_2 & b_3 \\ \alpha a_1 & \alpha a_2 & \alpha a_3 \end{pmatrix} = det \begin{pmatrix} a_1 & a_2 & a_3 \\ b_1 & b_2 & b_3 \\ \alpha a_1 + \beta b_1 & \alpha a_2 + \beta b_2 & \alpha a_3 + \beta b_3 \end{pmatrix}.$$

   Thus, if the third row is a linear combination of the first two rows, then the matrix has a determinant of zero.

(c)   Show that if the determinant of a matrix is nonzero, then the rows are inde-
      pendent. One way to do this is to use the fact that when a matrix is row
      reduced, this is done by applying the elementary row operations. These may
      change the value of the determinant but will not change whether the determi-
      nant is zero.

9. Show that if $\{\hat{v}_1, \hat{v}_2, \ldots, \hat{v}_n\}$ is a linearly dependent set of vectors, then $\{\hat{v}_1, \hat{v}_2, \ldots, \hat{v}_n, \hat{v}\}$,
   where $\hat{v}$ is any vector, is a linearly dependent set of vectors.

10. Let $\hat{v}_1 = (7, -4, 1, 0)$, $\hat{v}_2 = (6, -5, 0, 1))$. Find whether the following vectors are in the
    span $\{\hat{v}_1, \hat{v}_2\}$.

    (a) $(20, -13, 2, 1)$

    (b) $(-1, -1, -1, 1)$,

11. Let $\hat{v}_1 = (3, 1, 0)$, $\hat{v}_2 = (6, 0, -1))$. Find whether the following vectors are in the span
    $\{\hat{v}_1, \hat{v}_2\}$.

    (a) $(9, 4, 0)$

    (b) $(12, 2, -1)$

    (c) $(15, 1, -2)$

12. Let $\hat{v}_1 = (1, -1, 2)$, $\hat{v}_2 = (2, 4, 5)$. Find whether the following vectors are in the span
    $\{\hat{v}_1, \hat{v}_2\}$.

    (a) $(1, 0, 2)$

    (b) $(9, -9, 21)$

    (c) $(-3, -9, -8)$

13. Describe the values that a, b, and c must satisfy for the vector

$$\begin{pmatrix} a \\ b \\ c \end{pmatrix}$$

to be in the span of the following sets:

(a) $\left\{ \begin{pmatrix} 1 \\ 2 \\ 0 \end{pmatrix}, \begin{pmatrix} 3 \\ -2 \\ 0 \end{pmatrix} \right\}$

(b) $\left\{ \begin{pmatrix} 1 \\ -6 \\ 4 \end{pmatrix}, \begin{pmatrix} 3 \\ 2 \\ 0 \end{pmatrix} \right\}$

(c) $\left\{ \begin{pmatrix} 1 \\ 2 \\ 3 \end{pmatrix}, \begin{pmatrix} 2 \\ 4 \\ 6 \end{pmatrix} \right\}$

14. If V is a vector space that has a basis $\{\hat{v}_1, \ldots, \hat{v}_n\}$, what can you say about a set of
    vectors in V that have

    (a)   More than n vectors?

    (b)   Fewer than n vectors?

    (c)   Exactly n vectors?

15. Show that a set of vectors that contains the zero vector is linearly dependent.

16. Show that if V is a vector space of dimension n and U is a subspace of V of dimension n, then $U = V$.

17. Give an example of vectors $\hat{u}$, $\hat{v}$, and $\hat{w}$ for which $\{\hat{u}, \hat{v}\}, \{\hat{v}, \hat{w}\}$, and $\{\hat{u}, \hat{w}\}$ are linearly independent sets but $\{\hat{u}, \hat{v}, \hat{w}\}$ is not a linearly independent set.

18. Determine which of the sets is a basis for $\mathbb{R}^3$.

   (a) $(2,0,1),(5,-1,3)$

   (b) $(4,1,1),(3,-2,5),(1,1,1)$

   (c) $(2,3,6),(1,3,1),(6,12,14)$

   (d) $(2,0,1),(-6,9,-4),(0,0,0)$

19. Find a basis for the set of vectors of the form

$$\begin{bmatrix} a+2b-c \\ 3a+4b \\ 6a-2c \end{bmatrix}.$$

3.5 Converting a Set of Vectors to a Basis

Our next results say that a set of linearly independent vectors that is not a basis can be expanded to be a basis and a spanning set of vectors that is not a basis can be shrunk to be a basis.

In Theorem 16, we will show that it is possible to add vectors to a linearly independent set until a basis is obtained, and in Theorem 17 we will show that it is possible to delete vectors from a spanning set until a basis is obtained. However, the proofs do not give computationally simple ways to do this. After Theorem 16, we give an example of how to add vectors to get a basis, and after Theorem 17 we give an example of how to delete vectors to get a basis. In the next chapter, we give the justification for why the techniques are valid.

Theorem 16

Suppose that V is a finite dimensional vector space and $\{\hat{v}_1, \ldots, \hat{v}_n\}$ is a linearly independent set of vectors. If $\{\hat{v}_1, \ldots, \hat{v}_n\}$ is not a basis of V, then a finite number of vectors may be added to $\{\hat{v}_1, \ldots, \hat{v}_n\}$ so that the enlarged set will be a basis.

Proof

Since $\{\hat{v}_1, \ldots, \hat{v}_n\}$ is linearly independent but is not a basis, it does not span V. Thus, there is a $\hat{v} \in V, \hat{v} \neq \hat{0}$ that is not in the span of $\{\hat{v}_1, \ldots, \hat{v}_n\}$. We claim $\{\hat{v}_1, \ldots, \hat{v}_n, \hat{v}\}$ is a linearly independent set. Suppose

$$a_0\hat{v} + a_1\hat{v}_1 + \cdots + a_n\hat{v}_n = \hat{0}.$$

If $a_0 \neq 0$, then $\hat{v}$ is in the span of $\{\hat{v}_1, \ldots, \hat{v}_n\}$, which is a contradiction. Thus $a_0 = 0$, so

$$a_1 \hat{v}_1 + \cdots + a_n \hat{v}_n = \hat{0}$$

and $\{\hat{v}_1, \ldots, \hat{v}_n\}$ is linearly independent, so $a_i = 0$, $i = 1, \ldots, n$. Thus, $\{\hat{v}_1, \ldots, \hat{v}_n, \hat{v}\}$ is a linearly independent set. If this set spans V, then it is a basis. Otherwise repeat the process a finite number of times with the larger sets until a basis is obtained. If the dimension of V is k, there will be $k - n$ such additions.

Theorem 16 shows that a linearly independent set can be expanded to a basis and gives a theoretical method of how to accomplish this but does not give a method for finding the additional vectors. The example below gives one method.

Example

The vectors

$$\begin{pmatrix} 3 \\ -2 \\ 0 \\ 5 \end{pmatrix}, \begin{pmatrix} 1 \\ 4 \\ 7 \\ -9 \end{pmatrix}$$

are linearly independent. Complete this set of vectors to form a basis.

The technique we exploit is to choose a basis of the vector space—and it is usually simplest to use the standard basis—and attach vectors of the chosen basis to the given vectors. In this case, we would have

$$\begin{pmatrix} 3 \\ -2 \\ 0 \\ 5 \end{pmatrix}, \begin{pmatrix} 1 \\ 4 \\ 7 \\ -9 \end{pmatrix}, \begin{pmatrix} 1 \\ 0 \\ 0 \\ 0 \end{pmatrix}, \begin{pmatrix} 0 \\ 1 \\ 0 \\ 0 \end{pmatrix}, \begin{pmatrix} 0 \\ 0 \\ 1 \\ 0 \end{pmatrix}, \begin{pmatrix} 0 \\ 0 \\ 0 \\ 1 \end{pmatrix}.$$

Next, create a matrix whose columns are the vectors given earlier. Be sure to list the vectors that are to be part of the basis in the first columns. In this case, we have

$$\begin{pmatrix} 3 & 1 & 1 & 0 & 0 & 0 \\ -2 & 4 & 0 & 1 & 0 & 0 \\ 0 & 7 & 0 & 0 & 1 & 0 \\ 5 & -9 & 0 & 0 & 0 & 1 \end{pmatrix}.$$

Row reduce this matrix. In this case, the result is

$$\begin{pmatrix} 1 & 0 & 0 & 0 & 9/35 & 1/5 \\ 0 & 1 & 0 & 0 & 1/7 & 0 \\ 0 & 0 & 1 & 0 & -32/35 & -3/5 \\ 0 & 0 & 0 & 1 & -2/35 & 2/5 \end{pmatrix}.$$

There will be (in the case where the vector space is $\mathbb{R}^4$) four leading 1's. A basis that includes the two given vectors will be the columns in the original matrix that correspond to the columns in the row reduced matrix where the leading 1's occur (we will prove this later). So in this case, a basis that contains the two given vectors is

$$\begin{pmatrix} 3 \\ -2 \\ 0 \\ 5 \end{pmatrix}, \begin{pmatrix} 1 \\ 4 \\ 7 \\ -9 \end{pmatrix}, \begin{pmatrix} 1 \\ 0 \\ 0 \\ 0 \end{pmatrix}, \begin{pmatrix} 0 \\ 1 \\ 0 \\ 0 \end{pmatrix}.$$

Theorem 17

Suppose that V is a finite dimensional vector space and $\{\hat{v}_1, \ldots, \hat{v}_n\}$ is a spanning set of vectors. If $\{\hat{v}_1, \ldots, \hat{v}_n\}$ is not a basis, then a finite number of vectors may be deleted from $\{\hat{v}_1, \ldots, \hat{v}_n\}$ until the diminished set is a basis.

Proof

Since $\{\hat{v}_1, \ldots, \hat{v}_n\}$ is a spanning set but not a basis, then it is not linearly independent, and one of the vectors can be written as a linear combination of the others. Suppose

$$\hat{v}_1 = a_2\hat{v}_2 + \ldots + a_n\hat{v}_n.$$

Then $\{\hat{v}_2, \ldots, \hat{v}_n\}$ is a spanning set of vectors. If $\{\hat{v}_2, \ldots, \hat{v}_n\}$ is a linearly independent set of vectors, then it is a basis. If not, continue the process until a basis is obtained.

Theorem 17 does not show explicitly how to select the vectors that should be discarded from a linearly dependent set. There may be several choices that are valid. The next example gives a computational technique for discarding superfluous vectors. The technique will be justified in the next chapter.

Example

Select from the set of vectors

$$\left\{ \begin{pmatrix} 1 \\ 7 \\ -2 \end{pmatrix}, \begin{pmatrix} 4 \\ 3 \\ 2 \end{pmatrix}, \begin{pmatrix} 9 \\ 13 \\ 2 \end{pmatrix}, \begin{pmatrix} 10 \\ 29 \\ 0 \end{pmatrix}, \begin{pmatrix} 4 \\ 0 \\ 1 \end{pmatrix} \right\}$$

a maximal linearly independent set.

We form the matrix whose columns are the vectors, that is

$$\begin{pmatrix} 1 & 4 & 9 & 10 & 4 \\ 7 & 3 & 13 & 29 & 0 \\ -2 & 2 & 2 & 0 & 1 \end{pmatrix}.$$

When this matrix is row reduced, the result is

$$\begin{pmatrix} 1 & 0 & 1 & 0 & 73/45 \\ 0 & 1 & 2 & 0 & 191/90 \\ 0 & 0 & 0 & 1 & -11/18 \end{pmatrix}.$$

The leading 1's occur in the first, second, and fourth columns. Return to the original matrix, and those columns will be the maximal linearly independent set. This is

$$\left\{ \begin{pmatrix} 1 \\ 7 \\ -2 \end{pmatrix}, \begin{pmatrix} 4 \\ 3 \\ 2 \end{pmatrix}, \begin{pmatrix} 10 \\ 29 \\ 0 \end{pmatrix} \right\}.$$

The following example demonstrates that if $\hat{v}_1 . \hat{v}_2 . \hat{v}_3$ are vectors in $\mathbb{R}^3$ that do not span $\mathbb{R}^3$ then $\{\hat{v}_1 . \hat{v}_2 . \hat{v}_3\}$ is not linearly independent.

Example

Let

$$\hat{v}_1 = \begin{pmatrix} 1 \\ 2 \\ 3 \end{pmatrix}, \quad \hat{v}_2 = \begin{pmatrix} 1 \\ 1 \\ 1 \end{pmatrix}, \quad \hat{v}_3 = \begin{pmatrix} 2 \\ 3 \\ 4 \end{pmatrix}, \quad A = \begin{pmatrix} 1 & 1 & 2 \\ 2 & 1 & 3 \\ 3 & 1 & 4 \end{pmatrix}.$$

When A is row reduced, the result is

$$\begin{pmatrix} 1 & 0 & 1 \\ 0 & 1 & 1 \\ 0 & 0 & 0 \end{pmatrix}.$$

Thus, the solution to

$$\begin{pmatrix} 1 & 1 & 2 \\ 2 & 1 & 3 \\ 3 & 1 & 4 \end{pmatrix} \begin{pmatrix} x \\ y \\ z \end{pmatrix} = \begin{pmatrix} 0 \\ 0 \\ 0 \end{pmatrix}$$

is

$$x + z = 0,\ y + z = 0 \text{ so } z \text{ is a free variable} \quad \text{and} \quad x = -z,\ y = -z.$$

Thus, any vector

$$\begin{pmatrix} -t \\ -t \\ t \end{pmatrix}$$

has

$$\begin{pmatrix} 1 & 1 & 2 \\ 2 & 1 & 3 \\ 3 & 1 & 4 \end{pmatrix} \begin{pmatrix} -t \\ -t \\ t \end{pmatrix} = \begin{pmatrix} 0 \\ 0 \\ 0 \end{pmatrix}.$$

so

$$-t \begin{pmatrix} 1 \\ 2 \\ 3 \end{pmatrix} - t \begin{pmatrix} 1 \\ 1 \\ 1 \end{pmatrix} + t \begin{pmatrix} 2 \\ 3 \\ 4 \end{pmatrix} = \begin{pmatrix} 0 \\ 0 \\ 0 \end{pmatrix}$$

and thus $\{\hat{v}_1, \hat{v}_2, \hat{v}_3\}$ is not a linearly independent set of vectors.

The last row of zeros in the row reduced form of A suggests that any vector of the form

$$\begin{pmatrix} 0 \\ 0 \\ a \end{pmatrix} a \neq 0$$

will not be in the span of $\{\hat{v}_1, \hat{v}_2, \hat{v}_3\}$.

In fact, when

$$\begin{pmatrix} 1 & 1 & 2 & 0 \\ 2 & 1 & 3 & 0 \\ 3 & 1 & 4 & a \end{pmatrix}$$

is row reduced, the result is

$$\begin{pmatrix} 1 & 0 & 1 & 0 \\ 0 & 1 & 1 & 0 \\ 0 & 0 & 0 & 1 \end{pmatrix}$$

which confirms the conjecture.

These ideas can be modified to show that any set of n vectors in $\mathbb{R}^n$ that fails to be a basis is neither linearly independent nor a spanning set. The salient point being that when an $n \times n$ matrix is row reduced, the result is either the identity matrix (in which case the columns form a basis) or has a row of zeroes.

Theorem 18

Suppose that V is a vector space of dimension n.

(a) If $\{\hat{v}_1, \ldots, \hat{v}_n\}$ is a spanning set of vectors for V, then $\{\hat{v}_1, \ldots, \hat{v}_n\}$ is a basis for V.

(b) If $\{\hat{v}_1, \ldots, \hat{v}_n\}$ is a linearly independent set of vectors for V, then $\{\hat{v}_1, \ldots, \hat{v}_n\}$ is a basis for V.

Proof

(a) We give a proof by contradiction. Suppose that $\{\hat{v}_1, \ldots, \hat{v}_n\}$ is a spanning set but is not a basis for V. Since $\{\hat{v}_1, \ldots, \hat{v}_n\}$ spans V but is not a basis, it must be that $\{\hat{v}_1, \ldots, \hat{v}_n\}$ is linearly dependent. Thus, there is a vector in $\{\hat{v}_1, \ldots, \hat{v}_n\}$, say, $\hat{v}_1$, so that

$$\hat{v}_1 = a_2\hat{v}_2 + \cdots + a_n\hat{v}_n.$$

Thus, $\{\hat{v}_2, \ldots, \hat{v}_n\}$ is a spanning set and if it is not a basis, it can be shrunk to be a basis. Thus, there is a basis for V with fewer than n vectors, contradicting the assumption that V is a vector space of dimension n.

(b) Again, we give a proof by contradiction. Suppose that $\{\hat{v}_1, \ldots, \hat{v}_n\}$ is not a basis for V. Since $\{\hat{v}_1, \ldots, \hat{v}_n\}$ is linearly independent, then it must not span V. Thus, there is a vector $\hat{v} \in V$ with $\hat{v}$ not in the span of $\{\hat{v}_1, \ldots, \hat{v}_n\}$. We showed earlier that $\{\hat{v}_1, \ldots, \hat{v}_n, \hat{v}\}$ is a linearly independent set. Thus, we have a set of $n+1$ linearly independent vectors that can be expanded to a basis for V, contradicting the assumption that V is a vector space of dimension n.

3.6 A Synopsis of Sections 3.3.4 and 3.3.5

In these sections, we have derived several facts about bases of vector spaces. Here is a list of these results for the case of vector spaces that consist of more than the zero vector.

Suppose that V is a nontrivial vector space that has a basis of exactly k vectors. Then

1. Every basis of V has exactly k vectors.
2. Any subset of V that contains more than k vectors cannot be linearly independent.
3. Any subset of V that contains fewer than k vectors cannot span V.
4. Any subset of V that is linearly independent but does not span V can be expanded to form a basis.
5. Any subset of V that spans V but is not linearly independent can be contracted to form a basis.
6. Any linearly independent set of k vectors is a basis for V.
7. Any spanning set of k vectors is a basis for V.

We have excluded the trivial vector space because of what might be considered a philosophical point that we prefer to avoid, but it is common to say that the empty set is a basis for the trivial vector space.

Exercises

In Exercises 1 through 5, (i.) Tell whether the given set of vectors forms a basis. (ii.) If the set of vectors does not form a basis, select a maximal linearly independent set of vectors. (iii.) If the maximal linearly independent set of vectors does not form a basis, extend the set to form a basis.

1. $\left\{ \begin{pmatrix} 1 \\ -2 \\ 4 \end{pmatrix} \begin{pmatrix} 11 \\ 14 \\ 8 \end{pmatrix} \begin{pmatrix} 2 \\ 8 \\ -4 \end{pmatrix} \begin{pmatrix} 15 \\ 30 \\ 0 \end{pmatrix} \right\}$

2. $\left\{ \begin{pmatrix} 1 \\ -3 \\ -2 \end{pmatrix} \begin{pmatrix} -2 \\ 6 \\ 4 \end{pmatrix} \right\}$

3. $\left\{ \begin{pmatrix} 5 \\ 6 \end{pmatrix} \right\}$

4. $\left\{ \begin{pmatrix} 0 \\ 3 \\ 2 \\ 5 \end{pmatrix} \begin{pmatrix} 4 \\ -2 \\ 1 \\ 6 \end{pmatrix} \begin{pmatrix} 8 \\ 2 \\ 4 \\ 17 \end{pmatrix} \begin{pmatrix} 1 \\ 3 \\ 5 \\ 7 \end{pmatrix} \begin{pmatrix} 7 \\ -1 \\ -1 \\ 10 \end{pmatrix} \begin{pmatrix} 1 \\ 2 \\ 3 \\ 4 \end{pmatrix} \right\}$

5. $\left\{ \begin{pmatrix} 1 \\ 2 \\ 1 \end{pmatrix} \begin{pmatrix} -3 \\ 7 \\ 5 \end{pmatrix} \begin{pmatrix} 4 \\ 0 \\ 0 \end{pmatrix} \right\}$

6. Find a maximal linearly independent set in each of the following sets of vectors. If the maximal linearly independent set is not a basis, extend the maximal set to a basis.

    (a) $(2,1,6),(3,21,9),(1,7,3),(5,22,15)$

    (b) $(1,6,9),(4,6,2),(1,0,0),(6,2,9)$

    (c) $(1,5),(3,15),(2,10)$

7. Construct a basis of $\mathbb{R}^4$ that includes the vector $(1,2,0,5)$.

8. (a) Show that $\hat{v}_1$ and $\hat{v}_2$ can be written as a linear combination of $\{\hat{v}_1 + \hat{v}_2, \hat{v}_2 + \hat{v}_3, \hat{v}_{3,}\}$.

    (b) Show that if $\{\hat{v}_1, \hat{v}_2, \ldots, \hat{v}_n\}$ is a basis for the vector space V, then $\{\hat{v}_1 + \hat{v}_2, \hat{v}_2 + \hat{v}_3, \ldots, \hat{v}_{n-1} + \hat{v}_n, \hat{v}_n\}$ is also a basis for V.

9. Let

$$\hat{v}_1 = \begin{pmatrix} 1 \\ 0 \\ 2 \\ 1 \end{pmatrix}, \quad \hat{v}_2 = \begin{pmatrix} 1 \\ 1 \\ 0 \\ 0 \end{pmatrix}, \quad \hat{v}_3 = \begin{pmatrix} 3 \\ 2 \\ 2 \\ 1 \end{pmatrix}, \quad \hat{v}_4 = \begin{pmatrix} 6 \\ -1 \\ 2 \\ 3 \end{pmatrix}.$$

and let U be the subspace spanned by $\{\hat{v}_1, \hat{v}_2\}$ and let W be the subspace spanned by $\{\hat{v}_3, \hat{v}_4\}$.

Find a basis for $U + W$ and $U \cap W$.

10. The vectors

$$(1,6,3),(2,-3,7),(7,12,23)$$

are not linearly independent.

(a) Find a vector not in the span of $\{(1,6,3),(2,-3,7),(7,12,23)\}$

(b) Find a nonzero vector that can be expressed as a linear combination of

$$\{(1,6,3),(2,-3,7),(7,12,23)\}$$

in two different ways.

11. Determine h so that

$$\begin{pmatrix} 3 \\ 6 \\ 1 \end{pmatrix}, \begin{pmatrix} -2 \\ -5 \\ 0 \end{pmatrix}, \begin{pmatrix} 8 \\ 1 \\ h \end{pmatrix}$$

will not span $\mathbb{R}^3$.

12. Show that if S is a linearly independent set of vectors and S' is a proper subset of S, then span (S') is a proper subset of span (S).

13. In finding a basis for the row space of a matrix with the method described in the text, the basis vectors you obtain may not be the rows of the matrix. Describe how you could get a basis consisting of the rows of the matrix and apply your technique to the matrix

$$\begin{pmatrix} 2 & 0 & 3 & 1 & -2 & 5 \\ 4 & 1 & 6 & 3 & 3 & 9 \\ 8 & 1 & 12 & 5 & -1 & 0 \\ 0 & 7 & 2 & -4 & 4 & 0 \end{pmatrix}$$

3.7 Change of Bases

Different bases of a vector space describe the same vector in different ways.

There are two things we want to emphasize. The first is the relationship between a vector and a basis for the vector space in which the vector lies. Borrowing from Terrence Tao, the relationship of a vector to a basis is analogous to the relationship of an idea to words. The same idea can be described in many different languages. Likewise, a vector can be described by different bases.

The second point is that we have defined a basis to be a collection of vectors that are linearly independent and span a vector space. As sets, the order of appearance of elements in a set is immaterial; all that matters is the elements. For example, the set $\{a,b,c\}$ is the same as the set $\{c,a,b\}$. In what follows, we will describe the representation of a vector $\hat{v}$ with respect to a basis by the scalars that are the coefficients of the basis vectors in the linear combination that makes up the vector $\hat{v}$; the basis vectors themselves will not be listed. This type of representation assumes an order of the basis vectors. The effect of all this is that **when we list a set of basis vectors in this section, it will be assumed to be an ordered set.**

Thus, if it is understood that the basis of the vector space is the ordered set $\{\hat{v}_1, \ldots, \hat{v}_n\}$, we could describe

$$\hat{v} = a_1\hat{v}_1 + \cdots + a_n\hat{v}_n$$

by the ordered n-tuple of scalars $(a_1, \ldots, a_n)$.

We will call $(a_1, \ldots, a_n)$ the coordinates of $\hat{v}$ with respect to the basis $\{\hat{v}_1, \ldots, \hat{v}_n\}$.

Notation: if $\mathcal{B} = \{\hat{u}_1, \ldots, \hat{u}_n\}$ is a basis for the vector space V and $\hat{v}$ is a vector in V, then the ordered set of coordinates of $\hat{v}$ with respect to the basis $\mathcal{B}$ is denoted

$$\left[\hat{v}\right]_{\mathcal{B}}.$$

If we are in a situation where there is no ambiguity about the basis being used, it is usually the case that the subscript $\mathcal{B}$ will be suppressed. Also, if we are using the usual (standard) basis, the subscript will be suppressed.

We consider the following problem: given the coordinates of a vector in one basis, find the coordinates of the vector in another basis.

We first give a simple example, then give a description of the theory of the process in the more general case and then show how to solve the problem in a practical manner.

Example

Suppose that $\mathcal{B} = \{\hat{u}_1, \hat{u}_2\}$ and $\mathcal{C} = \{\hat{v}_1, \hat{v}_2\}$ are two bases for the vector space V. Suppose also that we know the representation of a vector $\hat{w}$ in the basis $\mathcal{B}$ and we want to find the representation of $\hat{w}$ in the basis $\mathcal{C}$.

We can express each vector $\hat{u}_i$ in terms of $\{\hat{v}_1, \hat{v}_2\}$. Suppose

$$\hat{u}_1 = a\hat{v}_1 + b\hat{v}_2$$
$$\hat{u}_2 = c\hat{v}_1 + d\hat{v}_2.$$

Suppose $\hat{w} \epsilon V$ and $\hat{w} = \alpha\hat{u}_1 + \beta\hat{u}_2$. That is,

$$\left[\hat{w}\right]_{\mathcal{B}} = \begin{pmatrix} \alpha \\ \beta \end{pmatrix} \quad \text{and} \quad \text{we want to find } \left[\hat{w}\right]_{\mathcal{C}}.$$

We have

$$\hat{w} = \alpha\hat{u}_1 + \beta\hat{u}_2 = \alpha\left(a\hat{v}_1 + b\hat{v}_2\right) + \beta\left(c\hat{v}_1 + d\hat{v}_2\right) = \left(\alpha a + \beta c\right)\hat{v}_1 + \left(\alpha b + \beta d\right)\hat{v}_2$$

so $\left[\hat{w}\right]_{\mathcal{C}} = \begin{pmatrix} \alpha a + \beta c \\ \alpha b + \beta d \end{pmatrix} = \begin{pmatrix} a & c \\ b & d \end{pmatrix}\begin{pmatrix} \alpha \\ \beta \end{pmatrix}.$

We want to determine how we can easily go from $\left[\hat{w}\right]_{\mathcal{B}}$ to $\left[\hat{w}\right]_{\mathcal{C}}$. If we analyze

$$\begin{pmatrix} a & c \\ b & d \end{pmatrix}\begin{pmatrix} \alpha \\ \beta \end{pmatrix}$$

we note that

$$\begin{pmatrix} \alpha \\ \beta \end{pmatrix} = \left[\hat{w}\right]_{\mathcal{B}}$$

and

$$\begin{pmatrix} a & c \\ b & d \end{pmatrix}$$

is the matrix whose first column is the representation of $\hat{u}_1$ in the $\mathcal{C}$ basis and whose second column is the representation of $\hat{u}_2$ in the $\mathcal{C}$ basis.

We extend these ideas to the general case.

Suppose that $\mathcal{B} = \{\hat{u}_1, \ldots, \hat{u}_n\}$ and $\mathcal{C} = \{\hat{v}_1, \ldots, \hat{v}_n\}$ are two bases for the vector space V. Suppose also that we know the representation of a vector $\hat{w}$ in the basis $\mathcal{B}$ and we want to find the representation of v in the basis $\mathcal{C}$.

We can express each vector $\hat{u}_i$ in terms of $\{\hat{v}_1, \ldots, \hat{v}_n\}$. Suppose

$$\hat{u}_1 = p_{11}\hat{v}_1 + p_{21}\hat{v}_2 + \cdots + p_{n1}\hat{v}_n$$
$$\hat{u}_2 = p_{12}\hat{v}_1 + p_{22}\hat{v}_2 + \cdots + p_{n2}\hat{v}_n$$
$$\vdots$$
$$\hat{u}_n = p_{1n}\hat{v}_1 + p_{2n}\hat{v}_2 + \cdots + p_{nn}\hat{v}_n$$

and suppose that $\hat{w}$ in the $\mathcal{B}$ basis is

$$\hat{w} = b_1\hat{u}_1 + \cdots + b_n\hat{u}_n.$$

That is,

$$\left[\hat{w}\right]_{\mathcal{B}} = \begin{pmatrix} b_1 \\ \vdots \\ b_n \end{pmatrix}.$$

Then we have

$$v = b_1\hat{u}_1 + \cdots + b_n\hat{u}_n = b_1\left(p_{11}\hat{v}_1 + p_{21}\hat{v}_2 + \cdots + p_{n1}\hat{v}_n\right) +$$
$$b_2\left(p_{12}\hat{v}_1 + p_{22}\hat{v}_2 + \cdots + p_{n2}\hat{v}_n\right) + \cdots +$$
$$b_n\left(p_{1n}\hat{v}_1 + p_{2n}\hat{v}_2 + \cdots + p_{nn}\hat{v}_n\right)$$
$$= \left(b_1p_{11} + b_2p_{21} + \cdots + b_np_{n1}\right)\hat{v}_1 +$$
$$\left(b_1p_{12} + b_2p_{22} + \cdots + b_np_{n2}\right)\hat{v}_2 +$$
$$\vdots$$
$$\left(b_1p_{1n} + b_2p_{2n} + \cdots + b_np_{nn}\right)\hat{v}_n.$$

This means

$$
[\hat{v}]_{\mathcal{C}} = \begin{pmatrix} b_1 p_{11} + b_2 p_{12} + \cdots + b_n p_{1n} \\ b_1 p_{21} + b_2 p_{22} + \cdots + b_n p_{2n} \\ \vdots \\ b_1 p_{n1} + b_2 p_{n2} + \cdots + b_n p_{nn} \end{pmatrix} = \begin{pmatrix} p_{11} & p_{12} & \cdots & p_{1n} \\ p_{21} & p_{22} & \cdots & p_{2n} \\ \vdots & \vdots & & \vdots \\ p_{n1} & p_{n2} & \cdots & p_{nn} \end{pmatrix} \begin{pmatrix} b_1 \\ b_2 \\ \vdots \\ b_n \end{pmatrix}. \tag{3.1}
$$

Now

$$
[\hat{v}]_{\mathcal{B}} = \begin{pmatrix} b_1 \\ b_2 \\ \vdots \\ b_n \end{pmatrix} \quad \text{and} \quad [\hat{v}]_{\mathcal{C}} = \begin{pmatrix} p_{11} & p_{12} & \cdots & p_{1n} \\ p_{21} & p_{22} & \cdots & p_{2n} \\ \vdots & \vdots & & \vdots \\ p_{n1} & p_{n2} & \cdots & p_{nn} \end{pmatrix} \begin{pmatrix} b_1 \\ b_2 \\ \vdots \\ b_n \end{pmatrix}
$$

so if

$$
P = \begin{pmatrix} p_{11} & p_{12} & \cdots & p_{1n} \\ p_{21} & p_{22} & \cdots & p_{2n} \\ \vdots & \vdots & & \vdots \\ p_{n1} & p_{n2} & \cdots & p_{nn} \end{pmatrix}
$$

then we have

$$
[\hat{v}]_{\mathcal{C}} = P[\hat{v}]_{\mathcal{B}},
$$

where P is the matrix whose ith column is the representation of $\hat{u}_i$ in the $\mathcal{C}$ basis.
The matrix P is called the transition matrix for the basis $\mathcal{C}$.

Algorithm for converting a vector whose representation in the $\mathcal{C}$ basis is $\begin{pmatrix} \alpha \\ \beta \end{pmatrix}$ to the representation of the vector in the standard basis

1. Form the matrix $P_{\mathcal{C}}$ whose columns are the vectors that make up the basis of $\mathcal{C}$.
2. Multiply the vector's representation in the $\mathcal{C}$ basis on the left by the matrix $P_{\mathcal{C}}$. The result is the representation of the vector in the standard basis.

Symbolically, this is expressed

$$
[\hat{v}]_{\mathcal{B}} = P_{\mathcal{C}}[\hat{v}]_{\mathcal{C}}.
$$

When $\mathcal{B}$ is the standard basis, the $\mathcal{B}$ is normally suppressed, and the expression is written

$$
[\hat{v}] = P_{\mathcal{C}}[\hat{v}]_{\mathcal{C}}. \tag{3.2}
$$

If we know the expression of the vector in the standard basis and we want to find the expression of the vector in the $\mathcal{C}$ basis, then we solve Equation 3.2 for $\left[\hat{v}\right]_c$ to get

$$\left[\hat{v}\right]_c = P_C^{-1}\left[\hat{v}\right].$$

We know that P_C is invertible because its columns are the vectors of a basis.

Example

Let $\mathcal{C}$ be the basis

$$\mathcal{C} = \left\{ \begin{pmatrix} 1 \\ -2 \\ 2 \end{pmatrix}, \begin{pmatrix} 3 \\ 4 \\ 0 \end{pmatrix}, \begin{pmatrix} 1 \\ 1 \\ -1 \end{pmatrix} \right\}.$$

(a) Find the representation of

$$\left[\hat{v}\right]_c = \begin{pmatrix} 6 \\ -9 \\ 8 \end{pmatrix}$$

in the standard basis.
   We have

$$P_C = \begin{pmatrix} 1 & 3 & 1 \\ -2 & 4 & 1 \\ 2 & 0 & -1 \end{pmatrix}$$

so

$$\left[\hat{v}\right] = P_C\left[\hat{v}\right]_c = \begin{pmatrix} 1 & 3 & 1 \\ -2 & 4 & 1 \\ 2 & 0 & -1 \end{pmatrix}\begin{pmatrix} 6 \\ -9 \\ 8 \end{pmatrix} = \begin{pmatrix} -13 \\ -40 \\ 4 \end{pmatrix}.$$

(b) Find the representation of

$$\left[\hat{v}\right] = \begin{pmatrix} 3 \\ 1 \\ 2 \end{pmatrix}$$

in the $\mathcal{C}$ basis.
   We have

$$\left[\hat{v}\right]_c = P_C^{-1}\left[\hat{v}\right] = \begin{pmatrix} 1 & 3 & 1 \\ -2 & 4 & 1 \\ 2 & 0 & -1 \end{pmatrix}^{-1}\begin{pmatrix} 3 \\ 1 \\ 2 \end{pmatrix} = \begin{pmatrix} 11/12 \\ 3/4 \\ -1/6 \end{pmatrix}.$$

Converting between Two Nonstandard Bases

Suppose that

$$C = \left\{ \begin{pmatrix} 1 \\ 3 \end{pmatrix}, \begin{pmatrix} -2 \\ 5 \end{pmatrix} \right\} \quad \text{and} \quad D = \left\{ \begin{pmatrix} 0 \\ 2 \end{pmatrix}, \begin{pmatrix} 1 \\ 6 \end{pmatrix} \right\}$$

and we want to convert the representation of a vector $\hat{v}$ in one basis to a representation in the other basis.

Let P_C be the transition matrix for C and P_D be the transition matrix for D, that is,

$$P_C = \begin{pmatrix} 1 & -2 \\ 3 & 5 \end{pmatrix} \quad \text{and} \quad P_D = \begin{pmatrix} 0 & 1 \\ 2 & 6 \end{pmatrix}.$$

Then

$$P_C\left[\hat{v}\right]_C = \left[\hat{v}\right] = P_D\left[\hat{v}\right]_D$$

and the equation

$$P_C\left[\hat{v}\right]_C = P_D\left[\hat{v}\right]_D$$

provides the mechanism for transition between bases. So if

$$\left[\hat{v}\right]_C = \begin{pmatrix} 9 \\ 4 \end{pmatrix}$$

then

$$\left[\hat{v}\right]_D = P_D^{-1}P_C\left[\hat{v}\right]_C = \begin{pmatrix} 0 & 1 \\ 2 & 6 \end{pmatrix}^{-1}\begin{pmatrix} 1 & -2 \\ 3 & 5 \end{pmatrix}\begin{pmatrix} 9 \\ 4 \end{pmatrix} = \begin{pmatrix} 41/2 \\ 1 \end{pmatrix}.$$

Example

The standard basis of $\mathcal{P}_2(x)$ is $\{1, x, x^2\}$. Find the representation of

$$3 - 2x + 5x^2$$

in the basis

$$B = \left\{ 1 + x, 1 - x, x^2 \right\}.$$

We have

$$P_B = \begin{pmatrix} 1 & 1 & 0 \\ 1 & -1 & 0 \\ 0 & 0 & 1 \end{pmatrix}, \quad \left[\hat{v}\right] = \begin{pmatrix} 3 \\ -2 \\ 5 \end{pmatrix}$$

so

$$\left[\hat{v}\right]_B = P_B^{-1}\left[\hat{v}\right] = \begin{pmatrix} 1 & 1 & 0 \\ 1 & -1 & 0 \\ 0 & 0 & 1 \end{pmatrix}^{-1}\begin{pmatrix} 3 \\ -2 \\ 5 \end{pmatrix} = \begin{pmatrix} 1/2 \\ 5/2 \\ 5 \end{pmatrix}.$$

One can check that

$$\frac{1}{2}(1+x)+\frac{5}{2}(1-x)+5x^2 = 3-2x+5x^2.$$

Example

The standard basis for $\mathcal{M}_{2\times2}(\mathbb{R})$ is

$$\left\{ \begin{pmatrix} 1 & 0 \\ 0 & 0 \end{pmatrix}, \begin{pmatrix} 0 & 1 \\ 0 & 0 \end{pmatrix}, \begin{pmatrix} 0 & 0 \\ 1 & 0 \end{pmatrix}, \begin{pmatrix} 0 & 0 \\ 0 & 1 \end{pmatrix} \right\}.$$

Find the transition matrix for the basis

$$B=\left\{ \begin{pmatrix} 1 & 0 \\ 0 & 1 \end{pmatrix}, \begin{pmatrix} 2 & 1 \\ 0 & 0 \end{pmatrix}, \begin{pmatrix} 0 & 0 \\ 1 & 1 \end{pmatrix}, \begin{pmatrix} 0 & 0 \\ 0 & 1 \end{pmatrix} \right\}.$$

We have

$$\begin{pmatrix} 1 & 0 \\ 0 & 1 \end{pmatrix}=1\begin{pmatrix} 1 & 0 \\ 0 & 0 \end{pmatrix}+0\begin{pmatrix} 0 & 1 \\ 0 & 0 \end{pmatrix}+0\begin{pmatrix} 0 & 0 \\ 1 & 0 \end{pmatrix}+1\begin{pmatrix} 0 & 0 \\ 0 & 1 \end{pmatrix}$$

$$\begin{pmatrix} 2 & 1 \\ 0 & 0 \end{pmatrix}=2\begin{pmatrix} 1 & 0 \\ 0 & 0 \end{pmatrix}+1\begin{pmatrix} 0 & 1 \\ 0 & 0 \end{pmatrix}+0\begin{pmatrix} 0 & 0 \\ 1 & 0 \end{pmatrix}+0\begin{pmatrix} 0 & 0 \\ 0 & 1 \end{pmatrix}$$

$$\begin{pmatrix} 0 & 0 \\ 1 & 1 \end{pmatrix}=0\begin{pmatrix} 1 & 0 \\ 0 & 0 \end{pmatrix}+0\begin{pmatrix} 0 & 1 \\ 0 & 0 \end{pmatrix}+1\begin{pmatrix} 0 & 0 \\ 1 & 0 \end{pmatrix}+1\begin{pmatrix} 0 & 0 \\ 0 & 1 \end{pmatrix}$$

$$\begin{pmatrix} 0 & 0 \\ 0 & 1 \end{pmatrix}=0\begin{pmatrix} 1 & 0 \\ 0 & 0 \end{pmatrix}+0\begin{pmatrix} 0 & 1 \\ 0 & 0 \end{pmatrix}+0\begin{pmatrix} 0 & 0 \\ 1 & 0 \end{pmatrix}+1\begin{pmatrix} 0 & 0 \\ 0 & 1 \end{pmatrix}$$

so

$$\mathcal{P}_B = \begin{pmatrix} 1 & 2 & 0 & 0 \\ 0 & 1 & 0 & 0 \\ 0 & 0 & 1 & 0 \\ 1 & 0 & 1 & 1 \end{pmatrix}.$$

If

$$A=\begin{pmatrix} a & b \\ c & d \end{pmatrix}=a\begin{pmatrix} 1 & 0 \\ 0 & 0 \end{pmatrix}+b\begin{pmatrix} 0 & 1 \\ 0 & 0 \end{pmatrix}+c\begin{pmatrix} 0 & 0 \\ 1 & 0 \end{pmatrix}+d\begin{pmatrix} 0 & 0 \\ 0 & 1 \end{pmatrix},$$

then

$$[A]_B = \mathcal{P}_B^{-1}\begin{pmatrix} a \\ b \\ c \\ d \end{pmatrix}=\begin{pmatrix} 1 & 2 & 0 & 0 \\ 0 & 1 & 0 & 0 \\ 0 & 0 & 1 & 0 \\ 1 & 0 & 1 & 1 \end{pmatrix}^{-1}\begin{pmatrix} a \\ b \\ c \\ d \end{pmatrix}=\begin{pmatrix} a-2b \\ b \\ c \\ -a+2b-c+d \end{pmatrix}.$$

Exercises

In Exercises 1 through 10, find the coordinates of the vector $\hat{u}$ relative to the given basis.

1. $V = \mathbb{R}^2$, $\quad \mathcal{B} = \left\{ \begin{pmatrix} 1 \\ 4 \end{pmatrix}, \begin{pmatrix} 0 \\ 3 \end{pmatrix} \right\}$, $\quad \hat{u} = \begin{pmatrix} 6 \\ 2 \end{pmatrix}$

2. $V = \mathbb{R}^2$, $\quad \mathcal{B} = \left\{ \begin{pmatrix} -2 \\ 1 \end{pmatrix}, \begin{pmatrix} 1 \\ 1 \end{pmatrix} \right\}$, $\quad \hat{u} = \begin{pmatrix} 0 \\ 5 \end{pmatrix}$

3. $V = \mathbb{R}^3$, $\quad \mathcal{B} = \left\{ \begin{pmatrix} 0 \\ 3 \\ 1 \end{pmatrix}, \begin{pmatrix} -2 \\ 1 \\ 4 \end{pmatrix}, \begin{pmatrix} 8 \\ 1 \\ 3 \end{pmatrix} \right\}$, $\quad \hat{u} = \begin{pmatrix} 2 \\ 2 \\ 5 \end{pmatrix}$

4. $V = \mathbb{R}^3$, $\quad \mathcal{B} = \left\{ \begin{pmatrix} 4 \\ 7 \\ -2 \end{pmatrix}, \begin{pmatrix} 0 \\ 0 \\ 1 \end{pmatrix}, \begin{pmatrix} -4 \\ 1 \\ 6 \end{pmatrix} \right\}$, $\quad \hat{u} = \begin{pmatrix} -1 \\ 0 \\ 3 \end{pmatrix}$

5. $V = \mathbb{R}^4$, $\quad \mathcal{B} = \left\{ \begin{pmatrix} 1 \\ 0 \\ 0 \\ 2 \end{pmatrix}, \begin{pmatrix} 2 \\ 1 \\ 1 \\ 0 \end{pmatrix}, \begin{pmatrix} 3 \\ -1 \\ 4 \\ 2 \end{pmatrix}, \begin{pmatrix} 0 \\ 0 \\ 1 \\ -2 \end{pmatrix}, \begin{pmatrix} 0 \\ 0 \\ 1 \\ -2 \end{pmatrix} \right\}$, $\quad \hat{u} = \begin{pmatrix} 4 \\ -2 \\ 0 \\ 1 \end{pmatrix}$

6. $V = \mathbb{R}^4$, $\quad \mathcal{B} = \left\{ \begin{pmatrix} -1 \\ 4 \\ 1 \\ 1 \end{pmatrix}, \begin{pmatrix} 1 \\ -1 \\ 2 \\ 3 \end{pmatrix}, \begin{pmatrix} 0 \\ 1 \\ 0 \\ 1 \end{pmatrix}, \begin{pmatrix} 1 \\ 0 \\ 1 \\ 1 \end{pmatrix} \right\}$, $\quad \hat{u} = \begin{pmatrix} 5 \\ 4 \\ 1 \\ 1 \end{pmatrix}$

7. $V = P_2(x)$, $\quad \mathcal{B} = \left\{ 1 + 2x, x - x^2, 3 + x^2 \right\}$, $\quad \hat{u} = 3 - 4x + x^2$

8. $V = P_2(x)$, $\quad \mathcal{B} = \left\{ 3 - x, 1 + 5x - x^2, 2x + x^2 \right\}$, $\quad \hat{u} = 1 + 4x - 3x^2$

9. $V = M_{2\times2}(\mathbb{R})$, $\quad \mathcal{B} = \left\{ \begin{pmatrix} 1 & 1 \\ 1 & 1 \end{pmatrix}, \begin{pmatrix} 1 & 1 \\ 1 & 0 \end{pmatrix}, \begin{pmatrix} 1 & 0 \\ 0 & 1 \end{pmatrix}, \begin{pmatrix} 0 & 2 \\ 1 & 1 \end{pmatrix} \right\}$, $\quad \hat{u} = \begin{pmatrix} 3 & -1 \\ 0 & 6 \end{pmatrix}$

10. $V = M_{2\times2}(\mathbb{R})$, $\quad \mathcal{B} = \left\{ \begin{pmatrix} 2 & 0 \\ 1 & 3 \end{pmatrix}, \begin{pmatrix} 0 & 4 \\ 1 & -1 \end{pmatrix}, \begin{pmatrix} 3 & 1 \\ 0 & 1 \end{pmatrix}, \begin{pmatrix} 1 & 0 \\ -2 & 1 \end{pmatrix} \right\}$, $\quad \hat{u} = \begin{pmatrix} 2 & 3 \\ 7 & 3 \end{pmatrix}$

11. Find a basis for the symmetric 3×3 matrices.

12. (a) Show that the set of $n \times n$ matrices whose trace is 0 is a vector space.
    (b) Find a basis for 2×2 matrices whose trace is 0.
    (c) Find a basis for 3×3 matrices whose trace is 0.

    In Exercises 13 through 16, find
    (a) The transition matrix from $\mathcal{B}_1$ to $\mathcal{B}_2$.
    (b) The transition matrix from $\mathcal{B}_2$ to $\mathcal{B}_1$.
    (c) For the given $[\hat{v}]_{\mathcal{B}_1}$ find $[\hat{v}]_{\mathcal{B}_2}$.
    (d) For the given $[\hat{v}]_{\mathcal{B}_2}$ find $[\hat{v}]_{\mathcal{B}_1}$.

13. $\mathcal{B}_1 = \left\{ \begin{pmatrix} 0 \\ 3 \end{pmatrix}, \begin{pmatrix} 1 \\ 2 \end{pmatrix} \right\}$, $\mathcal{B}_2 = \left\{ \begin{pmatrix} 4 \\ -1 \end{pmatrix}, \begin{pmatrix} 1 \\ 1 \end{pmatrix} \right\}$

$\left[\hat{v} \right]_{\mathcal{B}_1} = \begin{pmatrix} 2 \\ 5 \end{pmatrix}$, $\left[\hat{v} \right]_{\mathcal{B}_2} = \begin{pmatrix} 6 \\ -2 \end{pmatrix}$

14. $\mathcal{B}_1 = \left\{ \begin{pmatrix} 1 \\ -1 \end{pmatrix}, \begin{pmatrix} 2 \\ 3 \end{pmatrix} \right\}$, $\mathcal{B}_2 = \left\{ \begin{pmatrix} 1 \\ 5 \end{pmatrix}, \begin{pmatrix} 2 \\ -3 \end{pmatrix} \right\}$

$\left[\hat{v} \right]_{\mathcal{B}_1} = \begin{pmatrix} 4 \\ 3 \end{pmatrix}$, $\left[\hat{v} \right]_{\mathcal{B}_2} = \begin{pmatrix} 3 \\ 8 \end{pmatrix}$

15. $\mathcal{B}_1 = \left\{ \begin{pmatrix} 1 \\ 5 \\ -2 \end{pmatrix}, \begin{pmatrix} -3 \\ 0 \\ 2 \end{pmatrix}, \begin{pmatrix} 4 \\ 1 \\ 4 \end{pmatrix} \right\}$, $\mathcal{B}_2 = \left\{ \begin{pmatrix} 2 \\ 2 \\ -2 \end{pmatrix}, \begin{pmatrix} 1 \\ 6 \\ 3 \end{pmatrix}, \begin{pmatrix} 0 \\ 1 \\ 0 \end{pmatrix} \right\}$

$\left[\hat{v} \right]_{\mathcal{B}_1} = \begin{pmatrix} 2 \\ -3 \\ 1 \end{pmatrix}$, $\left[\hat{v} \right]_{\mathcal{B}_2} = \begin{pmatrix} 2 \\ 0 \\ 5 \end{pmatrix}$

16. $\mathcal{B}_1 = \left\{ \begin{pmatrix} 0 \\ 3 \\ 1 \end{pmatrix}, \begin{pmatrix} 4 \\ 5 \\ 2 \end{pmatrix}, \begin{pmatrix} 1 \\ 1 \\ 1 \end{pmatrix} \right\}$, $\mathcal{B}_2 = \left\{ \begin{pmatrix} 3 \\ 1 \\ 0 \end{pmatrix}, \begin{pmatrix} 2 \\ -1 \\ -1 \end{pmatrix}, \begin{pmatrix} 1 \\ 1 \\ 2 \end{pmatrix} \right\}$

$\left[\hat{v} \right]_{\mathcal{B}_1} = \begin{pmatrix} 6 \\ 1 \\ 1 \end{pmatrix}$, $\left[\hat{v} \right]_{\mathcal{B}_2} = \begin{pmatrix} 0 \\ 4 \\ -2 \end{pmatrix}$

3.8 Null Space, Row Space, and Column Space of a Matrix

Let A be an $m \times n$ matrix. Associated with each matrix are three vector spaces.

(1) The null space of A defined by

$$N(A) = \left\{ \hat{x} \in \mathcal{F}^n \mid A\hat{x} = \hat{0} \right\}.$$

(2) The row space of A, which is the vector space of linear combinations of the rows of A. If A is an $m \times n$ matrix, then the row space of A is a subspace of $\mathcal{F}^n$.

(3) The column space of A, which is the vector space of linear combinations of the columns of A. If A is an $m \times n$ matrix, then the column space of A is a subspace of $\mathcal{F}^m$.

In this section, we will show that for a given $m \times n$ matrix A

1. The number of vectors in a basis for the column space of A is equal to the number of vectors in a basis for the row space of A. This number is called the rank of A.
2. The number of vectors in a basis for the null space of A is called the nullity of A. We will show that

$$\text{Nullity of } A + \text{rank of } A = \text{number of columns of } A.$$

Furthermore, the nullity of A is the number of free variables in the row reduced form of A and the rank of A is the number of leading variables in the row reduced form of A.
Among other things, in this section, we will validate the algorithms we used in Section 3.5 to reduce/expand a given set of vectors to a basis.

Theorem 19

Let A be an $m \times n$ matrix. A vector can be written as a linear combination of the columns of A if and only if that vector can be written as $A\hat{c}$ for some

$$\hat{c} \in \mathcal{F}^n.$$

Proof

Let

$$\hat{v} = A\hat{c} = \begin{pmatrix} a_{11} & a_{12} & \cdots & a_{1n} \\ a_{21} & a_{22} & \cdots & a_{2n} \\ \vdots & \vdots & & \vdots \\ a_{m1} & a_{m2} & \cdots & a_{mn} \end{pmatrix} \begin{pmatrix} c_1 \\ c_2 \\ \vdots \\ c_n \end{pmatrix}$$

$$= \begin{pmatrix} a_{11}c_1 + a_{12}c_2 + \cdots + a_{1n}c_n \\ a_{21}c_1 + a_{22}c_2 + \cdots + a_{2n}c_n \\ \vdots \\ a_{m1}c_1 + a_{m2}c_2 + \cdots + a_{mn}c_n \end{pmatrix}$$

$$= \begin{pmatrix} a_{11}c_1 \\ a_{21}c_1 \\ \vdots \\ a_{m1}c_1 \end{pmatrix} + \cdots + \begin{pmatrix} a_{1n}c_n \\ a_{2n}c_n \\ \vdots \\ a_{mn}c_n \end{pmatrix}$$

$$= c_1 \begin{pmatrix} a_{11} \\ a_{21} \\ \vdots \\ a_{m1} \end{pmatrix} + \cdots + c_n \begin{pmatrix} a_{1n} \\ a_{2n} \\ \vdots \\ a_{mn} \end{pmatrix}.$$

Corollary

The equation $A\hat{x} = \hat{b}$ has a solution if and only if $\hat{b}$ is in the column space of A.
   Recall that for any function

$$f : X \rightarrow Y$$

the range of f is $\{y \in Y \mid y = f(x) \text{ for some } x \in X\}$. For

$$A : \mathbb{R}^n \rightarrow \mathbb{R}^m$$

Theorem 19 says the range of A is the column space of A.

Example

Let

$$A = \begin{pmatrix} 1 & -3 & 0 \\ 2 & 2 & -1 \\ 6 & 0 & 4 \\ 5 & 5 & 1 \end{pmatrix}.$$

Determine if there is a vector

$$\hat{x} = \begin{pmatrix} x_1 \\ x_2 \\ x_3 \end{pmatrix}$$

for which

$$A\hat{x} = \hat{y} = \begin{pmatrix} y_1 \\ y_2 \\ y_3 \\ y_4 \end{pmatrix}$$

if

(i) $\hat{y} = \begin{pmatrix} -10 \\ 0 \\ 10 \\ 14 \end{pmatrix}$

(ii) $\hat{y} = \begin{pmatrix} 2 \\ 1 \\ 7 \\ -6 \end{pmatrix}.$

A useful way to think of this problem is that we are asking if $\hat{y}$ is in the column space of A. Thus, we want to know if there is a vector

$$\hat{x} = \begin{pmatrix} x_1 \\ x_2 \\ x_3 \end{pmatrix}$$

for which

$$A\hat{x} = \begin{pmatrix} 1 & -3 & 0 \\ 2 & 2 & -1 \\ 6 & 0 & 4 \\ 5 & 5 & 1 \end{pmatrix} \begin{pmatrix} x_1 \\ x_2 \\ x_3 \end{pmatrix} = x_1 \begin{pmatrix} 1 \\ 2 \\ 6 \\ 5 \end{pmatrix} + x_2 \begin{pmatrix} -3 \\ 2 \\ 0 \\ 5 \end{pmatrix} + x_3 \begin{pmatrix} 0 \\ -1 \\ 4 \\ 1 \end{pmatrix} = \hat{y} = \begin{pmatrix} y_1 \\ y_2 \\ y_3 \\ y_4 \end{pmatrix}.$$

This is most conveniently solved by converting the vector equation to the system of linear equations

$$x_1 - 3x_2 = y_1$$
$$2x_1 + 2x_2 - x_3 = y_2$$
$$6x_1 + 4x_3 = y_3$$
$$5x_1 + 5x_2 + x_3 = y_4$$

(i) For

$$\hat{y} = \begin{pmatrix} -10 \\ 0 \\ 10 \\ 14 \end{pmatrix}$$

the system has the augmented matrix

$$\begin{pmatrix} 1 & -3 & 0 & -10 \\ 2 & 2 & -1 & 0 \\ 6 & 0 & 4 & 10 \\ 5 & 5 & 1 & 14 \end{pmatrix}.$$

When row reduced, the result is

$$\begin{pmatrix} 1 & 0 & 0 & -1 \\ 0 & 1 & 0 & 3 \\ 0 & 0 & 1 & 4 \\ 0 & 0 & 0 & 0 \end{pmatrix}$$

and we have $x_1 = -1, x_2 = 3, x_3 = 4$ so if

$$\hat{x} = \begin{pmatrix} -1 \\ 3 \\ 4 \end{pmatrix}$$

we have

$$A\hat{x} = \begin{pmatrix} 1 & -3 & 0 \\ 2 & 2 & -1 \\ 6 & 0 & 4 \\ 5 & 5 & 1 \end{pmatrix} \begin{pmatrix} -1 \\ 3 \\ 4 \end{pmatrix} = \begin{pmatrix} -10 \\ 0 \\ 10 \\ 14 \end{pmatrix}.$$

(ii) For

$$\hat{y} = \begin{pmatrix} 2 \\ 1 \\ 7 \\ -6 \end{pmatrix}$$

the system has the augmented matrix

$$\begin{pmatrix} 1 & -3 & 0 & 2 \\ 2 & 2 & -1 & 1 \\ 6 & 0 & 4 & 7 \\ 5 & 5 & 1 & -6 \end{pmatrix}.$$

When row reduced, the result is

$$\begin{pmatrix} 1 & 0 & 0 & 0 \\ 0 & 1 & 0 & 0 \\ 0 & 0 & 1 & 0 \\ 0 & 0 & 0 & 1 \end{pmatrix}$$

and there is no solution.

The major goal of this section is, given a matrix A, find bases for the null space of A, the row space of A, and the column space of A. We will first demonstrate how this is done when the matrix is in row reduced form and then see what changes occur in going from a matrix that is not in row reduced form to a matrix that is in row reduced form. We will see that the row space and null space are unchanged but there is a change in the column space.

Example

Consider the matrix A that is in row reduced form:

$$A = \begin{pmatrix} 1 & 4 & 0 & 0 & 0 \\ 0 & 0 & 1 & 0 & 5 \\ 0 & 0 & 0 & 1 & 0 \\ 0 & 0 & 0 & 0 & 0 \end{pmatrix}.$$

The nonzero rows all have leading 1's and these rows are linearly independent. This is because the leading 1's occur in columns where there is only one nonzero entry. In this case, the leading 1's occur at the first, third, and fourth components.

To expand what we have said, in this example, the first three rows have nonzero elements. Let

$$\hat{v}_1 = (1,4,0,0,0), \quad \hat{v}_2 = (0,0,1,0,5), \quad \hat{v}_3 = (0,0,0,1,0).$$

So if

$$a\hat{v}_1 + b\hat{v}_2 + c\hat{v}_3 = (0,0,0,0,0)$$

then $a=b=c=0$, so $\{\hat{v}_1, \hat{v}_2, \hat{v}_3\}$ is a linearly independent set.

Likewise, the columns where leading 1's occur form a basis for the column space. In the example, these columns are

$$\begin{pmatrix} 1 \\ 0 \\ 0 \\ 0 \end{pmatrix}, \begin{pmatrix} 0 \\ 1 \\ 0 \\ 0 \end{pmatrix}, \begin{pmatrix} 0 \\ 0 \\ 1 \\ 0 \end{pmatrix}.$$

In the matrix, the second column is a multiple of the first column, and the fifth column is a multiple of the third column. For all matrices that are in row reduced form, a column that does not have a leading 1 is a multiple of a column that has a leading 1.

The vectors mentioned earlier form a basis for the subspace

$$\left\{ \begin{pmatrix} x \\ y \\ z \\ 0 \end{pmatrix} \middle| \, x, y, z \in \mathbb{R} \right\}.$$

Thus, in the case of a row reduced matrix, the number of leading 1's is equal to the dimension of the row space and is also equal to the dimension of the column space.

For the null space of this example, the free variables are x_2 and x_5, and we have

$$x_1 + 4x_2 = 0 \quad \text{so } x_1 = -4x_2$$

$$x_3 + 5x_5 = 0, \quad \text{so } x_3 = -4x_5$$

$$x_4 = 0.$$

Thus, a vector in the null space of A is of the form

$$\hat{x} = \begin{pmatrix} x_1 \\ x_2 \\ x_3 \\ x_4 \\ x_5 \end{pmatrix} = \begin{pmatrix} -4x_2 \\ x_2 \\ -4x_5 \\ 0 \\ x_5 \end{pmatrix} = \begin{pmatrix} -4x_2 \\ x_2 \\ 0 \\ 0 \\ 0 \end{pmatrix} + \begin{pmatrix} 0 \\ 0 \\ -4x_5 \\ 0 \\ x_5 \end{pmatrix} = x_2 \begin{pmatrix} -4 \\ 1 \\ 0 \\ 0 \\ 0 \end{pmatrix} + x_5 \begin{pmatrix} 0 \\ 0 \\ -4 \\ 0 \\ 1 \end{pmatrix}$$

and a basis for the null space is

$$\left\{ \begin{pmatrix} -4 \\ 1 \\ 0 \\ 0 \\ 0 \end{pmatrix}, \begin{pmatrix} 0 \\ 0 \\ -4 \\ 0 \\ 1 \end{pmatrix} \right\}.$$

Note that each free variable gives rise to a basis vector of the null space.

Thus, if the matrix is row reduced, we have

Number of free variables=dimension of the null space.

We will later show this is true even if the matrix is not in row reduced form.

Since in row reduced form, each variable is a free variable or a leading variable, it follows that the number of leading variables plus the number of free variables are equal to the number of columns of the matrix.

We now explore the changes that occur in the row, column, and null spaces when elementary row operations are applied to a matrix. We found earlier that elementary row operations do not change the null space of a matrix.

Theorem 20

Applying an elementary row operation to a matrix does not change the row space of the matrix. That is, if E is a matrix for which multiplying the matrix A on the left by E results in having the matrix A changed by an elementary row operation, then the matrix EA has the same row space as A.

Note that if E is an elementary matrix, then it satisfies the hypothesis of the theorem.

Proof

We consider each of the elementary row operations.

(1) Interchanging two rows of a matrix only changes the order in the vectors whose linear combinations make up the row space.
(2) Multiplying a row of a matrix by a nonzero constant does not change the possible linear combinations of the rows.
(3) For the effect of replacing a row by itself plus a nonzero multiple of another row, suppose two rows are $\hat{r}_1$ and $\hat{r}_2$ and $\hat{r}_1$ is replaced by $\hat{r}_1 + a\hat{r}_2$. We show that any linear combination of $\hat{r}_1$ and $\hat{r}_2$ can be written as a linear combination of $\hat{r}_1 + a\hat{r}_2$ and $\hat{r}_2$.

We suppose that

$$c\hat{r}_1 + d\hat{r}_2 = e\left(\hat{r}_1 + a\hat{r}_2\right) + f\hat{r}_2.$$

We know a, c, and d, and want to find e and f.

Observe that if

$$e = c \quad \text{and} \quad f = d - ac$$

then we have

$$e\left(\hat{r}_1 + a\hat{r}_2\right) + f\hat{r}_2 = c\left(\hat{r}_1 + a\hat{r}_2\right) + (d - ac)\hat{r}_2 = c\hat{r}_1 + (ac + d - ac)\hat{r}_2 = c\hat{r}_1 + d\hat{r}_2.$$

Furthermore, any vector in the span of $\left\{\left(\hat{r}_1 + a\hat{r}_2\right), \hat{r}_2\right\}$ is in the span of $\left\{\hat{r}_1, \hat{r}_2\right\}$.

Corollary

If A is a matrix with row reduced form B, then A and B have the same row space.
  This result is useful because a basis of the row space of a row reduced matrix consists of the nonzero rows.

Example

Let

$$A = \begin{pmatrix} 5 & 10 & 2 & 8 & 0 & 12 \\ 1 & 2 & 0 & 0 & 0 & 2 \\ 2 & 4 & 1 & 4 & 0 & 5 \\ 1 & 2 & 0 & 0 & 1 & 5 \\ 2 & 4 & 1 & 4 & 2 & 2 \\ 3 & 6 & 1 & 4 & 0 & 7 \end{pmatrix}$$

When A is row reduced, the result is

$$B = \begin{pmatrix} 1 & 2 & 0 & 0 & 0 & 0 \\ 0 & 0 & 1 & 4 & 0 & 0 \\ 0 & 0 & 0 & 0 & 1 & 0 \\ 0 & 0 & 0 & 0 & 0 & 1 \\ 0 & 0 & 0 & 0 & 0 & 0 \\ 0 & 0 & 0 & 0 & 0 & 0 \end{pmatrix}$$

so a basis for the row space of A is

$$\{(1\ \ 2\ \ 0\ \ 0\ \ 0\ \ 0),\ (0\ \ 0\ \ 1\ \ 4\ \ 0\ \ 0),\ (0\ \ 0\ \ 0\ \ 0\ \ 1\ \ 0),\ (0\ \ 0\ \ 0\ \ 0\ \ 1\ \ 0)\}.$$

While elementary row operations do not change the row space or null space of a matrix, they can change the column space of a matrix as the next example shows.

Example

Consider a 2×2 matrix where the second row is a nonzero multiple of the first, say

$$A = \begin{pmatrix} 1 & 2 \\ 4 & 8 \end{pmatrix}.$$

The rows and columns are each multiples of the other. The column space is spanned by the vector

$$\begin{pmatrix} 1 \\ 4 \end{pmatrix}$$

and the row space is spanned by the vector $(1, 2)$.

When the matrix A is row reduced, the result is

$$\begin{pmatrix} 1 & 2 \\ 0 & 0 \end{pmatrix}.$$

The column space of the row reduced matrix is spanned by the vector

$$\begin{pmatrix} 1 \\ 0 \end{pmatrix}$$

which is different from the space spanned by

$$\begin{pmatrix} 1 \\ 4 \end{pmatrix}$$

So while the elementary row operations do not change the space spanned by the rows of a matrix, the span of the column space can change. In this example, while the column space changed with the elementary row operations, the *dimension* of the column space did not. The next theorem shows that this is true in general.

Theorem 21

If A and B are row equivalent matrices, then the columns of A are linearly independent if and only if the corresponding columns of B are linearly independent. In particular, a set of columns of A forms a basis for the column space of A if and only if the corresponding columns of B form a basis for the column space of B.

Proof

We use the fact that if A and B are row equivalent matrices, then the null space of A is equal to the null space of B. We also use the fact that the columns of A are linearly dependent if and only if there is a nonzero vector $\hat{x}$ for which $A\hat{x} = \hat{0}$.

Thus, we have

$$A\hat{x} = \hat{0} \quad \text{if and only if } B\hat{x} = \hat{0} \quad \text{for some } \hat{x} \neq \hat{0}$$

if and only if the columns of A are dependent, if and only if the columns of B are dependent.

This gives a way to find a basis for the column space of a matrix. Namely,

(1) Put the matrix in row reduced form.
(2) Identify the columns of the row reduced form where the leading 1's occur.
(3) The corresponding columns in the original matrix will be a basis for the column space of the original matrix.

Example

We find a basis for the column space of the matrix

$$A = \begin{pmatrix} 5 & 10 & 2 & 8 & 0 & 12 \\ 1 & 2 & 0 & 0 & 0 & 2 \\ 2 & 4 & 1 & 4 & 0 & 5 \\ 1 & 2 & 0 & 0 & 1 & 5 \\ 2 & 4 & 1 & 4 & 2 & 2 \\ 3 & 6 & 1 & 4 & 0 & 7 \end{pmatrix}.$$

When A is row reduced, the result is

$$B = \begin{pmatrix} 1 & 2 & 0 & 0 & 0 & 0 \\ 0 & 0 & 1 & 4 & 0 & 0 \\ 0 & 0 & 0 & 0 & 1 & 0 \\ 0 & 0 & 0 & 0 & 0 & 1 \\ 0 & 0 & 0 & 0 & 0 & 0 \\ 0 & 0 & 0 & 0 & 0 & 0 \end{pmatrix}.$$

In the row reduced form, the leading 1's occurred in the first, third, fifth, and sixth columns. These do not constitute a basis for the column space of A, but if **we take the corresponding columns from the original matrix, these will be a basis for the column space.** Thus, a basis for the column space of the matrix A is

$$\left\{ \begin{pmatrix} 5 \\ 1 \\ 2 \\ 1 \\ 2 \\ 3 \end{pmatrix}, \begin{pmatrix} 2 \\ 0 \\ 1 \\ 0 \\ 1 \\ 1 \end{pmatrix}, \begin{pmatrix} 0 \\ 0 \\ 0 \\ 1 \\ 2 \\ 0 \end{pmatrix}, \begin{pmatrix} 12 \\ 2 \\ 5 \\ 5 \\ 2 \\ 7 \end{pmatrix} \right\}.$$

Summarizing our results, we have that for any matrix $A: \mathbb{R}^n \to \mathbb{R}^m$

Each nonzero row in row reduced form gives rise to a distinct vector in a basis for the row space. Each nonzero row in row reduced form is headed by a leading 1. Thus,

The number of leading variables in row reduced form = dimension of the row space = dimension of the column space

so

number of free variables + number of leading variables = number of variables in the domain = number of columns of the matrix.

Another method to determine a basis for the column space of a matrix is to take the transpose of the matrix, and a basis for the row space of the transposed matrix will be a basis for the column space of the original matrix (when the rows are converted to columns).

Definition

The rank of a matrix is the dimension of the row space (which is also the dimension of the column space).

Definition

The dimension of the null space of a matrix is called the nullity of the matrix.
   In Theorem 22 we reiterate a result that was given earlier in a slightly different context.

Theorem 22

For a matrix A

$$\text{Rank of } A + \text{nullity of } A = \text{Number of columns of } A.$$

Example

Suppose that A is a 3×8 matrix. Find the minimum possible value of the nullity of A.
   Since A is a 3×8 matrix, the maximum possible value of the rank of A is of 3. Since

$$\text{Rank of } A + \text{nullity of } A = \text{Number of columns of } A = 8$$

and the minimum value of the nullity of A occurs when the rank of A is the largest, the minimum value of the nullity of A is

$$8 - 3 = 5.$$

Exercises

1. Express the product $A\hat{u}$ as a linear combination of the columns of A.

(a) $\begin{pmatrix} 3 & -1 \\ 4 & 6 \end{pmatrix}\begin{pmatrix} -2 \\ 5 \end{pmatrix}$

(b) $\begin{pmatrix} 1 & 0 \\ -3 & 2 \end{pmatrix}\begin{pmatrix} 3 \\ 6 \end{pmatrix}$

(c) $\begin{pmatrix} 2 & 2 & 0 \\ -1 & -3 & -6 \\ 1 & -2 & 4 \end{pmatrix}\begin{pmatrix} -2 \\ 0 \\ 3 \end{pmatrix}$

(d) $\begin{pmatrix} 8 & -2 & 2 \\ 0 & 3 & 1 \\ 5 & 4 & -4 \end{pmatrix}\begin{pmatrix} 4 \\ -1 \\ 1 \end{pmatrix}$

(e) $\begin{pmatrix} 2 & 1 & 1 \\ 0 & -2 & 3 \end{pmatrix} \begin{pmatrix} 2 \\ 3 \\ -5 \end{pmatrix}$

(f) $\begin{pmatrix} 1 & 6 \\ -2 & 2 \\ 3 & 4 \end{pmatrix} \begin{pmatrix} 2 \\ -5 \end{pmatrix}$

2. Determine whether $\hat{b}$ is in the column space of A. If that is the case, then express $\hat{b}$ as a linear combination of the columns of A.

(a) $A = \begin{pmatrix} 2 & 3 \\ -1 & 0 \end{pmatrix}$, $\hat{b} = \begin{pmatrix} 0 \\ 5 \end{pmatrix}$

(b) $A = \begin{pmatrix} 1 & 2 \\ 3 & 6 \end{pmatrix}$, $\hat{b} = \begin{pmatrix} 1 \\ -2 \end{pmatrix}$

(c) $A = \begin{pmatrix} 1 & 0 & 2 \\ 3 & 3 & 1 \\ 0 & -2 & 4 \end{pmatrix}$, $\hat{b} = \begin{pmatrix} 1 \\ 1 \\ 0 \end{pmatrix}$

(d) $A = \begin{pmatrix} 1 & 3 & 5 \\ -1 & 4 & 2 \\ 2 & 3 & 7 \end{pmatrix}$, $\hat{b} = \begin{pmatrix} 4 \\ 3 \\ -9 \end{pmatrix}$

3. Determine bases for the row space, column space, and null space for the matrices below.

(a) $\begin{pmatrix} 1 & 3 \\ 2 & 6 \end{pmatrix}$

(b) $\begin{pmatrix} 2 & 0 & -1 \\ 3 & 1 & 1 \end{pmatrix}$

(c) $\begin{pmatrix} 0 & 2 & 1 \\ 3 & -4 & 3 \\ 6 & -6 & 7 \end{pmatrix}$

(d) $\begin{pmatrix} 3 & 2 & 0 & 2 & -3 & 1 \\ -1 & 2 & 0 & 1 & -1 & 4 \\ 0 & 1 & 3 & 1 & -2 & 0 \\ 1 & 4 & 1 & 5 & 0 & 3 \end{pmatrix}$

4. Let A and B be square matrices so that AB is defined.

   (a) Show that the null space of AB contains the null space of B. Is it necessarily true that the null space of AB contains the null space of A? Can you add a hypothesis that will make that true?

   (b) Show that the row space of B contains the row space of AB.

5. Show that if A and B are matrices for which $AB = 0$, then the column space of B is contained in the null space of A.

3.9 Sums and Direct Sums of Vector Spaces (Optional)

The topics covered in this section deal with ways to decompose a vector space and ways to put vector spaces together to form a new vector space.

Suppose that V is a vector space and A and B are nonempty subsets of V. We define

$$A + B = \{v \in V \mid v = a + b, a \in A, b \in B\}.$$

It is not necessarily the case that $A + B$ is a subspace of V, but if A and B are subspaces of V, then $A + B$ is a subspace of V as we show in Theorem 23.

Examples

Let $V = \mathbb{R}^3, U_1 = \{(x, 0, 0) \mid x \in \mathbb{R}\}$, and $U_2 = \{(y, y, 0) \mid y \in \mathbb{R}\}$. Then

$$U_1 + U_2 = \{(w, z, 0) \mid, w, z \in \mathbb{R}\}.$$

This is because we can take $y = z$ and $x = w - z$, so that

$$(w - z, 0, 0) + (z, z, 0) = (w, z, 0).$$

Theorem 23

Suppose that V is a vector space and $U_1, \ldots, U_n$ are subspaces of V. Then $U_1 + \cdots + U_n$ is the smallest subspace of V containing $U_1, \ldots, U_n$.

Proof

We first show $U_1 + \cdots + U_n$ is a subspace of V. Let $\hat{v}, \hat{w} \in U_1 + \cdots + U_n$. So

$$\hat{v} = \hat{v}_1 + \cdots \hat{v}_n, \quad \hat{w} = \hat{w}_1 + \cdots \hat{w}_n; \quad \hat{v}_i, \hat{w}_i \in U_i, \quad i = 1, \ldots, n.$$

Then

$$\hat{v} + \hat{w} = (\hat{v}_1 + \cdots \hat{v}_n) + (\hat{w}_1 + \cdots \hat{w}_n) = (\hat{v}_1 + \hat{w}_1) + \cdots + (\hat{v}_n + \hat{w}_n) \varepsilon U_1 + \cdots + U_n$$

and

$$\alpha\hat{v} = \alpha(\hat{v}_1 + \cdots \hat{v}_n) = \alpha\hat{v}_1 + \cdots + \alpha\hat{v}_n \, \varepsilon \, U_1 + \cdots + U_n$$

so $U_1 + \cdots + U_n$ is a subspace of V.

Note that $U_1 + \cdots + U_n$ contains each U_i, because if $\hat{u}_i \in U_i$, then

$$\hat{u}_i = \hat{0} + \cdots + \hat{0} + \hat{u}_i + \hat{0} + \cdots + \hat{0} \, \varepsilon \, U_1 + \cdots + U_n.$$

Moreover, any subspace containing $U_1, \ldots, U_n$ contains all finite sums of elements from those sets and thus contains $U_1 + \cdots + U_n$.

More important than sums of subspaces to applications in linear algebra is the idea of the direct sum of subspaces, because it provides a way of decomposing a vector space.

Definition

Suppose that V is a vector space and $U_1, \ldots, U_n$ are subspaces of V. We say that V is the direct sum of $U_1, \ldots, U_n$ if each vector in V can be written as the sum of elements from $U_1, \ldots, U_n$ in exactly one way. In this case, we write

$$V = U_1 \oplus \cdots \oplus U_n.$$

Example

Suppose that V is a vector space with basis $\{\hat{v}_1, \ldots, \hat{v}_k\}$. Let

$$U_i = \{\alpha \hat{v}_i \mid \alpha \in \mathcal{F}\}.$$

Then U_i is a subspace of V and

$$V = U_1 \oplus \cdots \oplus U_n.$$

This is because if $\hat{v} \in V$ and if $\{\hat{v}_1, \ldots, \hat{v}_k\}$ is a basis for $\{\hat{v}_1, \ldots, \hat{v}_k\}$ then there is exactly one way to write $\hat{v}$ as a linear combination of $\{\hat{v}_1, \ldots, \hat{v}_k\}$.

Theorem 24

Suppose that V is a vector space and $U_1, \ldots, U_n$ are subspaces of V. We have

$$V = U_1 \oplus \cdots \oplus U_n$$

if and only if both of the following conditions hold:

(a) $V = U_1 + \cdots + U_n$
(b) If $\hat{u}_i \in U_i, i = 1, \ldots, n$ and $\hat{u}_1 + \cdots + \hat{u}_n = \hat{0}$, then $\hat{u}_i = 0, i = 1, \ldots, n$.

Proof

Suppose that

$$V = U_1 \oplus \cdots \oplus U_n.$$

Then $V = U_1 + \cdots + U_n$. Suppose that $\hat{u}_1 + \cdots + \hat{u}_n = \hat{0}$.
One way this can happen is $\hat{u}_i = \hat{0}$ for $i = 1, \ldots, n$. Since

$$V = U_1 \oplus \cdots \oplus U_n,$$

this is the only way that it can happen.

Suppose that conditions (*a.*) and (*b.*) hold. By condition (*a.*), every vector in V can be expressed as a sum of vectors $\hat{u}_1 + \cdots + \hat{u}_n$ with $\hat{u}_i \in U_i$. We must show the expression is unique. Suppose there is a vector $\hat{v} \in V$ with

$$\hat{v} = \hat{u}_1 + \cdots + \hat{u}_n, \quad \hat{u}_i \in U_i$$

and

$$\hat{v} = \hat{w}_1 + \cdots + \hat{w}_n, \quad \hat{w}_i \in U_i.$$

Then

$$\hat{0} = \hat{v} - \hat{v} = \left(\hat{u}_1 + \cdots + \hat{u}_n \right) - \left(\hat{w}_1 + \cdots + \hat{w}_n \right) = \left(\hat{u}_1 - \hat{w}_1 \right) + \cdots + \left(\hat{u}_n - \hat{w}_n \right)$$

and $\left(\hat{u}_i - \hat{w}_i \right) \in U_i$, so by condition (*b.*) $\hat{u}_i - \hat{w}_i = \hat{0}$ or $\hat{u}_i = \hat{w}_i$.

Corollary

Suppose that V is a vector space and U and W are subspaces of V. Then

$$V = U \oplus W$$

if and only if

$$V = U + W \quad \text{and} \quad U \cap W = \{\hat{0}\}$$

Proof

We need to show that $U \cap W = \{\hat{0}\}$ is equivalent to the statement if $\hat{u} \in U$, $\hat{w} \in W$ and

$$\hat{u} + \hat{w} = \hat{0}$$

then $\hat{u} = \hat{w} = \hat{0}$.

Now since $\hat{u} + \hat{w} = \hat{0}$, we have $\hat{u} = -\hat{w}$, so $-\hat{w} \in U$ and thus $\hat{w} \in U$. Since $U \cap W = \{\hat{0}\}$ and $\hat{w} \in U \cap W$ we must have $\hat{w} = \hat{0}$. Thus

$$\hat{u} = -\hat{w} = \hat{0}.$$

Theorem 25

Suppose that V is a vector space of dimension n and U is a subspace of V of dimension k where $k < n$. Then there is a subspace W of V of dimension $n - k$ with $V = U \oplus W$.

Proof

Let $\{\hat{v}_1,..,\hat{v}_k\}$ be a basis for U. Extend $\{\hat{v}_1,..,\hat{v}_k\}$ to be a basis for V, say $\{\hat{v}_1,..,\hat{v}_k,\hat{v}_{k+1},...,\hat{v}_n\}$. Let W be the subspace of V formed by linear combinations of $\{\hat{v}_{k+1},...,\hat{v}_n\}$. If $\hat{v}\in U\cap W$, then

$$\hat{v}=c_1\hat{v}_1+\cdots c_k\hat{v}_k \in U \quad\text{and}\quad \hat{v}=c_{k+1}\hat{v}_{k+1}+\cdots c_n\hat{v}_n \in W.$$

Thus

$$\hat{0}=\hat{v}-\hat{v}=\left(c_1\hat{v}_1+\cdots c_k\hat{v}_k\right)-\left(c_{k+1}\hat{v}_{k+1}+\cdots c_n\hat{v}_n\right)$$

and since $\{\hat{v}_1,..,\hat{v}_k,\hat{v}_{k+1},...,\hat{v}_n\}$ is a basis, $c_i=0$, $i=1,\ldots,n$ so $\hat{v}=\hat{0}$.

Also $U+W$ contains all linear combinations of $\{\hat{v}_1,..,\hat{v}_k,\hat{v}_{k+1},...,\hat{v}_n\}$ so $U\oplus W=V$.

Corollary

If $V=U\oplus W$, then $dim\,V=dim\,U+dim\,W$.

Proof

Suppose that $\{\hat{u}_1,..,\hat{u}_k\}$ is a basis for U and $\{\hat{w}_1,..,\hat{w}_j\}$ is a basis for W.

Since $V=U+W$, then $\{\hat{u}_1,..,\hat{u}_k,\hat{w}_1,..,\hat{w}_j\}$ spans V. Also

$$\{\hat{u}_1,..,\hat{u}_k\}\cap\{\hat{w}_1,..,\hat{w}_j\}=\varnothing.$$

We show $\{\hat{u}_1,..,\hat{u}_k,\hat{w}_1,..,\hat{w}_j\}$ is a linearly independent set. Suppose

$$c_1\hat{u}_1+\cdots+c_k\hat{u}_k+d_1\hat{w}_1+\cdots+d_j\hat{w}_j=\hat{0}.$$

Now

$$c_1\hat{u}_1+\cdots+c_k\hat{u}_k=\hat{u}\in U \quad\text{and}\quad d_1\hat{w}_1+\cdots+d_j\hat{w}_j=\hat{w}\in W$$

and

$$\hat{u}+\hat{w}=\hat{0} \quad\text{so}\,\hat{u}=-\hat{w}.$$

But U and W are subspaces of with $U\cap W=\{\hat{0}\}$, so $\hat{u}=\hat{w}=\hat{0}$.

Thus

$$c_1\hat{u}_1+\cdots+c_k\hat{u}_k=\hat{0} \quad\text{and}\quad d_1\hat{w}_1+\cdots+d_j\hat{w}_j=\hat{0}$$

and since $\{\hat{u}_1,..,\hat{u}_k\}$ and $\{\hat{w}_1,..,\hat{w}_j\}$ are linearly independent sets,

$$c_1=\cdots=c_k=0, \quad\text{and}\quad d_1=\cdots=d_j=0.$$

Exercises

1. Let U be the subspace of $\mathbb{R}^4$ spanned by $\hat{v}_1 = (3,0,1,1)$ and $\hat{v}_2 = (2,2,2,1)$. Find a basis for a subspace W of $\mathbb{R}^4$ for which

$$\mathbb{R}^4 = U \oplus W.$$

2. Let $\mathcal{P}_n(\mathbb{R})$ be the polynomials of degree n or less with coefficients in $\mathbb{R}$. Let $\mathcal{E}(\mathbb{R})$ be the polynomials in $\mathcal{P}_n(\mathbb{R})$ for which the odd powers of x have coefficients equal to 0 and $\mathcal{O}(\mathbb{R})$ be the polynomials in $\mathcal{P}_n(\mathbb{R})$ for which the even powers of x have coefficients equal to 0. Show that

$$\mathcal{P}_n(\mathbb{R}) = \mathcal{E}(\mathbb{R}) \oplus \mathcal{O}(\mathbb{R}).$$

3. Let S be the set of symmetric 2×2 matrices, that is, matrices for which $A = A^T$. These are matrices of the form

$$\begin{pmatrix} a & c \\ c & b \end{pmatrix}.$$

Let T be the set of antisymmetric 2×2 matrices, that is, matrices for which $A = -A^T$. These are matrices of the form

$$\begin{pmatrix} 0 & c \\ -c & 0 \end{pmatrix}.$$

    (a) Show that T and S are subspaces of $M_{2 \times 2}(\mathbb{R})$.

    (b) Show that $M_{2 \times 2}(\mathbb{R}) = S \oplus T$.

4. (a) Show that if U and W are subspaces of the vector space V, then $U + W$ is a subspace of V.

    (b) Give an example of nonempty subsets U and W of the vector space V for which $U + W$ is not a subspace of V.

5. Suppose that V is a vector space and A and B are subspaces of V with bases $\{\hat{a}_1, \ldots, \hat{a}_j\}$ and $\{\hat{b}_1, \ldots, \hat{b}_k\}$. Show that

$$V = A \oplus B$$

If and only if $\{\hat{a}_1, \ldots, \hat{a}_j, \hat{b}_1, \ldots, \hat{b}_k\}$ is a basis for V.

6. Use the result of Exercise 5 to determine whether $V = A \oplus B$. If $V \neq A \oplus B$, find a nonzero vector in $A \cap B$ or a vector in V that is not in $A + B$

    (a) $V = \mathbb{R}^4$, A is the subspace spanned by

$$\{(1,-2,0,6), (3,7,1,-1)\}$$

and B is the subspace spanned by

$$\{(1,0,0,0),(5,3,1,11)\}.$$

(b) $V = \mathbb{R}^5$, A is the subspace spanned by

$$\{(3,-2,6,6,4),(2,5,1,-1,0)\}$$

and B is the subspace spanned by

$$\{(1,2,3,0,0),(6,5,-3,1,1)\}.$$

(c) $V = \mathbb{R}^4$, A is the subspace spanned by

$$\{(1,-2,0,6),(3,7,1,-1)\}$$

and B is the subspace spanned by

$$\{(1,0,0,0),(3,2,1,5)\}.$$

7. Show that if V is a vector space with subspaces W_1 and W_2, then $W_1 \cup W_2$ is a subspace if and only if $W_1 \subset W_2$ or $W_2 \subset W_1$.

4

Linear Transformations

4.1 Properties of a Linear Transformation

Definition

Suppose that U and V are vector spaces over the same scalar field F. A linear transformation (linear function) T from U to V is a function

$$T : U \to V$$

that satisfies

$$T\left(\hat{u}_1 + \hat{u}_2\right) = T\left(\hat{u}_1\right) + T\left(\hat{u}_2\right) \quad \text{for all } \hat{u}_1, \hat{u}_2 \in U$$

and

$$T\left(a\hat{u}\right) = aT\left(\hat{u}\right) \quad \text{for all } \hat{u} \in U, a \in F.$$

These two conditions are often combined into the single condition

$$T\left(a_1\hat{u}_1 + a_2\hat{u}_2\right) = a_1T\left(\hat{u}_1\right) + a_2T\left(\hat{u}_2\right) \quad \text{for all } \hat{u}_1, \hat{u}_2 \in U, a_1, a_2 \in F.$$

If

$$T : V \to V$$

is a linear transformation, then T is said to be a linear operator on V.

Notation

The set of linear transformations from U to V is denoted $L(U, V)$, and the set of linear operators from V to V is denoted $L(V)$.

The properties described in the definition are saying that linear transformations preserve vector addition and scalar multiplication.

The diagram in Figure 4.1a illustrates what the definition is saying for vector addition.

One could begin with $\hat{u}_1, \hat{u}_2 \in U$ and first apply T, to each vector obtaining the vectors $T\left(\hat{u}_1\right), T\left(\hat{u}_2\right) \in V$.

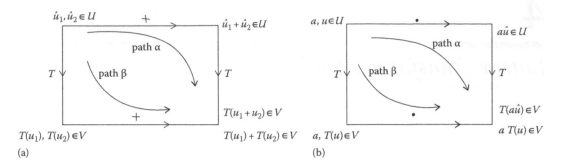

FIGURE 4.1

(a) The difference between $T\left(\hat{u}_1 + \hat{u}_2\right)$ and $T\left(\hat{u}_1\right) + T\left(\hat{u}_2\right)$. (b) The difference between $T\left(a\hat{u}\right)$ and $aT\left(\hat{u}\right)$.

One could then add $T\left(\hat{u}_1\right),$ and $T\left(\hat{u}_2\right)$ to get $T\left(\hat{u}_1\right) + T\left(\hat{u}_2\right) \in V$. This is the path α in Figure 4.1a.

A second choice would be to first add $\hat{u}_1, \hat{u}_2 \in U$, to obtain $\hat{u}_1 + \hat{u}_2 \in U$, then apply T to $\hat{u}_1 + \hat{u}_2$ obtaining $T(\hat{u}_1 + \hat{u}_2) \in V$. This is the path β in Figure 4.1a.

Depending on the function, one could get different answers following the different paths. For example, if

$$f\left(x\right) = x^2$$

then

$$f\left(x_1\right) + f\left(x_2\right) = x_1^2 + x_2^2$$

and

$$f\left(x_1 + x_2\right) = \left(x_1 + x_2\right)^2 = x_1^2 + 2x_1x_2 + x_2^2$$

We could follow the same ideas with scalar multiplication. If a is a scalar and $\hat{u} \in U$, one could form the vector $a\hat{u} \in U$, then apply T to $a\hat{u}$ to obtain $T\left(a\hat{u}\right) \in V$. This is the path α in Figure 4.1b.

One could have first applied T to $\hat{u} \in U$ to obtain $T\left(\hat{u}\right) \in V$, then multiplied by a to get $aT\left(\hat{u}\right) \in V$. This is the path β in Figure 4.1b.

The function

$$f\left(x\right) = x^2$$

again provides an example where the different paths give different answers. We have

$$f\left(2x\right) = \left(2x\right)^2 = 4x^2$$

and

$$2f\left(x\right) = 2\left(x^2\right) = 2x^2.$$

The definition says that if T is a linear transformation, then we end up in the same place with either path.

Definition

If $T_1, T_2 \in L(U, V)$ and $\alpha \in F$, we define $T_1 + T_2$ and αT_1 by

$$(T_1 + T_2)(\hat{u}) = T_1(\hat{u}) + T_2(\hat{u}),$$
$$(\alpha T_1)(\hat{u}) = T_1(\alpha\hat{u}) \quad \text{for all } \hat{u} \in U, \alpha \in F.$$

We note that $T_1 + T_2$ and αT_1 are in $L(U, V)$. This is because if $\hat{u}_1, \hat{u}_2 \in U$ and $\alpha \in F$, then

$$(T_1 + T_2)(\hat{u}_1 + \hat{u}_2) = T_1(\hat{u}_1 + \hat{u}_2) + T_2(\hat{u}_1 + \hat{u}_2) = T_1(\hat{u}_1) + T_1(\hat{u}_2) + T_2(\hat{u}_1) + T_2(\hat{u}_2)$$
$$= T_1(\hat{u}_1) + T_2(\hat{u}_1) + T_1(\hat{u}_2) + T_2(\hat{u}_2)$$
$$= (T_1 + T_2)(\hat{u}_1) + (T_1 + T_2)(\hat{u}_2)$$

and

$$\left[\alpha(T_1 + T_2)\right](\hat{u}_1) = \left[(\alpha T_1) + (\alpha T_2)\right](\hat{u}_1) = (\alpha T_1)(\hat{u}_1) + (\alpha T_2)(\hat{u}_1)$$
$$= T_1(\alpha\hat{u}_1) + T_2(\alpha\hat{u}_1) = (T_1 + T_2)(\alpha\hat{u}_1).$$

so that $T_1 + T_2$ is in $L(U, V)$.
Showing that αT_1 is in $L(U, V)$ is left as an exercise.

Examples

The linear transformation $T: U \to V$ defined by $T(\hat{u}) = \hat{0}_v$ for every $\hat{u} \in U$ is the zero transformation, which we will denote by 0.
The linear operator $T: U \to U$ defined by $T(\hat{u}) = \hat{u}$ for every $\hat{u} \in U$ is the identity operator on U, which we will denote by I_U.

Theorem 1

With the notation mentioned earlier, $L(U, V)$ and $L(V)$ are vector spaces over F.
The proof is left as an exercise.

Theorem 2

Let $T \in L(U, V)$ and $\hat{0}_U$ and $\hat{0}_V$ be the zero vectors for U and V, respectively. Then

(1) $T(\hat{0}_U) = \hat{0}_V$.

(2) For all $\hat{u} \in U$, $T(-\hat{u}) = -T(\hat{u})$.

Proof

(1) We have

$$T\left(\hat{0}_U\right) = T\left(0\hat{0}_U\right) = 0T\left(\hat{0}_U\right) = \hat{0}_V$$

since $T\left(\hat{0}_U\right)$ is a vector in V and $0v = \hat{0}_V$ for all $v \in V$.

(2) We have

$$\hat{0}_V = T\left(\hat{0}_U\right) = T\left[\hat{u} + \left(-\hat{u}\right)\right] = T\left(\hat{u}\right) + T\left(-\hat{u}\right),$$

so $T\left(-\hat{u}\right)$ is the additive inverse of $T\left(\hat{u}\right)$, and since additive inverses are unique, $T\left(-\hat{u}\right) = -T\left(\hat{u}\right)$.

Example

We show that the function

$$T : \mathbb{R}^2 \to \mathbb{R}^3$$

defined by

$$T\begin{pmatrix} x \\ y \end{pmatrix} = \begin{pmatrix} 2x + y \\ x - y \\ 4x \end{pmatrix}$$

is a linear transformation.

Let

$$\hat{u}_1 = \begin{pmatrix} x_1 \\ y_1 \end{pmatrix}, \quad \hat{u}_2 = \begin{pmatrix} x_2 \\ y_2 \end{pmatrix}.$$

Then

$$T\left(\hat{u}_1 + \hat{u}_2\right) = T\left[\begin{pmatrix} x_1 \\ y_1 \end{pmatrix} + \begin{pmatrix} x_2 \\ y_2 \end{pmatrix}\right] = T\begin{pmatrix} x_1 + x_2 \\ y_1 + y_2 \end{pmatrix} = \begin{pmatrix} 2\left(x_1 + x_2\right) + \left(y_1 + y_2\right) \\ \left(x_1 + x_2\right) - \left(y_1 + y_2\right) \\ 4\left(x_1 + x_2\right) \end{pmatrix} = \begin{pmatrix} \left(2x_1 + y_1\right) + \left(2x_2 + y_2\right) \\ \left(x_1 - y_1\right) + \left(x_2 - y_2\right) \\ 4x_1 + 4x_2 \end{pmatrix}$$

$$= \begin{pmatrix} 2x_1 + y_1 \\ x_1 - y_1 \\ 4x_1 \end{pmatrix} + \begin{pmatrix} 2x_2 + y_2 \\ x_2 - y_2 \\ 4x_2 \end{pmatrix} = T\begin{pmatrix} x_1 \\ y_1 \end{pmatrix} + T\begin{pmatrix} x_2 \\ y_2 \end{pmatrix} = T\left(\hat{u}_1\right) + T\left(\hat{u}_2\right).$$

Also,

$$T\left(a\hat{u}_1\right) = T\left[a\begin{pmatrix} x_1 \\ y_1 \end{pmatrix}\right] = T\begin{pmatrix} ax_1 \\ ay_1 \end{pmatrix} = \begin{pmatrix} 2ax_1 + ay_1 \\ ax_1 - ay_1 \\ 4ax_1 \end{pmatrix} = a\begin{pmatrix} 2x_1 + y_1 \\ x_1 - y_1 \\ 4x_1 \end{pmatrix} = aT\left(\hat{u}_1\right).$$

Example

We show the function

$$T : \mathbb{R}^2 \to \mathbb{R}^3$$

defined by

$$T\begin{pmatrix} x \\ y \end{pmatrix} = \begin{pmatrix} x^2 \\ x - y \\ 4x \end{pmatrix}$$

is not a linear transformation.

Using the notation of the previous example,

$$T\left(\hat{u}_1 + \hat{u}_2\right) = T\left[\begin{pmatrix} x_1 \\ y_1 \end{pmatrix} + \begin{pmatrix} x_2 \\ y_2 \end{pmatrix}\right] = T\begin{pmatrix} x_1 + x_2 \\ y_1 + y_2 \end{pmatrix} = \begin{pmatrix} \left(x_1 + x_2\right)^2 \\ \left(x_1 + x_2\right) - \left(y_1 + y_2\right) \\ 4\left(x_1 + x_2\right) \end{pmatrix} \begin{pmatrix} x_1^2 + 2x_1 x_2 + x_1^2 \\ \left(x_1 + x_2\right) - \left(y_1 + y_2\right) \\ 4\left(x_1 + x_2\right) \end{pmatrix}$$

but

$$T\left(\hat{u}_1\right) + T\left(\hat{u}_2\right) = T\begin{pmatrix} x_1 \\ y_1 \end{pmatrix} + T\begin{pmatrix} x_2 \\ y_2 \end{pmatrix} = \begin{pmatrix} x_1^2 \\ x_1 - y_1 \\ 4x_1 \end{pmatrix} + \begin{pmatrix} x_2^2 \\ x_2 - y_2 \\ 4x_2 \end{pmatrix} = \begin{pmatrix} x_1^2 + x_2^2 \\ \left(x_1 + x_2\right) - \left(y_1 + y_2\right) \\ 4\left(x_1 + x_2\right) \end{pmatrix}$$

so

$$T\left(\hat{u}_1 + \hat{u}_2\right) \neq T\left(\hat{u}_1\right) + T\left(\hat{u}_2\right).$$

The two examples above illustrate the fact that any function

$$T : \mathbb{R}^m \to \mathbb{R}^n$$

can be represented as

$$T\begin{pmatrix} x_1 \\ \vdots \\ x_m \end{pmatrix} = \begin{pmatrix} T_1\left(x_1, .., x_m\right) \\ \vdots \\ T_n\left(x_1, .., x_m\right) \end{pmatrix}$$

where

$$T_i : \mathbb{R}^m \to \mathbb{R}.$$

The function T will be linear if and only if every T_i, $i = 1, \ldots, n$, is linear.

Example

Consider

$$\frac{d}{dx} : \mathcal{P}_n(\mathbb{R}) \to \mathcal{P}_{n-1}(\mathbb{R}).$$

We have

$$\frac{d}{dx}\big(af(x)+bg(x)\big) = \frac{d}{dx}\big(af(x)\big) + \frac{d}{dx}\big(bg(x)\big) = a\frac{d}{dx}f(x) + b\frac{d}{dx}g(x)$$

so $\dfrac{d}{dx}$ is a linear transformation.

Example

Consider

$$\int_\alpha^\beta : \mathcal{P}_n(\mathbb{R}) \to \mathbb{R}.$$

We have

$$\int_\alpha^\beta \big(af(x)+bg(x)\big)dx = \int_\alpha^\beta \big(af(x)\big)dx + \int_\alpha^\beta \big(bg(x)\big)dx = a\int_\alpha^\beta f(x)dx + b\int_\alpha^\beta g(x)dx$$

so $\displaystyle\int_\alpha^\beta$ is a linear transformation.

Theorem 3

If $T \in L(U,V)$ is a one-to-one and onto linear transformation, then $T^{-1} \in L(V,U)$.

Proof

For

$$T^{-1} : V \to U$$

to exist, it is necessary and sufficient for T to be a one-to-one and onto function.
We show that T^{-1} is a linear transformation.
Suppose $\hat{v}_1, \hat{v}_2 \in V$. Then there are unique $\hat{u}_1, \hat{u}_2 \in U$ with

$$T(\hat{u}_1) = \hat{v}_1 \quad \text{and} \quad T(\hat{u}_2) = \hat{v}_2.$$

Thus

$$\hat{u}_1 = T^{-1}(\hat{v}_1) \quad \text{and} \quad \hat{u}_2 = T^{-1}(\hat{v}_2)$$

So

$$T^{-1}\left(\hat{v}_1+\hat{v}_2\right)=T^{-1}\left(T\left(\hat{u}_1\right)+T\left(\hat{u}_2\right)\right)=T^{-1}\left(T\left(\hat{u}_1+\hat{u}_2\right)\right)$$
$$=\hat{u}_1+\hat{u}_2=T^{-1}\hat{v}_1+T^{-1}\hat{v}_2.$$

For α a scalar, we have

$$T^{-1}\left(\alpha\hat{v}_1\right)=T^{-1}\left(\alpha T\left(\hat{u}_1\right)\right)=T^{-1}\left(T\left(\alpha\hat{u}_1\right)\right)=\alpha\hat{u}_1=\alpha T^{-1}\left(\hat{v}_1\right).$$

4.1.1 Null Space and Range (Image) of a Linear Transformation

Definition

If

$$T:U\to V$$

is a linear transformation, then the null space of T (also called the kernel of T), denoted $\mathcal{N}\left(T\right)$, is defined by

$$\mathcal{N}\left(T\right)=\left\{\hat{u}\in U\mid T\left(\hat{u}\right)=\hat{0}_V\right\}.$$

Thus, $\mathcal{N}\left(T\right)\subset U$.

Each linear transformation T also has another subspace associated with it, called the range of T.

Definition

Let

$$T:U\to V$$

be a linear transformation. The range of T (also called the image of T), denoted $\mathcal{R}\left(T\right)$, is defined by

$$\mathcal{R}\left(T\right)=\left\{\hat{v}\in V\mid\hat{v}=T\left(\hat{u}\right)\text{ for some }\hat{u}\in U\right\}=\left\{T\left(\hat{u}\right)\mid\hat{u}\in U\right\}.$$

Theorem 4

Let

$$T:U\to V$$

be a linear transformation. Then the range of T is a subspace of V.

Proof

Since $T\left(\hat{0}_U\right)=\hat{0}_V$, we have $\hat{0}_V \in \mathcal{R}\left(T\right)$.

Suppose that $\hat{v}_1,\hat{v}_2 \in \mathcal{R}\left(T\right)$. Then there are $\hat{u}_1,\hat{u}_2 \in U$ with

$$T\left(\hat{u}_1\right)=\hat{v}_1,\quad T\left(\hat{u}_2\right)=\hat{v}_2.$$

Thus,

$$T\left(\hat{u}_1+\hat{u}_2\right)=T\left(\hat{u}_1\right)+T\left(\hat{u}_2\right)=\hat{v}_1+\hat{v}_2 \quad \text{and so} \quad \hat{v}_1+\hat{v}_2 \in \mathcal{R}\left(T\right).$$

If $\alpha \in \mathcal{F}$ and $\hat{v} \in \mathcal{R}\left(T\right)$, then there is a $\hat{u} \in U$ with $\hat{v}=T\left(\hat{u}\right)$.

Then

$$T\left(\alpha\hat{u}\right)=\alpha T\left(\hat{u}\right)=\alpha\hat{v} \in \mathcal{R}\left(T\right).$$

In Chapter 2, we learned how to determine whether a vector was in the range or null space of a linear transformation (although we did not use those terms). Later in this chapter, we will demonstrate how to easily find a basis for the null space and the range space of a linear transformation.

Exercises

1. For $T:\mathbb{R}^3\to\mathbb{R}^3$ defined as follows, tell whether T is a linear transformation.

(a) $T\begin{pmatrix} x \\ y \\ z \end{pmatrix}=\begin{pmatrix} 3y-z \\ 2x+y \\ 0 \end{pmatrix}$

(b) $T\begin{pmatrix} x \\ y \\ z \end{pmatrix}=\begin{pmatrix} 3y-z \\ 2x+y \\ 1 \end{pmatrix}$

(c) $T\begin{pmatrix} x \\ y \\ z \end{pmatrix}=\begin{pmatrix} 0 \\ 2x+y^2 \\ z \end{pmatrix}$

2. Is $T:\mathcal{P}_n\left(\mathbb{R}\right)\to\mathcal{P}_{n+2}\left(\mathbb{R}\right)$ defined by

$$T\left(f\left(x\right)\right)=x^2\left(f\left(x\right)\right)$$

a linear transformation?

3. Tell whether the following transformations

$$T: M_{2 \times 2}(\mathbb{R}) \rightarrow \mathbb{R}$$

are linear.

(a) $T\left(\begin{pmatrix} a & b \\ c & d \end{pmatrix}\right) = ad - bc$

(b) $T\left(\begin{pmatrix} a & b \\ c & d \end{pmatrix}\right) = 3a + 2b - c$

4. Show that $T: M_{n \times m}(\mathbb{R}) \rightarrow M_{n \times m}(\mathbb{R})$ defined by

$$T(A) = A^T$$

is a linear transformation.

5. Show that $T: \mathbb{R}^2 \rightarrow \mathbb{R}^2$ defined by

$$T\begin{pmatrix} x \\ y \end{pmatrix} = \begin{pmatrix} \cos\theta & -\sin\theta \\ \sin\theta & \cos\theta \end{pmatrix}\begin{pmatrix} x \\ y \end{pmatrix} = \begin{pmatrix} x\cos\theta - y\sin\theta \\ x\sin\theta + y\cos\theta \end{pmatrix}$$

is a linear transformation. This is the function that rotates a vector counterclockwise through the angle θ.

6. (a) Show that the function $T: C[x] \rightarrow C[x]$ defined by

$$T(f(x)) = f(x^2 + 1)$$

is a linear transformation.

(b) Show that the function $T: C[x] \rightarrow C[x]$ defined by

$$T(f(x)) = (f(x))^2 + 1$$

is not a linear transformation.

7. Let $T: U \rightarrow V$ be a linear transformation. Show that

(a) T is one-to-one if and only if the image of a linearly independent set is a linearly independent set.

(b) T is a 1–1 function if and only if $\ker(T) = \{\hat{0}\}$.

8. If T is a linear transformation with

$$T(\hat{v}_1) = (4,7,-1), \quad T(\hat{v}_2) = (3,0,5), \quad T(\hat{v}_3) = (1,1,1,),$$

find

$$T(2\hat{v}_1 - 7\hat{v}_2 + \hat{v}_3).$$

9. Suppose $T:V{\to}V$ is a linear transformation.
   (a) Show that $T^2=0$ if and only if $\mathcal{R}\left(T\right)\subset\mathcal{N}\left(T\right)$.
   (b) Give an example of such a $T\neq0$.
10. Show that the zero transformation and identity transformation are linear transformations.
11. Show that a reflection $(x,y){\to}(x,-y)$ in $\mathbb{R}^2$ is a linear transformation.
12. Show that permuations of components in $\mathbb{R}^2$ are a linear transformation.
13. Show that a projection $(x,y){\to}(x,0)$ in $\mathbb{R}^2$ is a linear transformation.
14. Show that $L(U,V)$ and $L(V)$ are vector spaces over F.
15. Prove if

$$T:V \to V$$

is a linear operator, then

   (a) T is invertible if and only if the matrix of T with respect to any basis is invertible.
   (b) If T is invertible, and A is the matrix representation of T with respect to a particular basis, then the matrix representation of T^{-1} with respect to the same basis is A^{-1}.

4.2 Representing a Linear Transformation

We now prove the fundamental fact that a linear transformation can be expressed by multiplication of a vector by a matrix. However, for any particular linear transformation, the matrix will be determined by both the linear transformation and the choice of bases for the domain vector space and the range vector space.

We return to the idea of a basis for a vector space. A vector space has many different bases. A fixed vector has a unique expression for a given basis, but the same vector will usually have different representations in different bases.

Theorem 5

A linear transformation is determined by its effect on the basis elements.
What this theorem is saying is if

$$T:U \to V$$

is a linear transformation and $\left\{\hat{u}_1,\dots,\hat{u}_n\right\}$ is a basis of U and we know $T\left(\hat{u}_1\right),\dots,T\left(\hat{u}_n\right)$, then we know $T\left(\hat{u}\right)$ for any vector $\hat{u}\in U$. This is not true for an arbitrary function. For example, if

$$f:\mathbb{R} \to \mathbb{R}$$

and we know the value of $f(x)$ for any finite set $\{x_1,x_2,\dots,x_n\}$ of x's that is not enough to determine the value of $f(x)$ for a value of x that is not in $\{x_1,x_2,\dots,x_n\}$.

Proof

Let

$$T \in L(U, V),$$

and let $\{\hat{e}_1, \ldots, \hat{e}_n\}$ be a basis for U. If $\hat{u} \in U$, then there are unique scalars $a_1, \ldots, a_n$ for which

$$\hat{u} = a_1\hat{e}_1 + \cdots + a_n\hat{e}_n.$$

Thus,

$$T(\hat{u}) = T(a_1\hat{e}_1 + \cdots + a_n\hat{e}_n) = a_1T(\hat{e}_1) + \cdots + a_nT(\hat{e}_n).$$

So if $\{\hat{e}_1, \ldots, \hat{e}_n\}$ is a basis for U, then any vector $T(\hat{u})$ in the range of T is a linear combination of $\{T(\hat{e}_1), \ldots, T(\hat{e}_n)\}$.

A slightly different way to express the previous sentence is $\{T(\hat{e}_1), \ldots, T(\hat{e}_n)\}$ spans the range of T. This does not say that $\{T(\hat{e}_1), \ldots, T(\hat{e}_n)\}$ is a basis for the range of T because $\{T(\hat{e}_1), \ldots, T(\hat{e}_n)\}$ is not necessarily linearly independent, as we show in the exercises.

4.2.1 Representation of a Linear Transformation in the Usual Basis

As we stated, the major result of this section is that a linear transformation can be expressed as multiplication by a matrix. The example below will show how to do this. *However, the matrix will depend on the basis, and this important fact may not be clear until later.*

Example

Let $T : \mathbb{R}^2 \to \mathbb{R}^3$ be defined by

$$T\begin{pmatrix} x \\ y \end{pmatrix} = \begin{pmatrix} 3x - 4y \\ x + y \\ 2y \end{pmatrix} \tag{4.1}$$

We will use the standard bases for $\mathbb{R}^2$ and $\mathbb{R}^3$. According to Theorem 5, if we know the effect of T on $\hat{e}_1$ and $\hat{e}_2$ we know what T does to every vector.

Equation 4.1 can be written as

$$T\begin{pmatrix} x \\ y \end{pmatrix} = \begin{pmatrix} 3x - 4y \\ x + y \\ 2y \end{pmatrix} = \begin{pmatrix} 3x \\ x \\ 0 \end{pmatrix} + \begin{pmatrix} -4y \\ y \\ 2y \end{pmatrix} = x\begin{pmatrix} 3 \\ 1 \\ 0 \end{pmatrix} + y\begin{pmatrix} -4 \\ 1 \\ 2 \end{pmatrix}$$

and in Chapter 1, we learned the right-hand side can be expressed as the matrix equation

$$x\begin{pmatrix} 3 \\ 1 \\ 0 \end{pmatrix} + y\begin{pmatrix} -4 \\ 1 \\ 2 \end{pmatrix} = \begin{pmatrix} 3 & -4 \\ 1 & 1 \\ 0 & 2 \end{pmatrix}\begin{pmatrix} x \\ y \end{pmatrix}.$$

So we have

$$T\begin{pmatrix} x \\ y \end{pmatrix} = \begin{pmatrix} 3 & -4 \\ 1 & 1 \\ 0 & 2 \end{pmatrix} \begin{pmatrix} x \\ y \end{pmatrix}$$

and we have shown that evaluating T at the vector $\begin{pmatrix} x \\ y \end{pmatrix}$ is the same as multiplying the vector $\begin{pmatrix} x \\ y \end{pmatrix}$ by the matrix

$$\begin{pmatrix} 3 & -4 \\ 1 & 1 \\ 0 & 2 \end{pmatrix}.$$

We next want to determine how to easily construct the matrix associated with a linear transformation with respect to the standard bases. Consider again

$$T\begin{pmatrix} x \\ y \end{pmatrix} = \begin{pmatrix} 3 & -4 \\ 1 & 1 \\ 0 & 2 \end{pmatrix} \begin{pmatrix} x \\ y \end{pmatrix}$$

If we let $\begin{pmatrix} x \\ y \end{pmatrix}$ be the first basis vector, that is, $\begin{pmatrix} 1 \\ 0 \end{pmatrix}$, then we get

$$T\begin{pmatrix} 1 \\ 0 \end{pmatrix} = \begin{pmatrix} 3 & -4 \\ 1 & 1 \\ 0 & 2 \end{pmatrix} \begin{pmatrix} 1 \\ 0 \end{pmatrix} = \begin{pmatrix} 3 \\ 1 \\ 0 \end{pmatrix},$$

which is the first column of the matrix. Likewise, $T\begin{pmatrix} 0 \\ 1 \end{pmatrix}$ gives the second column of the matrix. So, at least when we use the standard basis vectors, we have found the following useful principle.

Theorem 6

Given a linear transformation $T : \mathbb{R}^n \to \mathbb{R}^m$, the matrix associated with T with the standard bases will have $T(\hat{e}_k)$ as its kth column where $\hat{e}_k$ is the kth standard basis vector in $\mathbb{R}^n$.
  This is saying that the matrix of T with respect to the usual bases is

$$\left[T(\hat{e}_1), T(\hat{e}_2), \cdots , T(\hat{e}_n) \right].$$

Examples

Find the matrix A for the following linear transformations with respect to the standard bases.

(a) $T\begin{pmatrix} x \\ y \\ z \end{pmatrix} = \begin{pmatrix} 4y - 6z \\ 2x + y \\ 5z \end{pmatrix}$

We have

$$T\begin{pmatrix} 1 \\ 0 \\ 0 \end{pmatrix} = \begin{pmatrix} 0 \\ 2 \\ 0 \end{pmatrix}; \quad T\begin{pmatrix} 0 \\ 1 \\ 0 \end{pmatrix} = \begin{pmatrix} 4 \\ 1 \\ 0 \end{pmatrix}; \quad T\begin{pmatrix} 0 \\ 0 \\ 1 \end{pmatrix} = \begin{pmatrix} -6 \\ 0 \\ 5 \end{pmatrix}$$

so

$$A = \begin{pmatrix} 0 & 4 & -6 \\ 2 & 1 & 0 \\ 0 & 0 & 5 \end{pmatrix}.$$

(b) T rotates each vector in $\mathbb{R}^2$ through the angle θ. See Figure 4.2. The vector $(1, 0)$ is moved to $(\cos\theta, \sin\theta)$, and $(0, 1)$ is moved to $(-\sin\theta, \cos\theta)$, so

$$T\begin{pmatrix} 1 \\ 0 \end{pmatrix} = \begin{pmatrix} \cos\theta \\ \sin\theta \end{pmatrix}; \quad T\begin{pmatrix} 0 \\ 1 \end{pmatrix} = \begin{pmatrix} -\sin\theta \\ \cos\theta \end{pmatrix}$$

so

$$A = \begin{pmatrix} \cos\theta & -\sin\theta \\ \sin\theta & \cos\theta \end{pmatrix}.$$

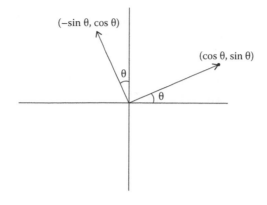

FIGURE 4.2
Rotating $\hat{e}_1$ and $\hat{e}_2$ through the angle θ.

(c) For the linear transformation

$$T\begin{pmatrix}x\\y\end{pmatrix}=\begin{pmatrix}x-y\\4x+3y\end{pmatrix}=x\begin{pmatrix}1\\4\end{pmatrix}+y\begin{pmatrix}-1\\3\end{pmatrix}$$

we might (correctly) infer from part (a.) that

$$A=\begin{pmatrix}1&-1\\4&3\end{pmatrix}$$

in the standard basis.

Example

Let $T:\mathbb{R}^3\to\mathbb{R}^2$ be defined by

$$T\begin{pmatrix}x\\y\\z\end{pmatrix}=\begin{pmatrix}2x-6y+2z\\3y+z\end{pmatrix}.$$

We find a basis for the range of T and the null space of T. We have

$$T\begin{pmatrix}1\\0\\0\end{pmatrix}=\begin{pmatrix}2\\0\end{pmatrix},\quad T\begin{pmatrix}0\\1\\0\end{pmatrix}=\begin{pmatrix}-6\\3\end{pmatrix},\quad T\begin{pmatrix}0\\0\\1\end{pmatrix}=\begin{pmatrix}2\\1\end{pmatrix}$$

so a spanning set for the range of T is

$$\left\{\begin{pmatrix}2\\0\end{pmatrix},\begin{pmatrix}-6\\3\end{pmatrix},\begin{pmatrix}2\\1\end{pmatrix}\right\}.$$

This cannot be a basis for the range of T (there are three vectors that have two components), so we use the methods of Chapter 3 to find a maximal linearly independent subset. With row reduction, we obtain

$$\begin{pmatrix}2&-6&2\\0&3&1\end{pmatrix}\to\begin{pmatrix}1&0&2\\0&1&1/3\end{pmatrix}$$

and so

$$\left\{\begin{pmatrix}2\\0\end{pmatrix},\begin{pmatrix}-6\\3\end{pmatrix}\right\}$$

is a basis for the range of T.

To obtain a basis for the null space of T, we note that the row reduced form of the matrix gives that x and y are leading variables and z is a free variable and

$$x = -2z, \quad y = -\frac{1}{3}z.$$

Thus, a vector in the null space of T is of the form

$$\begin{pmatrix} x \\ y \\ z \end{pmatrix} = \begin{pmatrix} -2z \\ -\dfrac{1}{3}z \\ z \end{pmatrix} = z \begin{pmatrix} -2 \\ -\dfrac{1}{3} \\ 1 \end{pmatrix}$$

so

$$\left\{ \begin{pmatrix} -2 \\ -\dfrac{1}{3} \\ 1 \end{pmatrix} \right\}$$

is a basis for the null space of T.

Theorem 7

Let

$$T : V \to V$$

be a linear operator. Then

(a) T is invertible if and only if the matrix of T with respect to any basis is invertible.
(b) If T is invertible, and A is the matrix representation of T with respect to a particular basis, then the matrix representation of T^{-1} with respect to the same basis is A^{-1}.

The proof is left to the exercises.
    The next theorem was given in a different context in Chapter 3.

Theorem 8

If V is a finite dimensional vector space and $T : V \to W$ is a linear transformation, then

$$\dim \mathcal{R}(T) + \dim \mathcal{N}(T) = \dim(V).$$

Exercises

In Exercises 1 through 4, give the matrix associated with the linear transformation.

1. $T\begin{pmatrix} x \\ y \end{pmatrix} = \begin{pmatrix} 4x - 3y \\ x \\ 5y \end{pmatrix}$

2. $T\begin{pmatrix} x \\ y \\ z \end{pmatrix} = \begin{pmatrix} x - y + 2z \\ 7x + 5z \end{pmatrix}$

3. $T(x) = (3x)$

4. $T\begin{pmatrix} z \\ y \\ z \end{pmatrix} = \begin{pmatrix} x - 2y + 9z \\ y + 4z \\ x - 5y \\ 0 \end{pmatrix}$

In Exercises 5 through 8, give the linear transformation associated with the matrix.

5. $\begin{pmatrix} 1 & -3 \\ 7 & 0 \\ 5 & 9 \end{pmatrix}$

6. $\begin{pmatrix} 2 & 0 & -3 & 1 \\ 5 & -2 & 4 & 0 \end{pmatrix}$

7. (2)

8. $\begin{pmatrix} 5 \\ 0 \\ 0 \\ 6 \\ 2 \end{pmatrix}$

9. Suppose $T : \mathbb{R}^2 \rightarrow \mathbb{R}^3$ is a linear transformation with

$$T\begin{pmatrix} 1 \\ -3 \end{pmatrix} = \begin{pmatrix} 3 \\ 7 \\ 2 \end{pmatrix} \quad \text{and} \quad T\begin{pmatrix} -2 \\ 2 \end{pmatrix} = \begin{pmatrix} -1 \\ -4 \\ 0 \end{pmatrix}.$$

   (a) Find $T\begin{pmatrix} 1 \\ 0 \end{pmatrix}$ and $T\begin{pmatrix} 0 \\ 1 \end{pmatrix}$

   (b) Find $T\begin{pmatrix} 5 \\ -6 \end{pmatrix}$

   (c) Find $T\begin{pmatrix} x \\ y \end{pmatrix}$

10. Suppose that $T_1 : U \rightarrow V$ and $T_2 : U \rightarrow V$ are linear transformations. Show $T_1 + T_2$ is a linear transformation.

11. The linear transformation T that reflects a vector about the line $y = mx$ is represented in the standard basis by the matrix

$$\begin{pmatrix} \dfrac{5}{13} & \dfrac{12}{13} \\ \dfrac{12}{13} & -\dfrac{5}{13} \end{pmatrix}.$$

In this problem, we want to find m.

(a) What are $T\begin{pmatrix} 1 \\ 0 \end{pmatrix}$ and $T\begin{pmatrix} 0 \\ 1 \end{pmatrix}$?

(b) What is the slope of the line between $\begin{pmatrix} 1 \\ 0 \end{pmatrix}$ and $T\begin{pmatrix} 1 \\ 0 \end{pmatrix}$?

(c) Use the answer to part (b.) to find m.

12. Find the value of m if the linear transformation T that reflects a vector about the line $y = mx$ is represented in the standard basis by the matrix

$$\begin{pmatrix} \dfrac{3}{5} & \dfrac{2}{5} \\ \dfrac{2}{5} & -\dfrac{3}{5} \end{pmatrix}.$$

13. Let $T : \mathbb{R}^3 \rightarrow \mathbb{R}^2$ be the linear transformation that reflects a vector about the line $y = x$.

(a) Find the matrix of T with respect to the standard bases.

(b) Find a basis for the null space (kernel) of T.

(c) Find a basis for the range of T.

14. Let $T : \mathbb{R}^3 \rightarrow \mathbb{R}^2$ be defined by

$$T\begin{pmatrix} x \\ y \\ z \end{pmatrix} = \begin{pmatrix} 3x + 2y - 4z \\ x - 5z \end{pmatrix}.$$

(a) Find the matrix of T with respect to the standard basis.

(b) Find a basis for the null space (kernel) of T.

(c) Find a basis for the range of T.

15. Let $T : P_n(\mathbb{R}) \rightarrow P_{n-1}(\mathbb{R})$ be defined by

$$T(f(x)) = \frac{d}{dx}(f(x)).$$

The standard basis for $P_k(\mathbb{R})$ is $\{1, x, x^2, \dots, x^k\}$.

(a) Find the matrix of T with respect to the standard bases.

(b) Find a basis for the null space (kernel) of T.

(c) Find a basis for the range of T.

16. (a) Give an example to demonstrate the following statement is false.

   If $T:\mathbb{R}^n \to \mathbb{R}^m$ is a linear transformation, and $\{\hat{v}_1,\ldots,\hat{v}_n\}$ is a basis for $\mathbb{R}^n$, then $\{T(\hat{v}_1),\ldots,T(\hat{v}_n)\}$ is a basis for the image of T.

(b) Add a hypothesis that will make the earlier statement true.

17. Let $T:P_2(\mathbb{R}) \to P_2(\mathbb{R})$ be defined by

$$T(a+bx+cx^2) = b+(a+2c)x+ax^2.$$

(a) Find the matrix representation of T with respect to the basis $\{1,x,x^2\}$.

(b) Find a basis for the null space of T.

(c) Find a basis for the image of T.

18. Let $T:P_2(\mathbb{R}) \to P_2(\mathbb{R})$ be defined by

$$T(a+bx+cx^2) = (a-2b)+(3a+c)x+(a+b)x^2.$$

(a) Find the matrix representation of T with respect to the basis $\{1,x,x^2\}$.

(b) Find a basis for the null space of T.

(c) Find a basis for the image of T.

19. Let V be the vector space spanned by $\{e^{3x}\sin x, e^{3x}\cos x\}$

Let $T:V \to V$ be defined by

$$T(f(x)) = 2f'(x)+f(x).$$

(a) Show that T is a linear transformation

(b) Give the matrix representation of V with respect to the given basis.

(c) Solve the equation

$$2y'+y = e^{3x}\sin x.$$

4.3 Finding the Representation of a Linear Operator with respect to Different Bases

Suppose we are given the representation of a linear transformation in the usual basis and a second basis as basis for the input vector space, and a third basis as a basis for the output vector space. We want to find the output vector of the linear transformation in the third basis if the input vector is given in the second basis.

Suppose

$$T : U \rightarrow V$$

is a linear transformation and we have a basis $\mathcal{B}_1 = \{\hat{u}_1, \ldots, \hat{u}_m\}$ for U and a basis $\mathcal{B}_2 = \{\hat{v}_1, \ldots, \hat{v}_n\}$ for V. Let T denote the matrix representation of T with respect to the standard basis. To find the matrix representation of T with respect to the bases $\mathcal{B}_1$ and $\mathcal{B}_2$, that is, where the input vector is expressed in the basis $\mathcal{B}_1$ and the output vector is expressed in the basis $\mathcal{B}_2$, we execute the following steps:

1. $\left[\hat{u}\right]_{\mathcal{B}_1} \rightarrow \hat{u}$ using $P_{\mathcal{B}_1}\left[\hat{u}\right]_{\mathcal{B}_1} = \hat{u}$ where $P_{\mathcal{B}_1}$ is the transition matrix defined in Chapter 3.
2. $\hat{u} \rightarrow T\hat{u}$
3. $T\hat{u} \rightarrow \left[T\hat{u}\right]_{\mathcal{B}_2}$ using $\left[T\hat{u}\right]_{\mathcal{B}_2} = P_{\mathcal{B}_2}^{-1}\left(T\hat{u}\right)$
4. If we let $\left[T\right]_{\mathcal{B}_1}^{\mathcal{B}_2}$ denote the matrix representation of the linear transformation T where the input vector is represented in the basis $\mathcal{B}_1$ and the output vector is represented in the basis $\mathcal{B}_2$, we have

$$\left[T\right]_{\mathcal{B}_1}^{\mathcal{B}_2} = P_{\mathcal{B}_2}^{-1} T P_{\mathcal{B}_1}$$

Example

Let $T : \mathbb{R}^2 \rightarrow \mathbb{R}^3$ be defined by

$$T\begin{pmatrix} x \\ y \end{pmatrix} = \begin{pmatrix} x + 2y \\ 3x - 4y \\ y \end{pmatrix}$$

and suppose the bases for $\mathbb{R}^2$ and $\mathbb{R}^3$ are

$$\mathcal{B}_1 = \left\{ \begin{pmatrix} 1 \\ 2 \end{pmatrix}, \begin{pmatrix} 3 \\ 0 \end{pmatrix} \right\} \quad \text{and} \quad \mathcal{B}_2 = \left\{ \begin{pmatrix} 1 \\ 1 \\ 0 \end{pmatrix}, \begin{pmatrix} 0 \\ 1 \\ 1 \end{pmatrix}, \begin{pmatrix} 1 \\ 0 \\ 1 \end{pmatrix} \right\}$$

respectively.

Here

$$P_{\mathcal{B}_1} = \begin{pmatrix} 1 & 3 \\ 2 & 0 \end{pmatrix} \quad P_{\mathcal{B}_2} = \begin{pmatrix} 1 & 0 & 1 \\ 1 & 1 & 0 \\ 0 & 1 & 1 \end{pmatrix}$$

and the matrix for T in the standard basis is

$$\begin{pmatrix} 1 & 2 \\ 3 & -4 \\ 0 & 1 \end{pmatrix}.$$

The matrix representation for the given bases is

$$(P_{\mathcal{B}_2})^{-1}TP_{\mathcal{B}_1} = \begin{pmatrix} 1 & 0 & 1 \\ 1 & 1 & 0 \\ 0 & 1 & 1 \end{pmatrix}^{-1} \begin{pmatrix} 1 & 2 \\ 3 & -4 \\ 0 & 1 \end{pmatrix} \begin{pmatrix} 1 & 3 \\ 2 & 0 \end{pmatrix} = \begin{pmatrix} -1 & 6 \\ -4 & 3 \\ 6 & -3 \end{pmatrix}.$$

Suppose the coordinates of $\hat{u}$ in the $\mathcal{B}_1$ basis are

$$[\hat{u}]_{\mathcal{B}_1} = \begin{pmatrix} 5 \\ 7 \end{pmatrix}$$

and we want to find the output in the $\mathcal{B}_2$ basis. We have

$$[T\hat{u}]_{\mathcal{B}_2} = ((P_{\mathcal{B}_2})^{-1}TP_{\mathcal{B}_1})[\hat{u}]_{\mathcal{B}_1} = [T]_{\mathcal{B}_1}^{\mathcal{B}_2}[\hat{u}]_{\mathcal{B}_1} = \begin{pmatrix} -1 & 6 \\ -4 & 3 \\ 6 & -3 \end{pmatrix} \begin{pmatrix} 5 \\ 7 \end{pmatrix} = \begin{pmatrix} 37 \\ 1 \\ 9 \end{pmatrix}.$$

Exercises

1. The set

$$\mathcal{B} = \left\{ \begin{pmatrix} 2 \\ 1 \end{pmatrix}, \begin{pmatrix} 3 \\ 5 \end{pmatrix} \right\}$$

is a basis for $\mathbb{R}^2$. Suppose T is a linear transformation with

$$T\begin{pmatrix} 1 \\ 0 \end{pmatrix} = \begin{pmatrix} 4 \\ 0 \end{pmatrix}, \quad T\begin{pmatrix} 0 \\ 1 \end{pmatrix} = \begin{pmatrix} 6 \\ -2 \end{pmatrix}.$$

(a) Find the matrix of T with respect to the basis $\mathcal{B}$.

(b) Find a basis for the null space (kernel) of T with respect to the basis $\mathcal{B}$.

(c) Find a basis for the range of T with respect to the basis $\mathcal{B}$.

2. For $T: \mathbb{R}^3 \to \mathbb{R}^3$ given by

$$T\begin{pmatrix} x \\ y \\ z \end{pmatrix} = \begin{pmatrix} 3x - 4z \\ 4x \\ x + y + z \end{pmatrix}$$

find the matrix representation of T with respect to the basis for the domain

$$\mathcal{B}_1 = \left\{ \begin{pmatrix} 1 \\ 0 \\ 0 \end{pmatrix}, \begin{pmatrix} -1 \\ 1 \\ 2 \end{pmatrix}, \begin{pmatrix} 3 \\ 2 \\ 0 \end{pmatrix} \right\}$$

and the basis for the range space

$$B_2 = \left\{ \begin{pmatrix} 1 \\ 1 \\ 0 \end{pmatrix}, \begin{pmatrix} 0 \\ 1 \\ 0 \end{pmatrix}, \begin{pmatrix} 0 \\ 1 \\ 1 \end{pmatrix} \right\}.$$

3. For $T:\mathbb{R}^3 \to \mathbb{R}^3$ given by

$$T\begin{pmatrix} x \\ y \\ z \end{pmatrix} = \begin{pmatrix} y - 3z \\ x - 4y \\ x + 2y - z \end{pmatrix}$$

find the matrix representation of T with respect to the basis for the domain

$$B_1 = \left\{ \begin{pmatrix} 1 \\ 0 \\ 1 \end{pmatrix}, \begin{pmatrix} 1 \\ 1 \\ 2 \end{pmatrix}, \begin{pmatrix} 0 \\ 2 \\ 1 \end{pmatrix} \right\}$$

and the basis for the range space

$$B_2 = \left\{ \begin{pmatrix} 1 \\ 1 \\ 0 \end{pmatrix}, \begin{pmatrix} 0 \\ 1 \\ 0 \end{pmatrix}, \begin{pmatrix} 0 \\ 1 \\ 1 \end{pmatrix} \right\}.$$

4. For $T:\mathbb{R}^3 \to \mathbb{R}^3$ given by

$$T\begin{pmatrix} x \\ y \\ z \end{pmatrix} = \begin{pmatrix} 3x - 4z \\ 4x \\ x + y + z \end{pmatrix}$$

find the matrix representation of T with respect to the basis for the domain

$$B_1 = \left\{ \begin{pmatrix} 1 \\ 0 \\ 1 \end{pmatrix}, \begin{pmatrix} -1 \\ 1 \\ 0 \end{pmatrix}, \begin{pmatrix} 1 \\ 2 \\ 0 \end{pmatrix} \right\}$$

and the basis for the range space

$$B_2 = \left\{ \begin{pmatrix} 1 \\ 1 \\ 0 \end{pmatrix}, \begin{pmatrix} 0 \\ 1 \\ 0 \end{pmatrix}, \begin{pmatrix} 0 \\ 1 \\ 1 \end{pmatrix} \right\}.$$

4.4 Composition (Multiplication) of Linear Transformations

The definition of matrix multiplication we gave earlier probably made little sense intuitively. Why was it so constructed? We will give a partial answer now.

Earlier, we showed there is an algebra connected with linear transformations. Two elements of $L(U, V)$ can be added to give an element of $L(U, V)$. That is, $L(U, V)$ is closed under addition. Also, if a is a scalar and $T \in L(U, V)$, then

$$aT \in L(U,V).$$

If we take the matrix representation of elements of $L(U, V)$, then the method of matrix addition gives the matrix of the sum of the corresponding linear transformations. Similar ideas hold for scalar multiples of linear transformations.

There is not a "multiplication" of linear transformations. Instead, linear transformations can be combined by function composition provided the domains and ranges are suitable. If $T \in L(V)$, then $T \circ T \in L(V)$ and we will use the notation

$$T \circ T = T^2$$

Similarly, for any positive integer n,

$$T^n = T \circ \cdots \circ T \quad (n\ factors).$$

In the next few paragraphs, we will represent a linear transformation T that has the matrix A as its representation by T_A.

If T_A and $T_B \in L(V)$, then $T_B \circ T_A \in L(V)$; and if $T_A \in L(U, V)$ and $T_B \in L(V, W)$, then $T_B \circ T_A \in L(U, W)$.

We would like to define matrix multiplication so that if A is the matrix associated with T_A and B is the matrix associated with T_B, then BA is the matrix associated with $T_B \circ T_A$.

The following example gives some insight into why the definition we gave for matrix multiplication is right for this purpose.

Example

Suppose that U, V, and W are vector spaces with bases $\{\hat{u}_1, \hat{u}_2, \hat{u}_3\}$, $\{\hat{v}_1, \hat{v}_2, \hat{v}_3\}$, and $\{\hat{w}_1, \hat{w}_2, \hat{w}_3\}$, respectively. Furthermore, suppose that

$$T_A : U \to V$$

with

$$T_A\left(\hat{u}_1\right) = a_{11}\hat{v}_1 + a_{21}\hat{v}_2 + a_{31}\hat{v}_3$$
$$T_A\left(\hat{u}_2\right) = a_{12}\hat{v}_1 + a_{22}\hat{v}_2 + a_{32}\hat{v}_3$$
$$T_A\left(\hat{u}_3\right) = a_{13}\hat{v}_1 + a_{23}\hat{v}_2 + a_{33}\hat{v}_3$$

so that the matrix of T_A with respect to the given bases is

$$A = \begin{pmatrix} a_{11} & a_{12} & a_{13} \\ a_{21} & a_{22} & a_{23} \\ a_{31} & a_{32} & a_{33} \end{pmatrix}.$$

Also, suppose that

$$T_B : V \to W$$

with

$$T_B\left(\hat{v}_1\right) = b_{11}\hat{w}_1 + b_{21}\hat{w}_2 + b_{31}\hat{w}_3$$
$$T_B\left(\hat{v}_2\right) = b_{12}\hat{w}_1 + b_{22}\hat{w}_2 + b_{32}\hat{w}_3$$
$$T_B\left(\hat{v}_3\right) = b_{13}\hat{w}_1 + b_{23}\hat{w}_2 + b_{33}\hat{w}_3$$

so that the matrix of T_B with respect to the given bases is

$$B = \begin{pmatrix} b_{11} & b_{12} & b_{13} \\ b_{21} & b_{22} & b_{23} \\ b_{31} & b_{32} & b_{33} \end{pmatrix}.$$

We want to find the matrix associated with $T_B \circ T_A$.
Now

$$u_1 \to a_{11}\hat{v}_1 + a_{21}\hat{v}_2 + a_{31}\hat{v}_3 \to a_{11}\left(b_{11}\hat{w}_1 + b_{21}\hat{w}_2 + b_{31}\hat{w}_3\right)$$

$$+ a_{21}\left(b_{12}\hat{w}_1 + b_{22}\hat{w}_2 + b_{32}\hat{w}_3\right) + a_{31}\left(b_{13}\hat{w}_1 + b_{23}\hat{w}_2 + b_{33}\hat{w}_3\right)$$

$$= \hat{w}_1\left(a_{11}b_{11} + a_{21}b_{12} + a_{31}b_{13}\right) + \hat{w}_2\left(a_{11}b_{21} + a_{21}b_{22} + a_{31}b_{23}\right) + \hat{w}_3\left(a_{11}b_{31} + a_{21}b_{32} + a_{31}b_{33}\right). \quad (4.2)$$

We rewrite the last terms in Equation 4.2 listing the b_{ij} factors first. This gives

$$\hat{w}_1\left(b_{11}a_{11} + b_{12}a_{21} + b_{13}a_{31}\right) + \hat{w}_2\left(b_{21}a_{11} + b_{22}a_{21} + b_{23}a_{31}\right) + \hat{w}_3\left(b_{31}a_{11} + b_{32}a_{21} + b_{33}a_{31}\right).$$

Thus, the first column of the matrix associated with $T_B \circ T_A$ is

$$\begin{pmatrix} b_{11}a_{11} + b_{12}a_{21} + b_{13}a_{31} \\ b_{21}a_{11} + b_{22}a_{21} + b_{23}a_{31} \\ b_{31}a_{11} + b_{32}a_{21} + b_{33}a_{31} \end{pmatrix} = \begin{pmatrix} \sum_{k=1}^{3} b_{1k}a_{k1} \\ \sum_{k=1}^{3} b_{2k}a_{k1} \\ \sum_{k=1}^{3} b_{3k}a_{k1} \end{pmatrix},$$

and the other columns are expressed in the analogous manner.

If we want to define the matrix product BA so that BA will be the matrix associated with $T_B \circ T_A$, the computations above suggest that

$$\left(BA\right)_{ij} = \sum_{k=1}^{3} b_{ik}a_{kj},$$

which is the formula we gave for matrix multiplication.

Exercises

1. Suppose

$$A = \begin{pmatrix} 1 & -1 \\ 2 & 3 \end{pmatrix} \quad \text{and} \quad B = \begin{pmatrix} 0 & 2 \\ -4 & 5 \end{pmatrix}$$

Let

$$T : \mathbb{R}^2 \to \mathbb{R}^2 \quad \text{and} \quad S : \mathbb{R}^2 \to \mathbb{R}^2$$

be defined by

$$T\begin{pmatrix} x \\ y \end{pmatrix} = A\begin{pmatrix} x \\ y \end{pmatrix} \quad \text{and} \quad S\begin{pmatrix} x \\ y \end{pmatrix} = B\begin{pmatrix} x \\ y \end{pmatrix}.$$

(a) Compute

$$T\begin{pmatrix} 1 \\ 0 \end{pmatrix}, \quad T\begin{pmatrix} 0 \\ 1 \end{pmatrix}, \quad S\left(T\begin{pmatrix} 1 \\ 0 \end{pmatrix}\right), \quad \text{and} \quad S\left(T\begin{pmatrix} 0 \\ 1 \end{pmatrix}\right).$$

(b) Compute BA.

(c) Compute $BA\begin{pmatrix} 1 \\ 0 \end{pmatrix}$ and $BA\begin{pmatrix} 0 \\ 1 \end{pmatrix}$.

2. Let $T : P_2(\mathbb{R}) \to P_2(\mathbb{R})$ be defined by $T(p(x)) = p(x-1)$ and

$$S : P_2(\mathbb{R}) \to P_2(\mathbb{R}) \quad \text{be defined by} \quad S\big(p(x)\big) = p(x+1).$$

(a) Find $(S \circ T)$ and $(T \circ S)$.

(b) The standard basis for $P_2(\mathbb{R})$ is $\{1, x, x^2\}$. Find the matrices of S and T with respect to the standard bases.

(c) Let A and B be the matrices for S and T, respectively.

Find AB and BA.

3. Let $S : \mathbb{R}^2 \to \mathbb{R}^2$ and $T : \mathbb{R}^2 \to \mathbb{R}^3$ be defined by

$$S\begin{pmatrix} a \\ b \end{pmatrix} = \begin{pmatrix} a+b \\ a-b \end{pmatrix}; \quad T\begin{pmatrix} a \\ b \end{pmatrix} = \begin{pmatrix} 2b \\ a+4b \\ -3a \end{pmatrix}.$$

(a) Find $(T \circ S)\begin{pmatrix} a \\ b \end{pmatrix}$.

(b) Find the matrix representation of $S, T,$ and $T \circ S$ with respect to the standard bases.

(c) Compare the product of the matrix representation of T and the matrix representation of S with the matrix representation $T \circ S$.

4. Show that if $T_1 : U \to V$ and $T_2 : V \to W$ are linear transformations, then $(T_2 \circ T_1) : U \to W$ is a linear transformation.

5. Recall that the matrix representation of the linear transformation that rotates a vector counterclockwise through an angle θ in $\mathbb{R}^2$ with respect to the standard basis is

$$\begin{pmatrix} \cos\theta & -\sin\theta \\ \sin\theta & \cos\theta \end{pmatrix}.$$

(a) Find the matrices of the linear transformations so that the first transformation rotates a vector counterclockwise through an angle θ and the second transformation rotates a vector counterclockwise through an angle α with respect to the standard basis.

(b) Should the product of these matrices commute?

(c) Use the product of these matrices to find a formula for $\sin(\theta + \alpha)$ and $\cos(\theta + \alpha)$.

(d) Show that $T(-\theta) = [T(\theta)]^{-1}$.

(e) Let

$$A(\theta) = \begin{pmatrix} \cos\theta & -\sin\theta \\ \sin\theta & \cos\theta \end{pmatrix}.$$

Give a geometric argument for why

$$A(2\theta) = \left[A(\theta) \right]^2$$

and use this to derive formulas for $\sin(2\theta)$ and $\cos(2\theta)$.

6. Let V be the vector space of functions that have continuous derivatives of all orders, and let

$$In : V \to V \quad \text{and} \quad D : V \to V \text{ be given by}$$

$$In(f) = \int_0^x f(t)\,dt; \quad D(f) = f'(x).$$

Find $(D \circ In)(f)$ and $(In \circ D)(f)$ for

(a) $f(x) = 5x^3 + 3x^2$

(b) $f(x) = 3x + 4$

(c) $f(x) = e^x$

5

Eigenvalues and Eigenvectors

5.1 Determining Eigenvalues and Eigenvectors

Suppose we have a finite dimensional vector space V and a linear operator

$$T : V \rightarrow V.$$

If V has dimension n, then T can be expressed as multiplication by an $n \times n$ matrix, but the matrix will depend on the basis for V. If the scalar field is $\mathbb{C}$, there will be certain "distinguished" vectors, that we call eigenvectors. If the scalar field is $\mathbb{R}$, this is a possibility, but not a certainty.

We begin our investigation where the scalar field is $\mathbb{R}$ and the vector space is $\mathbb{R}^n$.

Definition

Let V be the vector space $\mathbb{R}^n$ and let A be an $n \times n$ matrix. A nonzero vector $\hat{v} \in V$ is an eigenvector of A with eigenvalue λ if

$$A\hat{v} = \lambda\hat{v}.$$

Note that $\hat{0}$ cannot be an eigenvector, but 0 can be an eigenvalue. Also, if zero is an eigenvalue for A then A is not a one-to-one function. (Why?)

The effect of A on an eigenvector depends on the value of λ.

Value of λ	Effect of A on the Eigenvector $\hat{v}$
$0 < \lambda < 1$	Direction of $\hat{v}$ is unchanged. Length of $\hat{v}$ is decreased.
$\lambda = 1$	$\hat{v}$ is unchanged.
$1 < \lambda$	Direction of $\hat{v}$ is unchanged. Length of $\hat{v}$ is increased.
$\lambda = 0$	$\hat{v}$ becomes the zero vector.
$-1 < \lambda < 0$	Direction of $\hat{v}$ is reversed. Length of $\hat{v}$ is decreased.
$\lambda = -1$	Direction of $\hat{v}$ is reversed. Length of $\hat{v}$ is decreased.
$\lambda < -1$	Direction of $\hat{v}$ is reversed. Length of $\hat{v}$ is increased.

Our first task is given the matrix A, find the eigenvalues and eigenvectors of A.

Consider

$$A\hat{v} = \lambda\hat{v}, \ \hat{v} \neq \hat{0}.$$

Then

$$A\hat{v} - \lambda\hat{v} = \hat{0}. \tag{5.1}$$

We replace $\lambda\hat{v}$ by $\lambda I\hat{v}$ in Equation 5.1 to get

$$A\hat{v} - \lambda I\hat{v} = \hat{0}. \tag{5.2}$$

This is done so that factoring out the $\hat{v}$ will be legitimate. The expression $(A - \lambda I)$ makes sense, but $(A - \lambda)$ does not.

Now

$$A\hat{v} - \lambda I\hat{v} = (A - \lambda I)\hat{v} = \hat{0}$$

so we have the nonzero vector $\hat{v}$ in the null space of $(A - \lambda I)$. This occurs if and only if $(A - \lambda I)$ is not invertible, which occurs if and only if the determinant of $(A - \lambda I)$ is 0. Thus, we have the following result.

Theorem 1

The eigenvalues of the matrix A are the values of λ for which $det(A - \lambda I) = 0$.

Example

Find the eigenvalues for

$$A = \begin{pmatrix} 2 & 1 \\ 0 & -1 \end{pmatrix}.$$

We have

$$A - \lambda I = \begin{pmatrix} 2 - \lambda & 1 \\ 0 & -1 - \lambda \end{pmatrix}$$

$$det(A - \lambda I) = (2 - \lambda)(-1 - \lambda) - 0 = \lambda^2 - \lambda - 2.$$

The eigenvalues of this particular matrix are the values of λ for which

$$\lambda^2 - \lambda - 2 = (\lambda + 1)(\lambda - 2) = 0$$

which are $\lambda = 2$ and $\lambda = -1$.

The expression $det(A - \lambda I)$, when expanded, is called the characteristic polynomial of A, which is denoted $P_A(\lambda)$. If A is an $n \times n$ matrix, then the characteristic polynomial of A will have degree n. Expanding $det(A - \lambda I)$ and finding the values for which $det(A - \lambda I) = 0$ (i.e., finding the eigenvalues) can be computationally challenging and should be done using a computer in most cases. In the real numbers, the characteristic polynomial may not factor completely, but in the complex numbers, we have the following result.

Theorem 2

In the complex numbers, the characteristic polynomial factors into linear factors and

$$P_A(\lambda) = (\lambda - \lambda_1)(\lambda - \lambda_2)\cdots(\lambda - \lambda_n)$$

where λ_i are eigenvalues repeated according to their multiplicity.

Example

The matrix that rotates a vector through an angle θ in two dimensions is

$$A(\theta) = \begin{pmatrix} \cos\theta & -\sin\theta \\ \sin\theta & \cos\theta \end{pmatrix}.$$

If $\theta = 90°$, then there will not be an eigenvector. (What is the geometric reason there is not?)

To see what happens in this case, note that

$$A(90°) = \begin{pmatrix} 0 & -1 \\ 1 & 0 \end{pmatrix}$$

so

$$A(90°) - \lambda I = \begin{pmatrix} -\lambda & -1 \\ 1 & -\lambda \end{pmatrix}$$

and

$$det\left[A(90°) - \lambda I\right] = \lambda^2 + 1.$$

Thus, no eigenvalues of $A(90°)$ exist in the real numbers.

This example shows an important difference between the cases when the scalar field is $\mathbb{R}$ and when the scalar field is $\mathbb{C}$. If the scalar field is $\mathbb{C}$, the characteristic polynomial will always factor into linear factors.

5.1.1 Finding the Eigenvectors after the Eigenvalues Have Been Found

Suppose that we have found the eigenvalues for a matrix A. The next task is to find the eigenvectors associated with each eigenvalue. Suppose that λ_1 is an eigenvalue for A and we want to find the associated eigenvectors. That is, we want to find the nonzero vectors $\hat{v}$ for which

$$A\hat{v} = \lambda_1\hat{v} \quad \text{or} \quad (A - \lambda_1 I)\hat{v} = \hat{0}.$$

This is exactly the null space of $(A - \lambda_1 I)$ except for the zero vector.

Definition

If A is an $n \times n$ matrix with eigenvalue λ, the eigenspace of λ is the nullspace of $A - \lambda I$.
  We recall how we found the null space of a matrix B.
  Suppose

$$B = \begin{pmatrix} 1 & -1 & 3 \\ -10 & 8 & 8 \\ 14 & -12 & 4 \end{pmatrix}$$

and we want to solve $B\hat{x} = \hat{0}$.
  When B is row reduced, the result is

$$\begin{pmatrix} 1 & 0 & -16 \\ 0 & 1 & -19 \\ 0 & 0 & 0 \end{pmatrix}.$$

If we denote the variables as x_1, x_2, x_3, then we have the equations

$$x_1 - 16x_3 = 0, \quad \text{or} \quad x_1 = 16x_3$$
$$x_2 - 19x_3 = 0, \quad \text{or} \quad x_2 = 19x_3$$

So x_3 is the free variable and if $x_3 = t$, then $x_1 = 16t$ and $x_2 = 19t$.
  Thus, a vector in the null space of B is

$$\begin{pmatrix} x_1 \\ x_2 \\ x_3 \end{pmatrix} = \begin{pmatrix} 16t \\ 19t \\ t \end{pmatrix} = t \begin{pmatrix} 16 \\ 19 \\ 1 \end{pmatrix}$$

and

$$\left\{ \begin{pmatrix} 16 \\ 19 \\ 1 \end{pmatrix} \right\}$$

is a basis for the null space of B.
  We apply this technique to find a basis for each eigenspace. We continue with the example

$$A = \begin{pmatrix} 2 & 1 \\ 0 & -1 \end{pmatrix}.$$

To find a basis for the eigenspace for the eigenvalue $\lambda = 2$, we find

$$A - 2I = \begin{pmatrix} 2 & 1 \\ 0 & -1 \end{pmatrix} - 2 \begin{pmatrix} 1 & 0 \\ 0 & 1 \end{pmatrix} = \begin{pmatrix} 0 & 1 \\ 0 & -3 \end{pmatrix}.$$

When $A - 2I$ is row reduced, the result is

$$\begin{pmatrix} 0 & 1 \\ 0 & 0 \end{pmatrix}.$$

So $x_2 = 0$, and x_1 is the free variable. This gives the eigenvector

$$\begin{pmatrix} x_1 \\ 0 \end{pmatrix} = x_1 \begin{pmatrix} 1 \\ 0 \end{pmatrix}.$$

Putting this in standard form, we have the eigenvectors for $\lambda = 2$ are

$$t \begin{pmatrix} 1 \\ 0 \end{pmatrix}$$

and a basis for the eigenspace for $\lambda = 2$ is

$$\left\{ \begin{pmatrix} 1 \\ 0 \end{pmatrix} \right\}.$$

If we repeat this procedure for $\lambda = -1$, we get that a basis for the eigenspace for $\lambda = -1$ is

$$\left\{ \begin{pmatrix} 1 \\ -3 \end{pmatrix} \right\}.$$

Theorem 3

A set of eigenvectors, each of which has a different eigenvalue from the others, is a linearly independent set.

Proof

We give the proof in the case of three vectors and leave the general case as an exercise.

Suppose that $\hat{v}_1, \hat{v}_2,$ and $\hat{v}_3$ are eigenvectors for A with distinct eigenvalues $\lambda_1, \lambda_2, \lambda_3$ and suppose that

$$c_1 \hat{v}_1 + c_2 \hat{v}_2 + c_3 \hat{v}_3 = \hat{0}. \tag{5.3}$$

We will show that each $c_i = 0$.

Multiply Equation 5.3 by $(A - \lambda_3 I)(A - \lambda_2 I)$. Note that

$$\left(A - \lambda_2 I\right)\hat{v}_1 = A\hat{v}_1 - \lambda_2\hat{v}_1 = \lambda_1\hat{v}_1 - \lambda_2\hat{v}_1 = \left(\lambda_1 - \lambda_2\right)\hat{v}_1$$

so

$$\left(A - \lambda_3 I\right)\left(A - \lambda_2 I\right)\hat{v}_1 = \left(A - \lambda_3 I\right)\left[\left(A - \lambda_2 I\right)\hat{v}_1\right] = \left(A - \lambda_3 I\right)\left(\lambda_1 - \lambda_2\right)\hat{v}_1$$
$$= \left(\lambda_1 - \lambda_2\right)\left(A - \lambda_3 I\right)\hat{v}_1 = \left(\lambda_1 - \lambda_2\right)\left(\lambda_1 - \lambda_3\right)\hat{v}_1.$$

Similarly,

$$\left(A - \lambda_2 I\right)\hat{v}_2 = A\hat{v}_2 - \lambda_2\hat{v}_2 = \lambda_2\hat{v}_2 - \lambda_2\hat{v}_2 = \hat{0}$$

so

$$\left(A - \lambda_3 I\right)\left(A - \lambda_2 I\right)\hat{v}_2 = \left(A - \lambda_3 I\right)\hat{0} = \hat{0}.$$

Finally,

$$\left(A - \lambda_2 I\right)\hat{v}_3 = A\hat{v}_3 - \lambda_2\hat{v}_3 = \lambda_3\hat{v}_3 - \lambda_2\hat{v}_3 = \left(\lambda_3 - \lambda_2\right)\hat{v}_3$$

so

$$\left(A - \lambda_3 I\right)\left(A - \lambda_2 I\right)\hat{v}_3 = \left(A - \lambda_3 I\right)\left(\lambda_3 - \lambda_2\right)\hat{v}_3 = \left(\lambda_3 - \lambda_2\right)\left[\left(A - \lambda_3 I\right)\hat{v}_3\right]$$
$$= \left(\lambda_3 - \lambda_2\right)\left(\lambda_3 - \lambda_3\right)\hat{v}_3 = \hat{0}.$$

Thus,

$$\left(A - \lambda_3 I\right)\left(A - \lambda_2 I\right)\left[c_1\hat{v}_1 + c_2\hat{v}_2 + c_3\hat{v}_3\right] = c_1\left(\lambda_1 - \lambda_2\right)\left(\lambda_1 - \lambda_3\right)\hat{v}_1 + c_2\hat{0} + c_3\hat{0}$$
$$= c_1\left(\lambda_1 - \lambda_2\right)\left(\lambda_1 - \lambda_3\right)\hat{v}_1 = \hat{0}.$$

So,

$$c_1\left(\lambda_1 - \lambda_2\right)\left(\lambda_1 - \lambda_3\right)\hat{v}_1 = \hat{0}$$

and since $(\lambda_1 - \lambda_2)(\lambda_1 - \lambda_3) \neq 0$ and $\hat{v}_1 \neq \hat{0}$, it must be that $c_1 = 0$.

Similarly, multiplying Equation 5.3 by $(A - \lambda_3 I)(A - \lambda_1 I)$ yields $c_2 = 0$ and multiplying Equation 5.3 by $(A - \lambda_2 I)(A - \lambda_1 I)$ yields $c_3 = 0$.

Observations

(1) If A is a matrix with eigenvector $\hat{v}$ whose eigenvalue is λ, then

$$A^n \hat{v} = \lambda^n \hat{v}.$$

(2) If A is a matrix with eigenvector $\hat{v}$ whose eigenvalue is λ, then for any scalar $\alpha, \alpha\hat{v}$ is also an eigenvector of A with eigenvalue λ.

This is because

$$A(\alpha\hat{v}) = \alpha(A\hat{v}) = \alpha(\lambda\hat{v}) = \lambda(\alpha\hat{v}).$$

It is important to realize that a single eigenvalue may have more than one linearly independent eigenvector since the null space of $(A - \lambda I)$ may have dimension greater than one.

Example

Consider the two matrices

$$A = \begin{pmatrix} 2 & 0 \\ 0 & 2 \end{pmatrix} \quad \text{and} \quad B = \begin{pmatrix} 2 & 1 \\ 0 & 2 \end{pmatrix}.$$

We first find the eigenvalues and eigenvectors for A.
The characteristic polynomial for A is $(2-\lambda)^2$, so $\lambda=2$ is the only eigenvalue. Now

$$A - 2I = \begin{pmatrix} 0 & 0 \\ 0 & 0 \end{pmatrix}$$

So there are two free variables, x_1 and x_2, and there are two linearly independent eigenvectors

$$\begin{pmatrix} 1 \\ 0 \end{pmatrix} \quad \text{and} \quad \begin{pmatrix} 0 \\ 1 \end{pmatrix}.$$

The characteristic polynomial for B is also $(2-\lambda)^2$, so $\lambda=2$ is again the only eigenvalue. Now

$$B - 2I = \begin{pmatrix} 0 & 1 \\ 0 & 0 \end{pmatrix}$$

so $x_1=0$ and x_2 is free. Thus, there is only one linearly independent eigenvector

$$\begin{pmatrix} 0 \\ 1 \end{pmatrix}.$$

It is important to realize that even if the characteristic polynomials of different matrices are the same, their eigenspaces may be different.

Exercises

In Exercises 1 through 10, find (i) the characteristic polynomial, (ii) the eigenvalues, and (iii) a basis for each eigenspace.

1. $\begin{pmatrix} 3 & 0 & 0 \\ 0 & 2 & 1 \\ -1 & 0 & 2 \end{pmatrix}$

2. $\begin{pmatrix} 1 & 0 & 0 \\ -2 & -1 & -2 \\ 3 & 6 & 6 \end{pmatrix}$

3. $\begin{pmatrix} 2 & 1 & 0 \\ 0 & 2 & 1 \\ 0 & 0 & 2 \end{pmatrix}$

4. $\begin{pmatrix} 3 & 1 & 0 \\ 0 & 3 & 0 \\ 0 & 0 & 3 \end{pmatrix}$

5. $\begin{pmatrix} 3 & 3 & -2 \\ 2 & -3 & 2 \\ -4 & 1 & 1 \end{pmatrix}$

6. $\begin{pmatrix} 1 & -3 & 3 \\ 3 & -5 & 3 \\ 6 & -6 & 6 \end{pmatrix}$

7. $\begin{pmatrix} 0 & 0 & -2 \\ 1 & 2 & 1 \\ 0 & 0 & 3 \end{pmatrix}$

8. $\begin{pmatrix} 12 & 0 & 0 \\ -16 & -4 & 0 \\ -16 & 0 & -4 \end{pmatrix}$

9. $\begin{pmatrix} 3 & 6 & 0 & 0 \\ 1 & -2 & 0 & 0 \\ 0 & 0 & 3 & -1 \\ 0 & 0 & 0 & 3 \end{pmatrix}$

10. $\begin{pmatrix} 1 & 1 & 0 & -1 \\ 0 & -3 & 2 & 6 \\ 0 & 0 & 0 & 5 \\ 0 & 0 & 0 & 4 \end{pmatrix}$

11. Show that if A is an invertible matrix and λ is an eigenvalue of A with eigenvector $\hat{v}$, then $\dfrac{1}{\lambda}$ is an eigenvalue of A^{-1} with eigenvector $\hat{v}$.

12. (a) Show that if A is a square matrix, then A and A^T have the same eigenvalues.

   (b) Give an example of a 2×2 matrix A for which A and A^T have different eigenspaces.

13. (a) Suppose that A and B are $n \times n$ matrices with α an eigenvalue of A and β an eigenvalue of B. Must $\alpha + \beta$ be an eigenvalue of $A + B$?

   (b) Suppose that A and B are $n \times n$ matrices with $\hat{v}$ an eigenvector of A and B. Must $\hat{v}$ be an eigenvector of $A + B$?

14. Show that if α is an eigenvalue of A, then α^2 is an eigenvalue of A^2.

15. If A is a matrix for which A^k is the zero matrix for some positive integer k, show that every eigenvalue of A must be 0.

16. Show that if A is a square matrix then A is invertible if and only if 0 is not an eigenvalue of A.

17. Find the characteristic polynomial of

$$\begin{pmatrix} 3 & 9 & 5 & 1 \\ 0 & -4 & 0 & 0 \\ 0 & 0 & 2 & 4 \\ 0 & 0 & 0 & 6 \end{pmatrix}.$$

18. Give a geometric argument for finding two independent eigenvectors for the matrix that reflects a point about the line $y = mx$ in $\mathbb{R}^2$.

19. Without doing major calculations, find an eigenvector for the following matrices. Give the eigenvalue of the vector you selected.

(a) $\begin{pmatrix} 2 & 2 & 2 & 2 \\ 2 & 2 & 2 & 2 \\ 2 & 2 & 2 & 2 \\ 2 & 2 & 2 & 2 \end{pmatrix}$

(b) $\begin{pmatrix} 1 & 2 & 3 & 4 \\ 4 & 2 & 3 & 1 \\ 2 & 1 & 4 & 3 \\ 3 & 1 & 2 & 4 \end{pmatrix}$

(c) A^6 where $A = \begin{pmatrix} 2 & 2 \\ 2 & 2 \end{pmatrix}$.

20. Show that for a 2×2 matrix

$$A = \begin{pmatrix} a & b \\ c & d \end{pmatrix},$$

the characteristic polynomial is

$$\lambda^2 - \lambda Tr(A) + det(A),$$

where $Tr(A)$ is the trace of A.

This is a special case of the formula for the characteristic polynomial of an $n \times n$ matrix, which is

$$\lambda^n - a_1 \lambda^{n-1} + a_2 \lambda^{n-2} - \cdots + (-1)^n a_n,$$

where $a_1 = Tr(A)$ and $a_n = det(A)$.

21. If the characteristic polynomial of a matrix A is $(\lambda - 1)^2(\lambda - 3)$
    (a) What is the size of the matrix A?
    (b) What is $det(A)$?
    (c) What is $Tr(A)$?
    (d) Is A invertible?

22. Show that if λ is an eigenvalue of A with eigenvector $\hat{v}$, then for any number c, $(\lambda + c)$ is an eigenvalue of $A + cI$ with eigenvector $\hat{v}$.

5.2 Diagonalizing a Matrix

If we have a basis for $\mathcal{F}^n$ that consists of eigenvectors of an $n \times n$ matrix A, then the representation of A with respect to that basis is diagonal. The advantages of this from a computational viewpoint are huge.

For example, if

$$A = \begin{pmatrix} 0 & 0 & 0 \\ 0 & -1 & 0 \\ 0 & 0 & -.5 \end{pmatrix},$$

then

$$A^n = \begin{pmatrix} 0 & 0 & 0 \\ 0 & (-1)^n & 0 \\ 0 & 0 & (-.5)^n \end{pmatrix}.$$

It is sometimes possible to make sense of a transcendental function of a matrix using Taylor series expansions. We show in the exercises that for certain matrices, expressions such as e^A and $\sin A$ make sense.

The next result is immediate from Theorem 3.

Theorem 4

If the characteristic polynomial of A factors into distinct linear factors, then there is a basis of V that consists of eigenvectors of A.

5.2.1 Algebraic and Geometric Multiplicities of an Eigenvalue

Definition

The geometric multiplicity of an eigenvalue is the dimension of the eigenspace of the eigenvalue.

The algebraic multiplicity of an eigenvalue λ is the exponent of the factor $(x-\lambda)$ in the characteristic polynomial.

In Volume 2, we show that the geometric multiplicity of an eigenvalue is less than or equal to the algebraic multiplicity. As an earlier example shows, one can determine the algebraic multiplicity of an eigenvalue from the characteristic polynomial, but not the geometric multiplicity unless the algebraic multiplicity is one.

Example

Suppose the matrix A has characteristic polynomial

$$P_A(x) = (x)^3 (x-3)^4 (x+1)^2.$$

What conclusions can we draw?

(1) A is a 9×9 matrix, because the sum of the exponents is 9.
(2) The eigenvalues of A are $\lambda = 0, 3$, and -1.
(3) The algebraic multiplicity of $\lambda = 0$ is 3. All we can say about the geometric multiplicity of $\lambda = 0$ is that it is 1, 2, or 3.
(4) The algebraic multiplicity of $\lambda = 3$ is 4. All we can say about the geometric multiplicity of $\lambda = 3$ is that it is 1, 2, 3, or 4.
(5) The algebraic multiplicity of $\lambda = -1$ is 2. All we can say about the geometric multiplicity of $\lambda = -1$ is that it is 1 or 2.
(6) There will be a basis consisting of eigenvectors of A if and only if the geometric multiplicity of each eigenvalue is equal to the algebraic multiplicity of that eigenvalue.
(7) The matrix is not invertible because 0 is an eigenvalue.

5.2.2 Diagonalizing a Matrix

A crucial thing to remember in the next discussion is that if A and B are matrices so that AB is defined, then

$$AB = \left[A\hat{b}_1, \ldots, A\hat{b}_k \right],$$

where $A\hat{b}_i$ is the column vector obtained by multiplying the ith column of B on the left by A. We review this in the 2×2 case. If

$$A = \begin{pmatrix} a_{11} & a_{12} \\ a_{21} & a_{22} \end{pmatrix} \quad B = \begin{pmatrix} b_{11} & b_{12} \\ b_{21} & b_{22} \end{pmatrix},$$

then

$$A\hat{b}_1 = \begin{pmatrix} a_{11} & a_{12} \\ a_{21} & a_{22} \end{pmatrix} \begin{pmatrix} b_{11} \\ b_{21} \end{pmatrix} = \begin{pmatrix} a_{11}b_{11} + a_{12}b_{21} \\ a_{21}b_{11} + a_{22}b_{21} \end{pmatrix}$$

$$A\hat{b}_2 = \begin{pmatrix} a_{11} & a_{12} \\ a_{21} & a_{22} \end{pmatrix} \begin{pmatrix} b_{12} \\ b_{22} \end{pmatrix} = \begin{pmatrix} a_{11}b_{12} + a_{12}b_{22} \\ a_{21}b_{12} + a_{22}b_{22} \end{pmatrix}$$

$$AB = \begin{pmatrix} a_{11} & a_{12} \\ a_{21} & a_{22} \end{pmatrix} \begin{pmatrix} b_{11} & b_{12} \\ b_{21} & b_{22} \end{pmatrix} = \begin{pmatrix} a_{11}b_{11} + a_{12}b_{21} & a_{11}b_{12} + a_{12}b_{22} \\ a_{21}b_{11} + a_{22}b_{21} & a_{21}b_{12} + a_{22}b_{22} \end{pmatrix}.$$

Example

Suppose that A is an $n \times n$ matrix that has n linearly independent eigenvectors $\{\hat{v}_1,...,\hat{v}_n\}$ and λ_i is the eigenvalue of $\hat{v}_i$. We are not saying that all the λ_i's are distinct.

We have

$$A\begin{bmatrix} \hat{v}_1, \hat{v}_2, ..., \hat{v}_n \end{bmatrix} = \begin{bmatrix} A\hat{v}_1, A\hat{v}_2, .., A\hat{v}_n \end{bmatrix} = \begin{bmatrix} \lambda_1\hat{v}_1, \lambda_2\hat{v}_2, ..., \lambda_n\hat{v}_n \end{bmatrix} = \begin{bmatrix} \hat{v}_1, \hat{v}_2, ..., \hat{v}_n \end{bmatrix} \begin{pmatrix} \lambda_1 & 0 & \cdots & 0 \\ 0 & \lambda_2 & \cdots & 0 \\ \vdots & \vdots & \ddots & \vdots \\ 0 & 0 & \cdots & \lambda_n \end{pmatrix}.$$

Let P be the matrix whose columns are the eigenvectors $\hat{v}_1,...,\hat{v}_n$ and D the diagonal matrix

$$D = \begin{pmatrix} \lambda_1 & 0 & \cdots & 0 \\ 0 & \lambda_2 & \cdots & 0 \\ \vdots & \vdots & \cdots & \vdots \\ 0 & 0 & \cdots & \lambda_n \end{pmatrix}.$$

Then

$$AP = PD. \tag{5.4}$$

NOTE: If Equation 5.4 is not clear, consider the 2×2 case where A has eigenvectors

$$\hat{v}_1 = \begin{pmatrix} v_{11} \\ v_{21} \end{pmatrix} \quad \hat{v}_2 = \begin{pmatrix} v_{12} \\ v_{22} \end{pmatrix}$$

with eigenvalues λ_1 and λ_2, respectively. Then

$$AP = A\begin{bmatrix} \hat{v}_1 & \hat{v}_2 \end{bmatrix} = \begin{bmatrix} A\hat{v}_1 & A\hat{v}_2 \end{bmatrix} = \begin{bmatrix} \lambda_1\hat{v}_1 & \lambda_2\hat{v}_2 \end{bmatrix} = \begin{pmatrix} \lambda_1 v_{11} & \lambda_2 v_{12} \\ \lambda_1 v_{21} & \lambda_2 v_{22} \end{pmatrix}$$

and

$$PD = \begin{pmatrix} v_{11} & v_{12} \\ v_{21} & v_{22} \end{pmatrix} \begin{pmatrix} \lambda_1 & 0 \\ 0 & \lambda_2 \end{pmatrix} = \begin{pmatrix} v_{11}\lambda_1 + 0 & 0 + v_{12}\lambda_2 \\ v_{21}\lambda_1 + 0 & 0 + v_{22}\lambda_2 \end{pmatrix} = \begin{pmatrix} \lambda_1 v_{11} & \lambda_2 v_{12} \\ \lambda_1 v_{21} & \lambda_2 v_{22} \end{pmatrix}.$$

Typically, we will be given the matrix A and want to find the matrix D. Rearranging Equation 5.4 gives

$$D = P^{-1}AP.$$

We know that P^{-1} exists because $\{\hat{v}_1, \ldots, \hat{v}_n\}$ is a basis.

Recapping, we have the following result.

Theorem 5

Let A be a matrix for which there is a basis of eigenvectors $\{\hat{v}_1, \ldots, \hat{v}_n\}$. Then

$$P^{-1}AP = D,$$

where P is the matrix whose columns are the eigenvectors of A and D is the diagonal matrix whose diagonal entries are the eigenvalues listed in the same order as the corresponding eigenvectors in P.

Not every square matrix can be diagonalized. Theorem 5 gives the method of accomplishing a diagonalization when it can be done. In most cases, it is not immediately obvious when a matrix can be diagonalized. We give several examples and exercises showing matrices that can and matrices that cannot be diagonalized.

There is a special category of matrices that can always be diagonalized and are easily recognized. In $\mathbb{R}^n$, these are the symmetric matrices—matrices for which $A^T = A$. We will see in Chapter 7 that a matrix is symmetric if and only if there is a matrix P for which

$$P = P^T = P^{-1} \quad \text{and} \quad P^{-1}AP = D.$$

Such a matrix P is called an orthogonal matrix.

Example

Diagonalize, if possible, the matrix

$$A = \begin{pmatrix} 2 & 0 & 0 \\ -3 & 0 & 1 \\ 0 & 1 & 0 \end{pmatrix},$$

The characteristic polynomial is

$$(2-\lambda)(\lambda-1)(\lambda+1)$$

so there are three eigenvalues: $\lambda = 2$, $\lambda = 1$, and $\lambda = -1$.

Since the eigenvalues have algebraic dimension one, the matrix can be diagonalized. For

$$\lambda = 2, \text{eigenvector} = \begin{pmatrix} -1 \\ 2 \\ 1 \end{pmatrix}; \quad \lambda = 1, \text{eigenvector} = \begin{pmatrix} 0 \\ 1 \\ 1 \end{pmatrix}; \quad \lambda = -1, \text{eigenvector} = \begin{pmatrix} 0 \\ 1 \\ -1 \end{pmatrix}.$$

Then, we can take

$$P = \begin{pmatrix} -1 & 0 & 0 \\ 2 & 1 & 1 \\ 1 & 1 & -1 \end{pmatrix}$$

and

$$P^{-1}AP = \begin{pmatrix} -1 & 0 & 0 \\ 2 & 1 & 1 \\ 1 & 1 & -1 \end{pmatrix}^{-1} \begin{pmatrix} 2 & 0 & 0 \\ -3 & 0 & 1 \\ 0 & 1 & 0 \end{pmatrix} \begin{pmatrix} -1 & 0 & 0 \\ 2 & 1 & 1 \\ 1 & 1 & -1 \end{pmatrix} = \begin{pmatrix} 2 & 0 & 0 \\ 0 & 1 & 0 \\ 0 & 0 & -1 \end{pmatrix}.$$

Exercises

In Exercises 1 through 10, the matrices are the same as those from Section 7.1. If possible, diagonalize the matrix.

1. $\begin{pmatrix} 3 & 0 & 0 \\ 0 & 2 & 1 \\ -1 & 0 & 2 \end{pmatrix}$

2. $\begin{pmatrix} 1 & 0 & 0 \\ -2 & -1 & -2 \\ 3 & 6 & 6 \end{pmatrix}$

3. $\begin{pmatrix} 2 & 1 & 0 \\ 0 & 2 & 1 \\ 0 & 0 & 2 \end{pmatrix}$

4. $\begin{pmatrix} 3 & 1 & 0 \\ 0 & 3 & 0 \\ 0 & 0 & 3 \end{pmatrix}$

5. $\begin{pmatrix} 3 & 3 & -2 \\ 2 & -3 & 2 \\ -4 & 1 & 1 \end{pmatrix}$

6. $\begin{pmatrix} 1 & -3 & 3 \\ 3 & -5 & 3 \\ 6 & -6 & 6 \end{pmatrix}$

7. $\begin{pmatrix} 0 & 0 & -2 \\ 1 & 2 & 1 \\ 0 & 0 & 3 \end{pmatrix}$

8. $\begin{pmatrix} 12 & 0 & 0 \\ -16 & -4 & 0 \\ -16 & 0 & -4 \end{pmatrix}$

9. $\begin{pmatrix} 3 & 6 & 0 & 0 \\ 1 & -2 & 0 & 0 \\ 0 & 0 & 3 & -1 \\ 0 & 0 & 0 & 3 \end{pmatrix}$

10. $\begin{pmatrix} 1 & 1 & 0 & -1 \\ 0 & -3 & 2 & 6 \\ 0 & 0 & 0 & 5 \\ 0 & 0 & 0 & 4 \end{pmatrix}$

11. If the characteristic polynomial of a matrix A is $(\lambda+2)^3(\lambda-1)^4(\lambda)$

(a) What is the size of the matrix A?

(b) What are the algebraic and geometric dimensions of each eigenspace?

(c) Is A invertible?

5.3 Similar Matrices

If $T:V\to V$ is a linear operator and $\mathcal{B}$ is a basis for V, then there is a matrix A for which

$$T\left(\hat{x}\right) = A\hat{x},$$

where the matrix A depends on both T and $\mathcal{B}$. An important distinction between T and A is that T describes how a vector $\hat{x}$ is changed and is independent of the basis of V, whereas A gives the representation of $T\left(\hat{x}\right)$ in a given coordinate system, and is basis and T dependent.

In this section, we characterize the class of matrices that represent a particular linear operator.

Definition

If A and B are square matrices, then A is similar to B if there is an invertible matrix P with $B=P^{-1}AP$.

The main result of this section is that two matrices represent the same linear operator with respect to different bases if and only if the matrices are similar.

For the purposes of this section, it will be convenient to describe what is meant by an "equivalence relation," but that idea will not be used in the sequel. In the present setting, we consider a relation between $n\times n$ matrices. We will say that two such matrices are related if they are similar.

A relation is an equivalence relation if three conditions hold. These are

1. A matrix is related to itself. (This is called the reflexive property.)
2. If A is related to B, then B is related to A (symmetric property).
3. If A is related to B and B is related to C, then A is related to C (transitive property).

Theorem 6

Being similar is an equivalence relation on the set of $n \times n$ matrices.

Proof

A is similar to A since $A = I^{-1}AI$.
  If A is similar to B then B is similar to A since if $B = P^{-1}AP$, then

$$PBP^{-1} = P\left(P^{-1}AP\right)P^{-1} = \left(PP^{-1}\right)A\left(PP^{-1}\right) = A.$$

Said another way,

$$\left(P^{-1}\right)^{-1} BP^{-1} = A.$$

Also, if A is similar to B and B is similar to C, then A is similar to C because if

$$B = P^{-1}AP \quad \text{and} \quad C = Q^{-1}BQ,$$

then

$$C = Q^{-1}BQ = C = Q^{-1}\left(P^{-1}AP\right)Q = \left(Q^{-1}P^{-1}\right)A\left(PQ\right) = \left(PQ\right)^{-1}A\left(PQ\right).$$

An equivalence relation has the effect of partitioning a set. This means it divides a set into pieces, called equivalence classes, and each element in the set is in exactly one equivalence class. The equivalence classes are determined by the condition that A and B are in the same equivalence class if and only if they are related to each other.
  Similar matrices share several characteristics. The next results enumerate some of them.

Theorem 7

Similar matrices have the same characteristic polynomial.

Proof

Suppose A is similar to B. Then there is an invertible matrix P with

$$B = P^{-1}AP$$

so

$$B - \lambda I = P^{-1}AP - \lambda I = P^{-1}AP - P^{-1}\lambda IP = P^{-1}(A - \lambda I)P$$

and thus

$$det(B - \lambda I) = det\left[P^{-1}(A - \lambda I)P\right] = det(P^{-1})det(A - \lambda I)det(P)$$

$$= \left[det(P)\right]^{-1} det(A - \lambda I)det(P) = det(A - \lambda I).$$

Corollary

For similar matrices, the eigenvalues and the algebraic dimensions of their eigenspaces are the same.

While similar matrices have the same characteristic polynomial, two matrices that have the same characteristic polynomial are not necessarily similar.

Example

The matrices A and B where

$$A = \begin{pmatrix} 2 & 1 \\ 0 & 2 \end{pmatrix} \quad B = \begin{pmatrix} 2 & 0 \\ 0 & 2 \end{pmatrix}$$

have the same characteristic polynomial but are not similar.

We showed in a previous example that A has a basis of eigenvectors but B does not.

The next theorem states that two $n \times n$ matrices are similar if and only if they represent the same linear transformation with respect to different bases.

Theorem 8

Let

$$T : \mathcal{F}^n \to \mathcal{F}^n$$

be a linear transformation. Let A be the matrix representation of T with respect to the standard basis. The matrix A is similar to the matrix B if and only if there is a basis of $\mathcal{F}^n$ for which B is the representation of T with respect to that basis.

Proof

We have seen most of the ideas at the end of Section 4.2. The explanation and next example show why the result is true.

There is a one-to-one and onto correspondence between **ordered bases** of $\mathcal{F}^n$ and $n \times n$ invertible matrices with entries in $\mathcal{F}$. Let

$$T : V \to V$$

be a linear operator and let A be the matrix representation of T with respect to the standard basis. To be more concrete, we give the idea of the proof with an example. Let

$$T : \mathbb{R}^3 \to \mathbb{R}^3$$

and let A be the matrix representation of T with respect to the ordered usual basis, with

$$A = \begin{pmatrix} 1 & 2 & 5 \\ 3 & 0 & 3 \\ -2 & 4 & 6 \end{pmatrix}.$$

Choose the ordered basis

$$B = \left\{ \begin{pmatrix} 1 \\ 1 \\ 1 \end{pmatrix}, \begin{pmatrix} 2 \\ 2 \\ 0 \end{pmatrix}, \begin{pmatrix} 3 \\ 0 \\ 0 \end{pmatrix} \right\}.$$

Now B is uniquely associated with the invertible matrix

$$S = \begin{pmatrix} 1 & 2 & 3 \\ 1 & 2 & 0 \\ 1 & 0 & 0 \end{pmatrix}.$$

We have previously shown that

$$S^{-1}AS = \left[T \right]_{B}^{B}$$

so that $S^{-1}AS$ is the representation of T with respect to the basis B.

The theorems mentioned earlier say that each linear transformation on a vector space gives rise to an equivalence class of matrices, and that different linear transformations give rise to different equivalence classes of matrices.

We have shown that similar matrices have the same characteristic polynomial, and hence the same eigenvalues and the eigenvalues have the same algebraic multiplicity. We have now shown that similar matrices have the same geometric multiplicity.

The next example demonstrates that while similar matrices have the same correspondence of eigenvectors and eigenvalues, the representation of the eigenvectors will be different in different bases.

Example

Let $T : \mathbb{R}^3 \to \mathbb{R}^3$. Suppose the representation of T in the standard basis is

$$A = \begin{pmatrix} -2 & 3 & 1 \\ 0 & 1 & 1 \\ -3 & 4 & 1 \end{pmatrix}.$$

The eigenvalues of A are $\lambda = -2, 0$, and 2. Suppose that $\mathcal{B}$ is the basis

$$\mathcal{B} = \left\{ \begin{pmatrix} 1 \\ 1 \\ 1 \end{pmatrix}, \begin{pmatrix} 1 \\ 1 \\ 0 \end{pmatrix}, \begin{pmatrix} 1 \\ 0 \\ 0 \end{pmatrix} \right\}$$

so that the change of basis matrix is

$$P_{\mathcal{B}} = \begin{pmatrix} 1 & 1 & 1 \\ 1 & 1 & 0 \\ 1 & 0 & 0 \end{pmatrix}.$$

Now

$$B = P_{\mathcal{B}}^{-1} A P_{\mathcal{B}} = \begin{pmatrix} 2 & 1 & 3 \\ 0 & 0 & -3 \\ 0 & 0 & -2 \end{pmatrix}.$$

The eigenvalues of B are also $\lambda = -2, 0$, and 2.
The eigenvectors of A are

$$\lambda = -2, \quad \hat{v}_{-2} = \begin{pmatrix} 5 \\ -3 \\ 9 \end{pmatrix}; \quad \lambda = 0, \quad \hat{v}_0 = \begin{pmatrix} 1 \\ 1 \\ -1 \end{pmatrix}; \quad \lambda = 2, \quad \hat{v}_2 = \begin{pmatrix} 1 \\ 1 \\ 1 \end{pmatrix}.$$

The eigenvectors of B are

$$\lambda = -2, \quad \hat{w}_{-2} = \begin{pmatrix} 9 \\ -12 \\ 8 \end{pmatrix}_{\mathcal{B}}; \quad \lambda = 0, \quad \hat{w}_0 = \begin{pmatrix} -1 \\ 2 \\ 0 \end{pmatrix}_{\mathcal{B}}; \quad \lambda = 2, \quad \hat{w}_2 = \begin{pmatrix} 1 \\ 0 \\ 0 \end{pmatrix}_{\mathcal{B}}.$$

Note that the representations of the eigenvectors of B are in the $\mathcal{B}$ basis. We show that the eigenvectors are the same; it is just that the representations are different. We check this

$$\hat{w}_{-2} = \begin{pmatrix} 9 \\ -12 \\ 8 \end{pmatrix}_{\mathcal{B}} = 9 \begin{pmatrix} 1 \\ 1 \\ 1 \end{pmatrix} - 12 \begin{pmatrix} 1 \\ 1 \\ 0 \end{pmatrix} + 8 \begin{pmatrix} 1 \\ 0 \\ 0 \end{pmatrix} = \begin{pmatrix} 5 \\ -3 \\ 9 \end{pmatrix} = \hat{v}_{-2}$$

$$\hat{w}_0 = \begin{pmatrix} -1 \\ 2 \\ 0 \end{pmatrix}_{\mathcal{B}} = -1 \begin{pmatrix} 1 \\ 1 \\ 1 \end{pmatrix} + 2 \begin{pmatrix} 1 \\ 1 \\ 0 \end{pmatrix} + 0 \begin{pmatrix} 1 \\ 0 \\ 0 \end{pmatrix} = \begin{pmatrix} 1 \\ 1 \\ -1 \end{pmatrix} = \hat{v}_0$$

$$\hat{w}_2 = \begin{pmatrix} 1 \\ 0 \\ 0 \end{pmatrix}_{\mathcal{B}} = 1 \begin{pmatrix} 1 \\ 1 \\ 1 \end{pmatrix} + 0 \begin{pmatrix} 1 \\ 1 \\ 0 \end{pmatrix} + 0 \begin{pmatrix} 1 \\ 0 \\ 0 \end{pmatrix} = \begin{pmatrix} 1 \\ 1 \\ 1 \end{pmatrix} = \hat{v}_2.$$

If we visualize a vector as an arrow, the eigenvector of a linear operator is a vector that changes only its length (if the eigenvalue is negative, the direction is reversed) when the linear transformation acts on it and the basis in which the eigenvector is represented is immaterial.

Exercises

1. Suppose $T: \mathbb{R}^2 \to \mathbb{R}^2$ with

$$T\begin{pmatrix} 1 \\ 0 \end{pmatrix} = \begin{pmatrix} 2 \\ 5 \end{pmatrix}, \quad T\begin{pmatrix} 0 \\ 1 \end{pmatrix} = \begin{pmatrix} -1 \\ 3 \end{pmatrix}.$$

Consider the basis $\mathcal{B} = \left\{\hat{b}_1, \hat{b}_2\right\}$ with

$$\hat{b}_1 = \begin{pmatrix} 2 \\ 1 \end{pmatrix}, \quad \hat{b}_2 = \begin{pmatrix} 1 \\ 1 \end{pmatrix}.$$

Find $[T]_{\mathcal{B}}^{\mathcal{B}}$.

2. Suppose $T: \mathbb{R}^2 \to \mathbb{R}^2$ with

$$T\begin{pmatrix} 1 \\ 0 \end{pmatrix} = \begin{pmatrix} 3 \\ -2 \end{pmatrix}, \quad T\begin{pmatrix} 0 \\ 1 \end{pmatrix} = \begin{pmatrix} 6 \\ 4 \end{pmatrix}.$$

Consider the basis $\mathcal{B} = \left\{\hat{b}_1, \hat{b}_2\right\}$ with

$$\hat{b}_1 = \begin{pmatrix} 1 \\ -1 \end{pmatrix}, \quad \hat{b}_2 = \begin{pmatrix} -4 \\ 3 \end{pmatrix}.$$

Find $[T]_{\mathcal{B}}^{\mathcal{B}}$.

3. Suppose $T: \mathbb{R}^3 \to \mathbb{R}^3$ with

$$T\begin{pmatrix} 1 \\ 0 \\ 0 \end{pmatrix} = \begin{pmatrix} 1 \\ 1 \\ 2 \end{pmatrix}, \quad T\begin{pmatrix} 1 \\ 0 \\ 0 \end{pmatrix} = \begin{pmatrix} 0 \\ 2 \\ 3 \end{pmatrix}, \quad T\begin{pmatrix} 0 \\ 0 \\ 1 \end{pmatrix} = \begin{pmatrix} 1 \\ 1 \\ 0 \end{pmatrix}.$$

Consider the basis $\mathcal{B} = \left\{\hat{b}_1, \hat{b}_2, \hat{b}_3\right\}$ with

$$\hat{b}_1 = \begin{pmatrix} 2 \\ 0 \\ 1 \end{pmatrix}, \quad \hat{b}_2 = \begin{pmatrix} 0 \\ 4 \\ -1 \end{pmatrix}, \quad \hat{b}_3 = \begin{pmatrix} 3 \\ 3 \\ 0 \end{pmatrix}.$$

Find $[T]_{\mathcal{B}}^{\mathcal{B}}$.

4. Suppose $T: \mathbb{R}^2 \to \mathbb{R}^2$ with

$$T\begin{pmatrix} 1 \\ 0 \\ 0 \end{pmatrix} = \begin{pmatrix} 3 \\ 5 \\ -1 \end{pmatrix}, \quad T\begin{pmatrix} 1 \\ 0 \\ 0 \end{pmatrix} = \begin{pmatrix} 1 \\ 1 \\ 3 \end{pmatrix}, \quad T\begin{pmatrix} 0 \\ 0 \\ 1 \end{pmatrix} = \begin{pmatrix} -2 \\ -1 \\ 1 \end{pmatrix}.$$

Consider the basis $\mathcal{B} = \{\hat{b}_1, \hat{b}_2, \hat{b}_3\}$ with

$$\hat{b}_1 = \begin{pmatrix} 0 \\ 2 \\ 3 \end{pmatrix}, \quad \hat{b}_2 = \begin{pmatrix} 1 \\ 1 \\ -2 \end{pmatrix}, \quad \hat{b}_3 = \begin{pmatrix} 2 \\ 6 \\ 1 \end{pmatrix}.$$

Find $[T]_{\mathcal{B}}^{\mathcal{B}}$.

5. If A and B are similar matrices, list the properties that they share, for example, they have the same determinant.

6. Show that if A and B are similar matrices, then A^2 and B^2 are similar matrices.

7. If A is invertible, show that AB is similar to BA.

5.4 Eigenvalues and Eigenvectors in Systems of Differential Equations

One application of eigenvectors occurs in solving systems of differential equations.

Suppose that a_{ij}, $i, j = 1, \ldots, n$ are constant and $x_1(t), \ldots, x_n(t)$ are continuously differentiable functions and we have the system of n linear differential equations with constant coefficients

$$x_1'(t) = a_{11}x_1(t) + \cdots + a_{1n}x_n(t)$$
$$x_2'(t) = a_{21}x_1(t) + \cdots + a_{2n}x_n(t)$$
$$\vdots$$
$$x_n'(t) = a_{n1}x_1(t) + \cdots + a_{nn}x_n(t)$$

with initial conditions $x_i(0) = c_i$, $i = 1, \ldots, n$.

We can write this system as a vector equation

$$\hat{x}'(t) = A\hat{x}(t), \tag{5.5}$$

where

$$\hat{x}(t) = \begin{pmatrix} x_1(t) \\ \vdots \\ x_n(t) \end{pmatrix}, \quad \hat{x}'(t) = \begin{pmatrix} x_1'(t) \\ \vdots \\ x_n'(t) \end{pmatrix}, \quad \text{and} \quad A = \begin{pmatrix} a_{11} & \cdots & a_{1n} \\ \vdots & & \vdots \\ a_{n1} & \cdots & a_{nn} \end{pmatrix}.$$

We first consider an example where A is diagonal.

Example

Suppose that we have the matrix equation

$$\begin{pmatrix} x_1'(t) \\ x_2'(t) \\ x_3'(t) \end{pmatrix} = \begin{pmatrix} 2 & 0 & 0 \\ 0 & -3 & 0 \\ 0 & 0 & 5 \end{pmatrix} \begin{pmatrix} x_1(t) \\ x_2(t) \\ x_3(t) \end{pmatrix}$$

with initial conditions

$$x_1(0) = -1, \quad x_2(0) = 4, \quad x_3(0) = 9.$$

This yields the equivalent system

$$x_1'(t) = 2x_1(t), \quad x_1(0) = -1$$
$$x_2'(t) = -3x_2(t), \quad x_2(0) = 4$$
$$x_3'(t) = 5x_3(t), \quad x_3(0) = 9$$

The solutions to these three equations are

$$x_1(t) = -1e^{2t}, \quad x_2(t) = 4e^{-3t}, \quad x_3(t) = 9e^{5t}$$

So we could write the solution as

$$\hat{x}(t) = \begin{pmatrix} x_1(t) \\ x_2(t) \\ x_3(t) \end{pmatrix} = \begin{pmatrix} -1e^{2t} \\ 4e^{-3t} \\ 9e^{5t} \end{pmatrix}.$$

Suppose now that the matrix A in Equation 5.5 is not diagonal but can be diagonalized. Recall, this is the same thing as A having a basis of eigenvectors.

Suppose that P is a matrix that diagonalizes A. That is,

$$P^{-1}AP = D,$$

where D is diagonal. The hypothesis that A can be diagonalized means the individual equations are independent and there is enough information to ensure a unique solution to the system.

If we have

$$\hat{x}'(t) = A\hat{x}(t)$$

and let

$$\hat{y}(t) = P^{-1}\hat{x}(t),$$

then

$$P\hat{y}(t) = \hat{x}(t)$$

so

$$P\hat{y}'(t) = \hat{x}'(t) = A\hat{x}(t) = AP\hat{y}(t)$$

and we have

$$\hat{y}'(t) = P^{-1}AP\hat{y}(t) = D\hat{y}(t),$$

which we know how to solve.

We must convert the solution to $\hat{x}(t)$, but $\hat{x}(t) = P\hat{y}(t)$.

An algorithm to solve the matrix equation $\hat{x}'(t) = A\hat{x}(t)$ when A can be diagonalized.

1. Find the eigenvalues and eigenvectors of A. The matrix P is the matrix whose columns are the eigenvectors of A and the entries on the diagonal matrix are the eigenvalues of A.

2. Let

$$\hat{y}(t) = P^{-1}\hat{x}(t)$$

and solve

$$\hat{y}'(t) = P^{-1}AP\hat{y}(t) = D\hat{y}(t).$$

3. The solution is

$$\hat{y}(t) = \begin{pmatrix} c_1 e^{\lambda_1 t} \\ c_2 e^{\lambda_2 t} \\ \vdots \\ c_n e^{\lambda_n t} \end{pmatrix},$$

where λ_i are the eigenvalues of A and c_i are to be determined.

4. Convert the solution back to $\hat{x}(t)$ using $\hat{x}(t) = P\hat{y}(t)$.

It is easy to get lost in the explanation of this step and the next one, but this will be easier to follow after the next example.

The solution will be

$$\begin{pmatrix} x_1(t) \\ x_2(t) \\ \vdots \\ x_4(t) \end{pmatrix} = \begin{pmatrix} P_{11}c_1 e^{\lambda_1 t} + P_{12}c_2 e^{\lambda_2 t} + \cdots + P_{1n}c_n e^{\lambda_n t} \\ P_{21}c_1 e^{\lambda_1 t} + P_{22}c_2 e^{\lambda_2 t} + \cdots + P_{2n}c_n e^{\lambda_n t} \\ \vdots \\ P_{n1}c_1 e^{\lambda_1 t} + P_{n2}c_2 e^{\lambda_2 t} + \cdots + P_{nn}c_n e^{\lambda_n t} \end{pmatrix}$$

so

$$x_i(t) = P_{i1}c_1 e^{\lambda_1 t} + P_{i2}c_2 e^{\lambda_2 t} + \cdots + P_{in}c_n e^{\lambda_n t}.$$

5. We now solve for the c_is.

Setting $t = 0$ gives the system of equations

$$x_1(0) = P_{11}c_1 + P_{12}c_2 + \cdots + P_{1n}c_n$$
$$x_2(0) = P_{21}c_1 + P_{22}c_2 + \cdots + P_{2n}c_n$$
$$\vdots$$
$$x_n(0) = P_{n1}c_1 + P_{n2}c_2 + \cdots + P_{nn}c_n.$$

This can be expressed as the matrix equation

$$\hat{x}(0) = P\hat{c},$$

where

$$\hat{c} = \begin{pmatrix} c_1 \\ c_2 \\ \vdots \\ c_n \end{pmatrix}.$$

Thus

$$\hat{c} = P^{-1}\hat{x}(0).$$

The values $x_i(0)$ are given and the values P_{ij} have been determined. Thus, we can determine the values of c_i. We then know the solution, that is

$$x_i(t) = P_{i1}c_1 e^{\lambda_1 t} + P_{i2}c_2 e^{\lambda_2 t} + \cdots + P_{in}c_n e^{\lambda_n t}; \quad i = 1, \ldots, n.$$

Example

We solve the system of equations

$$x_1'(t) = -2x_1(t) - 4x_2(t) + 2x_3(t)$$
$$x_2'(t) = -2x_1(t) + x_2(t) + 2x_3(t)$$
$$x_3'(t) = 4x_1(t) + 2x_2(t) + 5x_3(t)$$

with initial conditions

$$x_1(0) = 3, \quad x_2(0) = -1, \quad x_3(0) = 0.$$

We write this in the form

$$\hat{x}'(t) = A\hat{x}(t),$$

where

$$A = \begin{pmatrix} -2 & -4 & 2 \\ -2 & 1 & 2 \\ 4 & 2 & 5 \end{pmatrix}.$$

1. We first find that the characteristic polynomial for A is

$$P_A(\lambda) = (\lambda - 3)(\lambda + 5)(\lambda - 6),$$

so the eigenvalues are $\lambda = 3$, $\lambda = -5$, $\lambda = 6$.

2. For each eigenvalue, we find an eigenvector.
   For

$$\lambda_1 = 3, \quad \hat{v}_1 = \begin{pmatrix} -2 \\ 3 \\ 1 \end{pmatrix}; \quad \lambda_2 = -5, \quad \hat{v}_2 = \begin{pmatrix} -2 \\ -1 \\ 1 \end{pmatrix}; \quad \lambda_3 = 6, \quad \hat{v}_3 = \begin{pmatrix} 1 \\ 6 \\ 16 \end{pmatrix}.$$

3. We then have that

$$P = \begin{pmatrix} -2 & -2 & 1 \\ 3 & -1 & 6 \\ 1 & 1 & 16 \end{pmatrix} \quad \text{and} \quad D = \begin{pmatrix} 3 & 0 & 0 \\ 0 & -5 & 0 \\ 0 & 0 & 6 \end{pmatrix}.$$

4. The solution to

$$y'(t) = \begin{pmatrix} 3 & 0 & 0 \\ 0 & -5 & 0 \\ 0 & 0 & 6 \end{pmatrix} \begin{pmatrix} y_1(t) \\ y_2(t) \\ y_3(t) \end{pmatrix}$$

is

$$y_1(t) = c_1 e^{3t}$$
$$y_2(t) = c_2 e^{-5t}$$
$$y_3(t) = c_3 e^{6t}.$$

So

$$\hat{x}(t) = P\hat{y}(t) = \begin{pmatrix} -2 & -2 & 1 \\ 3 & -1 & 6 \\ 1 & 1 & 16 \end{pmatrix} \begin{pmatrix} c_1 e^{3t} \\ c_2 e^{-5t} \\ c_3 e^{6t} \end{pmatrix} = \begin{pmatrix} -2c_1 e^{3t} - 2c_2 e^{-5t} + c_3 e^{6t} \\ 3c_1 e^{3t} - c_2 e^{-5t} + 6c_3 e^{6t} \\ c_1 e^{3t} + c_2 e^{-5t} + 16c_3 e^{6t} \end{pmatrix}.$$

Now

$$\hat{x}(0) = \begin{pmatrix} 3 \\ -1 \\ 0 \end{pmatrix}.$$

So

$$\hat{c} = \begin{pmatrix} c_1 \\ c_2 \\ c_3 \end{pmatrix} = P^{-1}\hat{x}(0) = \begin{pmatrix} -2 & -2 & 1 \\ 3 & -1 & 6 \\ 1 & 1 & 16 \end{pmatrix}^{-1} \begin{pmatrix} 3 \\ -1 \\ 0 \end{pmatrix} = \begin{pmatrix} -3/4 \\ -31/44 \\ 1/11 \end{pmatrix}$$

gives

$$c_1 = -\frac{3}{4}, \quad c_2 = -\frac{31}{44}, \quad c_3 = \frac{1}{11}.$$

Thus,

$$x_1(t) = -2\left(-\frac{3}{4}\right)e^{3t} - 2\left(-\frac{31}{44}\right)e^{-5t} + \left(\frac{1}{11}\right)e^{6t} = \frac{3}{2}e^{3t} + \frac{31}{22}e^{-5t} + \frac{1}{11}e^{6t}$$

$$x_2(t) = 3\left(-\frac{3}{4}\right)e^{3t} - \left(-\frac{31}{44}\right)e^{-5t} + 6\left(\frac{1}{11}\right)e^{6t} = -\frac{9}{4}e^{3t} + \frac{31}{44}e^{-5t} + \frac{6}{11}e^{6t}$$

$$x_3(t) = -\frac{3}{4}e^{3t} - \frac{31}{44}e^{-5t} + \frac{16}{11}e^{6t}.$$

We demonstrate that the first equation in the system of equations is satisfied. That is,

$$x_1'(t) = -2x_1(t) - 4x_2(t) + 2x_3(t).$$

Since

$$x_1(t) = \frac{3}{2}e^{3t} + \frac{31}{22}e^{-5t} + \frac{1}{11}e^{6t},$$

we have

$$x_1'(t) = \frac{9}{2}e^{3t} - \frac{155}{22}e^{-5t} + \frac{6}{11}e^{6t}$$

and

$$-2x_1(t) - 4x_2(t) + 2x_3(t) = -2\left[\frac{3}{2}e^{3t} + \frac{31}{22}e^{-5t} + \frac{1}{11}e^{6t}\right] - 4\left[-\frac{9}{4}e^{3t} + \frac{31}{44}e^{-5t} + \frac{6}{11}e^{6t}\right]$$

$$+ 2\left[-\frac{3}{4}e^{3t} - \frac{31}{44}e^{-5t} + \frac{16}{11}e^{6t}\right] = \left(-3 + 9 - \frac{3}{2}\right)e^{3t} + \left(-\frac{31}{11} - \frac{31}{11} - \frac{31}{22}\right)e^{-5t} + \left(-\frac{2}{11} - \frac{24}{11} + \frac{32}{11}\right)e^{6t}$$

$$= \frac{9}{2}e^{3t} - \frac{155}{22}e^{-5t} + \frac{6}{11}e^{6t}.$$

The other two equations can be checked in a similar manner.

Exercises

1. Solve the system of differential equations

$$x_1'(t) = 3x_1(t) + 3x_2(t) - 2x_3(t)$$
$$x_2'(t) = 2x_1(t) - 3x_2(t) + 2x_3(t)$$
$$x_3'(t) = -4x_1(t) + x_2(t) + x_3(t)$$

with initial conditions

$$x_1(0) = 2, \quad x_2(0) = 0, \quad x_3(0) = 1.$$

2. Solve the system of differential equations

$$x_1'(t) = 4x_1(t) + x_3(t)$$
$$x_2'(t) = -2x_1(t) + x_2(t)$$
$$x_3'(t) = -2x_1(t) + x_3(t)$$

   with initial conditions

$$x_1(0) = 1, \quad x_2(0) = 3, \quad x_3(0) = -1.$$

3. Solve the system of differential equations

$$x_1'(t) = x_1(t) + 2x_2(t) + x_3(t)$$
$$x_2'(t) = 6x_1(t) - x_2(t)$$
$$x_3'(t) = -x_1(t) - 2x_2(t) - x_3(t)$$

   with initial conditions

$$x_1(0) = 2, \quad x_2(0) = 0, \quad x_3(0) = 1.$$

4. Solve the system of differential equations

$$x_1'(t) = -x_1(t) + 2x_2(t) + 2x_3(t)$$
$$x_2'(t) = 2x_1(t) + 2x_2(t) - x_3(t)$$
$$x_3'(t) = 2x_1(t) - x_2(t) + 2x_3(t)$$

   with initial conditions

$$x_1(0) = 3, \quad x_2(0) = -2, \quad x_3(0) = 1.$$

6

*Inner Product Spaces**

Up to this point, we have made qualitative deductions about vector spaces, but we have not derived quantitative properties (except in Section 3.1). An example of a quantitative property that would almost certainly be of interest is the length of a vector. In this chapter, we introduce one way of forming the product of vectors, called the inner product, that associates a scalar with each pair of vectors. Among other things, an inner product enables us to generalize the notion of length that occurs in Euclidean spaces.

The terms "inner product" and "dot product" are often used synonymously, but they are not exactly the same. The dot product is used only when the scalar field is $\mathbb{R}$. The inner product generalizes the dot product and applies when the scalar field is $\mathbb{R}$ or $\mathbb{C}$.

In Section 3.1, we introduced vector spaces by analyzing $\mathbb{R}^2$ as an example. That vector space is a dot product space and provides some intuition for this chapter.

6.1 Some Facts about Complex Numbers

In this chapter, complex numbers come more to the forefront of our discussion. Here we give enough background in complex numbers to study vector spaces where the scalars are the complex numbers.

Complex numbers, denoted by $\mathbb{C}$, are defined by

$$\mathbb{C} = \{a + bi \mid a, b \in \mathbb{R}\},$$

where $i = \sqrt{-1}$.

Complex numbers are usually represented as z, where $z = a + bi$, with $a, b \in \mathbb{R}$. The arithmetic operations are what one would expect. For example, if

$$z_1 = a_1 + b_1 i \quad \text{and} \quad z_2 = a_2 + b_2 i,$$

then

$$z_1 + z_2 = (a_1 + b_1 i) + (a_2 + b_2 i) = (a_1 + a_2) + (b_1 + b_2)i$$

and

$$z_1 z_2 = (a_1 + b_1 i)(a_2 + b_2 i) = a_1 a_2 + (a_1 b_2 + b_1 a_2)i + b_1 b_2 i^2 = (a_1 a_2 - b_1 b_2) + (a_1 b_2 + b_1 a_2)i.$$

---

* *Note*: In this chapter, there are some subsections that are marked as optional. The material in these sections is beyond what is normally done in a first linear algebra course. Omitting them will not compromise later material.

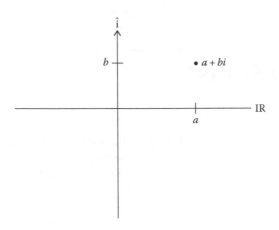

FIGURE 6.1

If $z=a+bi$, then the real part of z is a and the imaginary part of z is b. These are often denoted *Re z* and *Im z*, respectively.

If $z=a+bi$, then the complex conjugate of z, denoted $\bar{z}$, is defined by

$$\bar{z} = a - bi.$$

To plot complex numbers, one uses a two-dimensional grid. (See Figure 6.1.)

The modulus of a complex number z, denoted $|z|$, is the distance of the number z from the origin. For $z=a+bi$, we have

$$|z|^2 = a^2 + b^2 = z\bar{z}.$$

If the vector space is $\mathbb{C}^n$, the formula for the usual dot product breaks down in that it does not necessarily give the length of a vector. For example, with the definition of the Euclidean dot product, the length of the vector $(3i, 2i)$ would be $\sqrt{-13}$. In $\mathbb{C}^n$, the length of the vector $(a_1,\ldots,a_n)$ should be the distance from the origin to the terminal point of the vector, which is

$$\sqrt{a_1\bar{a_1} + \cdots + a_n\bar{a_n}} = \sqrt{|a_1|^2 + \cdots + |a_n|^2}.$$

The usual inner product in $\mathbb{C}^n$ is if $\hat{u} = (u_1,\ldots,u_n)$ and $\hat{v} = (v_1,\ldots,v_n)$, then

$$\langle \hat{u}, \hat{v} \rangle = \sum_{i=1}^{n} u_i \bar{v_i}.$$

Exercises

In Exercises 1 through 4, perform the arithmetic operations.

1. $(2+4i) - 3(-1+2i)$
2. $2(5-3i)(-6+i)$

3. $6(2+i)\overline{(3-i)}$

4. $|-2-3i|$

5. Show that if

$$z = a + bi$$

then

$$z^{-1} = \frac{1}{z} = \frac{1}{a^2 + b^2}(a - bi)$$

if a and b are not both 0.

6. Find the usual inner product, $\langle \hat{u}, \hat{v} \rangle$, for the following pairs of vectors.

(a) $\hat{u} = (1 - 2i, 3 + 6i, 2), \hat{v} = (i, -2i, 3 - 2i)$

(b) $\hat{u} = (i, 3, 2 - i), \hat{v} = (3, 4 + i, 7i)$

(c) $\hat{u} = (2i, 5i, -3i), \hat{v} = (1, 5, 3)$

(d) $\hat{u} = (1, 5, 3), \hat{v} = (2i, 5i, -3i)$

7. In this problem, we demonstrate a connection between complex numbers and 2×2 matrices. We associate

$$1 \text{ with } \begin{pmatrix} 1 & 0 \\ 0 & 1 \end{pmatrix} \quad \text{and} \quad i \text{ with } \begin{pmatrix} 0 & -1 \\ 1 & 0 \end{pmatrix}.$$

(a) Show that

$$\begin{pmatrix} 0 & -1 \\ 1 & 0 \end{pmatrix}^2 = \begin{pmatrix} -1 & 0 \\ 0 & -1 \end{pmatrix} = -\begin{pmatrix} 1 & 0 \\ 0 & 1 \end{pmatrix}$$

analogous to $i^2 = -1$.
   Identify

$$a + bi \sim \begin{pmatrix} a & -b \\ b & a \end{pmatrix}.$$

(b) Identify $z_1 \sim \begin{pmatrix} a & -b \\ b & a \end{pmatrix}$ and $z_2 \sim \begin{pmatrix} c & -d \\ d & c \end{pmatrix}$.

Show that

$$\begin{pmatrix} a & -b \\ b & a \end{pmatrix}\begin{pmatrix} c & -d \\ d & c \end{pmatrix} = \begin{pmatrix} c & -d \\ d & c \end{pmatrix}\begin{pmatrix} a & -b \\ b & a \end{pmatrix}$$

as $z_1 z_2 = z_2 z_1$ even though matrices do not typically commute in multiplication.

(c) Show that if $a+bi \sim \begin{pmatrix} a & -b \\ b & a \end{pmatrix}$, then $\overline{a+bi} \sim \begin{pmatrix} a & -b \\ b & a \end{pmatrix}^T$.

(d) Show that if $z=a+bi \sim \begin{pmatrix} a & -b \\ b & a \end{pmatrix}$, then

$$det\begin{pmatrix} a & -b \\ b & a \end{pmatrix} = a^2 + b^2 = |z|^2.$$

(e) In complex numbers, Euler's theorem states that

$$e^{i\theta} = \cos\theta + i\sin\theta.$$

In the matrix analogy, this says

$$\exp\begin{pmatrix} 0 & -\theta \\ \theta & 0 \end{pmatrix} = \begin{pmatrix} \cos\theta & -\sin\theta \\ \sin\theta & \cos\theta \end{pmatrix}.$$

Recall that $\begin{pmatrix} \cos\theta & -\sin\theta \\ \sin\theta & \cos\theta \end{pmatrix}$ is the rotation matrix in two dimensions.

Taking derivatives gives $\dfrac{d}{d\theta}e^{i\theta} = ie^{i\theta} = i\cos\theta - \sin\theta$ whose matrix representation is

$$\begin{pmatrix} -\sin\theta & -\cos\theta \\ \cos\theta & -\sin\theta \end{pmatrix} = \begin{pmatrix} 0 & -1 \\ 1 & 0 \end{pmatrix}\begin{pmatrix} \cos\theta & -\sin\theta \\ \sin\theta & \cos\theta \end{pmatrix} = i\exp\begin{pmatrix} 0 & -\theta \\ \theta & 0 \end{pmatrix}.$$

6.2 Inner Product Spaces

Definition

Let V be a vector space with scalar field $\mathcal{F}$ ($\mathbb{R}$ or $\mathbb{C}$). An inner product on V is a function

$$\langle,\rangle : V \times V \to \mathcal{F}$$

that satisfies the following conditions for all $\hat{u}$, $\hat{v}$, $\hat{w} \in V$, and $\lambda \in \mathcal{F}$:

1. $\langle \hat{u}, \hat{u}\rangle \geq 0$ and $\langle \hat{u}, \hat{u}\rangle = 0$ if and only if $\hat{u} = \hat{0}$.
2. $\langle \hat{u}+\hat{v}, \hat{w}\rangle = \langle \hat{u}, \hat{w}\rangle + \langle \hat{v}, \hat{w}\rangle$.
3. $\langle \lambda\hat{u}, \hat{v}\rangle = \lambda\langle \hat{u}, \hat{v}\rangle$.
4. $\langle \hat{u}, \hat{v}\rangle = \overline{\langle \hat{v}, \hat{u}\rangle}$.

If the scalar field is $\mathbb{R}$, then condition 4 becomes $\langle \hat{u}, \hat{v}\rangle = \langle \hat{v}, \hat{u}\rangle$ so the definition is the same as that for the dot product.

Definition

A vector space together with an inner product is called an inner product space.
   As a consequence of property 4,

$$\langle \hat{u}, \hat{v} \rangle \langle \hat{v}, \hat{u} \rangle = \langle \hat{u}, \hat{v} \rangle \overline{\langle \hat{u}, \hat{v} \rangle} = \left| \langle \hat{u}, \hat{v} \rangle \right|^2$$

and, as a consequence of properties 3 and 4,

$$\langle \hat{u}, \lambda \hat{v} \rangle = \overline{\lambda} \langle \hat{u}, \hat{v} \rangle.$$

Furthermore, $\langle \hat{v}, \hat{0} \rangle = \langle \hat{0}, \hat{v} \rangle = 0$ for every vector $\hat{v}$ since

$$\langle \hat{0}, \hat{v} \rangle = \langle \hat{0} + \hat{0}, \hat{v} \rangle = \langle \hat{0}, \hat{v} \rangle + \langle \hat{0}, \hat{v} \rangle = 2 \langle \hat{0}, \hat{v} \rangle.$$

Theorem 1: (Cauchy–Schwarz Inequality)

If $\hat{u}$ and $\hat{v}$ are vectors in an inner product space V, then

$$\left| \langle \hat{u}, \hat{v} \rangle \right| \leq \|\hat{u}\| \|\hat{v}\|.$$

Proof

The result is clear if $\hat{u} = \hat{0}$, $\hat{v} = \hat{0}$, or $\left| \langle \hat{u}, \hat{v} \rangle \right| = 0$, so suppose that none of these is the case.
   For any scalar α, we have

$$0 \leq \|\hat{u} - \alpha \hat{v}\|^2 = \langle \hat{u} - \alpha \hat{v}, \hat{u} - \alpha \hat{v} \rangle = \langle \hat{u}, \hat{u} \rangle - \overline{\alpha} \langle \hat{u}, \hat{v} \rangle - \alpha \langle \hat{v}, \hat{u} \rangle + \alpha \overline{\alpha} \langle \hat{v}, \hat{v} \rangle. \tag{6.1}$$

The rest of the proof amounts to making a judicious choice for α. Let

$$\alpha = \frac{\langle \hat{u}, \hat{v} \rangle}{\langle \hat{v}, \hat{v} \rangle}.$$

Then

$$\overline{\alpha} = \frac{\overline{\langle \hat{u}, \hat{v} \rangle}}{\langle \hat{v}, \hat{v} \rangle} \quad \text{and} \quad \alpha \overline{\alpha} = \frac{\left| \langle \hat{u}, \hat{v} \rangle \right|^2}{\langle \hat{v}, \hat{v} \rangle \langle \hat{v}, \hat{v} \rangle}$$

so we can rewrite (6.1) to get

$$0 \leq \|\hat{u}\|^2 - \frac{\overline{\langle \hat{u}, \hat{v} \rangle}}{\langle \hat{v}, \hat{v} \rangle} \langle \hat{u}, \hat{v} \rangle - \frac{\langle \hat{u}, \hat{v} \rangle}{\langle \hat{v}, \hat{v} \rangle} \langle \hat{v}, \hat{u} \rangle + \frac{\left| \langle \hat{u}, \hat{v} \rangle \right|^2}{\langle \hat{v}, \hat{v} \rangle \langle \hat{v}, \hat{v} \rangle} \langle \hat{v}, \hat{v} \rangle = \|\hat{u}\|^2 - \frac{\left| \langle \hat{u}, \hat{v} \rangle \right|^2}{\|\hat{v}\|^2} - \frac{\left| \langle \hat{u}, \hat{v} \rangle \right|^2}{\|\hat{v}\|^2} + \frac{\left| \langle \hat{u}, \hat{v} \rangle \right|^2}{\|\hat{v}\|^4} \|\hat{v}\|^2.$$

Thus,

$$\|\hat{u}\|^2 + \frac{\left|\langle\hat{u},\hat{v}\rangle\right|^2}{\|\hat{v}\|^2} \geq 2\frac{\left|\langle\hat{u},\hat{v}\rangle\right|^2}{\|\hat{v}\|^2}$$

so

$$\|\hat{u}\|^2 \geq \frac{\left|\langle\hat{u},\hat{v}\rangle\right|^2}{\|\hat{v}\|^2}$$

and

$$\|\hat{u}\|\|\hat{v}\| \geq \left|\langle\hat{u},\hat{v}\rangle\right|.$$

Exercises

1. (a) Show that for the vector space of real-valued continuous functions on $[0,1]$ the function $\langle,\rangle$ given by

$$\langle f,g\rangle = \int_0^1 f(x)g(x)dx$$

is an inner product.

(b) With the inner product above find $\|f\|$ if $f(x)=x^2+x$.

(c) With the inner product above find $\|f-g\|$ if

$$f(x)=x^2+x \quad \text{and} \quad g(x)=x-1.$$

2. (a) Show that the function

$$\left\langle \begin{pmatrix} x_1 \\ y_1 \end{pmatrix}, \begin{pmatrix} x_2 \\ y_2 \end{pmatrix} \right\rangle = 2x_1x_2 + y_1y_2$$

is an inner product on $\mathbb{R}^2$.

(b) With the inner product above find $\|\hat{u}\|$ if

$$\hat{u} = \begin{pmatrix} 3 \\ -4 \end{pmatrix}.$$

(c) With the inner product above find $\|\hat{u} - \hat{v}\|$ if

$$\hat{u} = \begin{pmatrix} 3 \\ -4 \end{pmatrix} \quad \text{and} \quad \hat{v} = \begin{pmatrix} 2 \\ 2 \end{pmatrix}.$$

3. Show why

$$\left\langle \begin{pmatrix} x_1 \\ y_1 \end{pmatrix}, \begin{pmatrix} x_2 \\ y_2 \end{pmatrix} \right\rangle = x_1{}^2 x_2 + y_1 y_2$$

is not an inner product on $\mathbb{R}^2$.

4. Show that for any inner product space

$$\|\hat{u} + \hat{v}\|^2 + \|\hat{u} - \hat{v}\|^2 = 2\|\hat{u}\|^2 + 2\|\hat{v}\|^2.$$

5. Show that for any inner product space

$$\|\hat{u} + \hat{v}\|^2 - \|\hat{u} - \hat{v}\|^2 = 4\langle \hat{u}, \hat{v} \rangle.$$

6. Show that in $\mathbb{R}^2$ with the usual inner product and A a 2×2 matrix

$$\langle \hat{u}, A\hat{v} \rangle = \langle A^T \hat{u}, \hat{v} \rangle.$$

6.3 Orthogonality

With the Euclidean dot product in $\mathbb{R}^2$ and $\mathbb{R}^3$, we had a method of defining the angle between two vectors, but with an inner product this cannot be done in $\mathbb{C}^n$ (with the usual interpretation of angle) because $\langle \hat{u}, \hat{v} \rangle$ can be complex. However, we will define what it means for vectors to be orthogonal.

Definition

A set of nonzero vectors $\{\hat{v}_1, \ldots, \hat{v}_n\}$ in the inner product space V is an orthogonal set if $\langle \hat{v}_i, \hat{v}_j \rangle = 0$ whenever $i \neq j$. An orthonormal set is an orthogonal set that has the additional property that $\langle \hat{v}_i, \hat{v}_i \rangle = 1$ for $i = 1, \ldots, n$.

If we have an orthogonal set $\{\hat{v}_1, \ldots, \hat{v}_n\}$, it can be converted to the orthonormal set

$$\left\{ \frac{\hat{v}_1}{\|\hat{v}_1\|}, \ldots, \frac{\hat{v}_n}{\|\hat{v}_n\|} \right\}.$$

Orthogonality generalizes the idea of being perpendicular to settings other than $\mathbb{R}^n$.

Theorem 2

An orthogonal set of vectors on an inner product space is a linearly independent set.

Proof

Let $\{\hat{v}_1,\ldots,\hat{v}_n\}$ be an orthogonal set of vectors in the inner product space V, and suppose that

$$a_1\hat{v}_1 + \cdots + a_n\hat{v}_n = \hat{0}.$$

Now

$$\langle a_1\hat{v}_1 + \cdots + a_n\hat{v}_n, \hat{v}_1 \rangle = a_1\langle \hat{v}_1, \hat{v}_1 \rangle + a_2\langle \hat{v}_2, \hat{v}_1 \rangle + \cdots + a_n\langle \hat{v}_n, \hat{v}_1 \rangle = a_1\langle \hat{v}_1, \hat{v}_1 \rangle$$

but

$$\langle a_1\hat{v}_1 + \cdots + a_n\hat{v}_n, \hat{v}_1 \rangle = \langle \hat{0}, \hat{v}_1 \rangle = 0$$

so

$$a_1\langle \hat{v}_1, \hat{v}_1 \rangle = 0.$$

Since $\langle \hat{v}_1, \hat{v}_1 \rangle \neq 0$, then $a_1 = 0$. Similarly, $a_i = 0$; $i = 2, \ldots, n$.

Corollary

An orthogonal spanning set of vectors for an inner product space is a basis.
The following example shows why orthogonal bases—and why, even more so, orthonormal bases—are convenient.

Example

Suppose that $\{\hat{v}_1,\ldots,\hat{v}_n\}$ is an orthogonal basis for the inner product space V. If $\hat{v}$ is a vector in V for which

$$\hat{v} = a_1\hat{v}_1 + \cdots + a_n\hat{v}_n,$$

we can find a formula for a_i.
　We find the formula for a_1. We have

$$\langle \hat{v}, \hat{v}_1 \rangle = \langle a_1\hat{v}_1 + \cdots + a_n\hat{v}_n, \hat{v}_1 \rangle = a_1\langle \hat{v}_1, \hat{v}_1 \rangle + a_2\langle \hat{v}_2, \hat{v}_1 \rangle + \cdots + a_n\langle \hat{v}_n, \hat{v}_1 \rangle = a_1\langle \hat{v}_1, \hat{v}_1 \rangle$$

$$a_1 = \frac{\langle \hat{v}, \hat{v}_1 \rangle}{\langle \hat{v}_1, \hat{v}_1 \rangle}.$$

Likewise,

$$a_i = \frac{\langle \hat{v}, \hat{v}_i \rangle}{\langle \hat{v}_i, \hat{v}_i \rangle}; \quad i = 1,\ldots,n.$$

If $\{\hat{v}_1,\ldots,\hat{v}_n\}$ is an orthonormal basis for the inner product space V, then we have the simpler formula:

$$a_i = \langle \hat{v}, \hat{v}_i \rangle; \quad i = 1,\ldots,n.$$

From this, we get the following important result.

Theorem 3

If $\{\hat{v}_1,\ldots,\hat{v}_n\}$ is an orthonormal basis for the inner product space V, then for $\hat{v}$ a vector in V we have

$$\hat{v} = \langle \hat{v}, \hat{v}_1 \rangle \hat{v}_1 + \cdots + \langle \hat{v}, \hat{v}_n \rangle \hat{v}_n.$$

Theorem 4

If $\{\hat{v}_1,\ldots,\hat{v}_n\}$ is an orthonormal basis for the inner product space V, then for $\hat{v}$ a vector in V we have

$$\|\hat{v}\|^2 = \sum_{i=1}^{n} \left| \langle \hat{v}, \hat{v}_i \rangle \right|^2.$$

Proof

By Theorem 3, we have

$$\|\hat{v}\|^2 = \langle \hat{v}, \hat{v} \rangle = \left\langle \left(\langle \hat{v}, \hat{v}_1 \rangle \hat{v}_1 + \cdots + \langle \hat{v}, \hat{v}_n \rangle \hat{v}_n, \langle \hat{v}, \hat{v}_1 \rangle \hat{v}_1 + \cdots + \langle \hat{v}, \hat{v}_n \rangle \hat{v}_n \right) \right\rangle. \quad (6.2)$$

Since $\{\hat{v}_1,\ldots,\hat{v}_n\}$ is an orthogonal set, expression (6.2) is equal to

$$\sum_{i=1}^{n} \left\langle \langle \hat{v}, \hat{v}_i \rangle \hat{v}_i, \langle \hat{v}, \hat{v}_i \rangle \hat{v}_i \right\rangle = \sum_{i=1}^{n} \langle \hat{v}, \hat{v}_i \rangle \overline{\langle \hat{v}, \hat{v}_i \rangle} \langle \hat{v}_i, \hat{v}_i \rangle = \sum_{i=1}^{n} \langle \hat{v}, \hat{v}_i \rangle \overline{\langle \hat{v}, \hat{v}_i \rangle} = \sum_{i=1}^{n} \left| \langle \hat{v}, \hat{v}_i \rangle \right|^2 \quad (6.3)$$

Example

The set of vectors

$$\left\{ \begin{pmatrix} 1 \\ 0 \\ 0 \end{pmatrix}, \begin{pmatrix} 0 \\ 1/\sqrt{2} \\ 1/\sqrt{2} \end{pmatrix}, \begin{pmatrix} 0 \\ -1/\sqrt{2} \\ 1/\sqrt{2} \end{pmatrix} \right\}$$

is an orthonormal basis of $\mathbb{R}^3$. We express the vector

$$\hat{v} = \begin{pmatrix} -3 \\ 4 \\ 7 \end{pmatrix}$$

in terms of this basis. We have

$$\langle \hat{v}, \hat{v}_1 \rangle = -3, \quad \langle \hat{v}, \hat{v}_2 \rangle = \frac{4}{\sqrt{2}} + \frac{7}{\sqrt{2}} = \frac{11}{\sqrt{2}}, \quad \langle \hat{v}, \hat{v}_3 \rangle = -\frac{4}{\sqrt{2}} + \frac{7}{\sqrt{2}} = \frac{3}{\sqrt{2}}$$

so

$$\hat{v} = \langle \hat{v}, \hat{v}_1 \rangle \hat{v}_1 + \langle \hat{v}, \hat{v}_2 \rangle \hat{v}_2 + \langle \hat{v}, \hat{v}_3 \rangle \hat{v}_3 = -3 \begin{pmatrix} 1 \\ 0 \\ 0 \end{pmatrix} + \frac{11}{\sqrt{2}} \begin{pmatrix} 0 \\ 1/\sqrt{2} \\ 1/\sqrt{2} \end{pmatrix} + \frac{3}{\sqrt{2}} \begin{pmatrix} 0 \\ -1/\sqrt{2} \\ 1/\sqrt{2} \end{pmatrix}.$$

Exercises

1. For the following sets of vectors, (i) tell whether the set of subscripted vectors is orthogonal, orthonormal, or neither. (ii) For those that are orthogonal or orthonormal, find the coordinates of the given vector $\hat{v}$ with respect to that basis:

(a) $\hat{v}_1 = \begin{pmatrix} 1 \\ -4 \end{pmatrix}, \quad \hat{v}_2 = \begin{pmatrix} 4 \\ 1 \end{pmatrix}, \quad \hat{v} = \begin{pmatrix} 7 \\ -3 \end{pmatrix}$

(b) $\hat{v}_1 = \begin{pmatrix} \dfrac{1}{\sqrt{2}} \\ \dfrac{1}{\sqrt{2}} \end{pmatrix}, \quad \hat{v}_2 = \begin{pmatrix} -\dfrac{1}{\sqrt{2}} \\ \dfrac{1}{\sqrt{2}} \end{pmatrix}, \quad \hat{v} = \begin{pmatrix} 2 \\ 5 \end{pmatrix}$

(c) $\hat{v}_1 = \begin{pmatrix} \dfrac{1}{\sqrt{5}} \\ -\dfrac{2}{\sqrt{5}} \end{pmatrix}, \quad \hat{v}_2 = \begin{pmatrix} -\dfrac{2}{\sqrt{5}} \\ -\dfrac{1}{\sqrt{5}} \end{pmatrix}, \quad \hat{v} = \begin{pmatrix} 0 \\ 3 \end{pmatrix}$

(d) $\hat{v}_1 = \begin{pmatrix} \dfrac{1}{\sqrt{3}} \\ \dfrac{1}{\sqrt{3}} \\ \dfrac{1}{\sqrt{3}} \end{pmatrix}, \quad \hat{v}_2 = \begin{pmatrix} \dfrac{-2}{\sqrt{6}} \\ \dfrac{1}{\sqrt{6}} \\ \dfrac{1}{\sqrt{6}} \end{pmatrix}, \quad \hat{v}_3 = \begin{pmatrix} 0 \\ \dfrac{-1}{\sqrt{2}} \\ \dfrac{1}{\sqrt{2}} \end{pmatrix}, \quad \hat{v} = \begin{pmatrix} 3 \\ 0 \\ -8 \end{pmatrix}$

(e) $\hat{v}_1 = \begin{pmatrix} -3 \\ -1 \\ 1 \end{pmatrix}, \quad \hat{v}_2 = \begin{pmatrix} \dfrac{1}{\sqrt{6}} \\ \dfrac{-2}{\sqrt{6}} \\ \dfrac{-1}{\sqrt{6}} \end{pmatrix}, \quad \hat{v}_3 = \begin{pmatrix} 1 \\ 4 \\ -7 \end{pmatrix}, \quad \hat{v} = \begin{pmatrix} 5 \\ -1 \\ 3 \end{pmatrix}$

6.4 The Gram–Schmidt Process

Our goal in this section is to show that if V is an inner product space, then there is an orthonormal basis for V and to determine an algorithm to construct such a basis.

The process we will use to convert a given set of vectors to an orthonormal set of vectors is called the Gram–Schmidt process. There are two approaches that can be used to apply the Gram–Schmidt process. One is to construct an orthogonal set of vectors and when that task is completed, convert the orthogonal set to an orthonormal set by normalizing the vectors. The other approach is to normalize the vectors at each step of the process. We will use the former method.

We first give an algorithm that converts a basis to an orthogonal basis, demonstrate the use of the algorithm, and then prove a theorem that validates the ideas.

6.4.1 Algorithm for the Gram–Schmidt Process

Let $\{\hat{x}_1,\ldots,\hat{x}_k\}$ be a basis for the subspace X of the vector space V. Define

$$\hat{v}_1 = \hat{x}_1$$

$$\hat{v}_2 = \hat{x}_2 - \frac{\hat{x}_2,\hat{v}_1}{\hat{v}_1,\hat{v}_1}\hat{v}_1$$

$$\hat{v}_3 = \hat{x}_3 - \frac{\hat{x}_3,\hat{v}_1}{\hat{v}_1,\hat{v}_1}\hat{v}_1 - \frac{\hat{x}_3,\hat{v}_2}{\hat{v}_2,\hat{v}_2}\hat{v}_2$$

$$\vdots$$

$$\hat{v}_k = \hat{x}_k - \frac{\hat{x}_k,\hat{v}_1}{\hat{v}_1,\hat{v}_1}\hat{v}_1 - \frac{\hat{x}_k,\hat{v}_2}{\hat{v}_2,\hat{v}_2}\hat{v}_2 - \cdots - \frac{\hat{x}_k,\hat{v}_{k-1}}{\hat{v}_{k-1},\hat{v}_{k-1}}\hat{v}_{k-1}.$$

After the next example, we will show that $\{\hat{v}_1,\ldots,\hat{v}_k\}$ is an orthogonal set and

$$\text{span}\{\hat{v}_1,\ldots,\hat{v}_k\} = \text{span}\{\hat{x}_1,\ldots,\hat{x}_k\}; \quad k = 1,\ldots,n.$$

Example

Find the orthogonal basis that is derived from the basis

$$\left\{ \begin{pmatrix} 1 \\ 1 \\ 1 \end{pmatrix}, \begin{pmatrix} 2 \\ 0 \\ 4 \end{pmatrix}, \begin{pmatrix} -3 \\ 1 \\ 2 \end{pmatrix} \right\}$$

using the Gram–Schmidt process.
   We have

$$\hat{x}_1 = \begin{pmatrix} 1 \\ 1 \\ 1 \end{pmatrix}, \quad \hat{x}_2 = \begin{pmatrix} 2 \\ 0 \\ 4 \end{pmatrix}, \quad \hat{x}_3 = \begin{pmatrix} -3 \\ 1 \\ 2 \end{pmatrix}$$

so

$$\hat{v}_1 = \hat{x}_1 = \begin{pmatrix} 1 \\ 1 \\ 1 \end{pmatrix}.$$

Furthermore,

$$\hat{v}_2 = \hat{x}_2 - \frac{\langle \hat{x}_2, \hat{v}_1 \rangle}{\langle \hat{v}_1, \hat{v}_1 \rangle} \hat{v}_1 = \begin{pmatrix} 2 \\ 0 \\ 4 \end{pmatrix} - \frac{\begin{pmatrix} 2 \\ 0 \\ 4 \end{pmatrix} \cdot \begin{pmatrix} 1 \\ 1 \\ 1 \end{pmatrix}}{\begin{pmatrix} 1 \\ 1 \\ 1 \end{pmatrix} \cdot \begin{pmatrix} 1 \\ 1 \\ 1 \end{pmatrix}} \begin{pmatrix} 1 \\ 1 \\ 1 \end{pmatrix} = \begin{pmatrix} 2 \\ 0 \\ 4 \end{pmatrix} - \frac{6}{3} \begin{pmatrix} 1 \\ 1 \\ 1 \end{pmatrix} = \begin{pmatrix} 0 \\ -2 \\ 2 \end{pmatrix}$$

and

$$\hat{v}_3 = \hat{x}_3 - \frac{\langle \hat{x}_3, \hat{v}_1 \rangle}{\langle \hat{v}_1, \hat{v}_1 \rangle} \hat{v}_1 - \frac{\langle \hat{x}_3, \hat{v}_2 \rangle}{\langle \hat{v}_2, \hat{v}_2 \rangle} \hat{v}_2 = \begin{pmatrix} -3 \\ 1 \\ 2 \end{pmatrix} - \frac{\begin{pmatrix} -3 \\ 1 \\ 2 \end{pmatrix} \cdot \begin{pmatrix} 1 \\ 1 \\ 1 \end{pmatrix}}{\begin{pmatrix} 1 \\ 1 \\ 1 \end{pmatrix} \cdot \begin{pmatrix} 1 \\ 1 \\ 1 \end{pmatrix}} \begin{pmatrix} 1 \\ 1 \\ 1 \end{pmatrix} - \frac{\begin{pmatrix} -3 \\ 1 \\ 2 \end{pmatrix} \cdot \begin{pmatrix} 0 \\ -2 \\ 2 \end{pmatrix}}{\begin{pmatrix} 0 \\ -2 \\ 2 \end{pmatrix} \cdot \begin{pmatrix} 0 \\ -2 \\ 2 \end{pmatrix}} \begin{pmatrix} 0 \\ -2 \\ 2 \end{pmatrix}$$

$$= \begin{pmatrix} -3 \\ 1 \\ 2 \end{pmatrix} - \frac{0}{3} \begin{pmatrix} 1 \\ 1 \\ 1 \end{pmatrix} - \frac{2}{8} \begin{pmatrix} 0 \\ -2 \\ 2 \end{pmatrix} = \begin{pmatrix} -3 \\ 1 \\ 2 \end{pmatrix} - \begin{pmatrix} 0 \\ -\frac{1}{2} \\ \frac{1}{2} \end{pmatrix} = \begin{pmatrix} -3 \\ \frac{3}{2} \\ \frac{3}{2} \end{pmatrix}.$$

Thus, the orthogonal basis generated by the Gram–Schmidt process is

$$\{\hat{v}_1, \hat{v}_2, \hat{v}_3\} = \left\{ \begin{pmatrix} 1 \\ 1 \\ 1 \end{pmatrix}, \begin{pmatrix} 0 \\ -2 \\ 2 \end{pmatrix}, \begin{pmatrix} -3 \\ \frac{3}{2} \\ \frac{3}{2} \end{pmatrix} \right\}.$$

It is often the case that one would prefer an orthonormal basis, which in this case is

$$\left\{ \frac{\hat{v}_1}{\|\hat{v}_1\|}, \frac{\hat{v}_2}{\|\hat{v}_2\|}, \frac{\hat{v}_3}{\|\hat{v}_3\|} \right\} = \left\{ \begin{pmatrix} \frac{1}{\sqrt{3}} \\ \frac{1}{\sqrt{3}} \\ \frac{1}{\sqrt{3}} \end{pmatrix}, \begin{pmatrix} 0 \\ \frac{-2}{\sqrt{8}} \\ \frac{2}{\sqrt{8}} \end{pmatrix}, \begin{pmatrix} \frac{-3}{\sqrt{27/2}} \\ \frac{3/2}{\sqrt{27/2}} \\ \frac{3/2}{\sqrt{27/2}} \end{pmatrix} \right\}.$$

The formulas in the Gram–Schmidt process may appear daunting, but the intuition is not. The process is based on the following fact from Fourier analysis.

Fact

Let $\{\hat{v}_1,\ldots,\hat{v}_n\}$ be an orthogonal basis for the inner product space V and suppose that $\hat{v} \in V$. Then for each $k=1,\ldots,n$

$$\left\| \hat{v} - \sum_{j=1}^{k} a_i \hat{v}_i \right\|$$

is minimized when

$$a_i = \frac{\langle \hat{v}, \hat{v}_i \rangle}{\langle \hat{v}_i, \hat{v}_i \rangle}; \quad i = 1,\ldots,n.$$

For a proof of this fact, see Kirkwood, *An Introduction to Analysis*. Later, we will call

$$\sum_{i=1}^{k} \frac{\langle \hat{v}, \hat{v}_i \rangle}{\langle \hat{v}_i, \hat{v}_i \rangle} \hat{v}_i$$

the projection of $\hat{v}$ onto the subspace generated by $\{\hat{v}_1, \ldots, \hat{v}_n\}$. In terms of distance, this is the closest vector in the subspace to the vector $\hat{v}$.

What we are doing in the Gram–Schmidt process is this: We begin by letting $\hat{v}_1 = \hat{x}_1$. Because of linear independence

$$\hat{x}_k \notin \mathrm{span}\{\hat{x}_1, \ldots, \hat{x}_{k-1}\}; \quad k = 2,\ldots,n.$$

So $\hat{x}_2 \notin \mathrm{span}\{\hat{x}_1\} = \mathrm{span}\{\hat{v}_1\}$ and

$$\frac{\langle \hat{x}_2, \hat{v}_1 \rangle}{\langle \hat{v}_1, \hat{v}_1 \rangle} \hat{v}_1$$

is the best approximation that $\mathrm{span}\{\hat{v}_1\}$ can provide to $\hat{x}_2$. The vector

$$\hat{v}_2 = \hat{x}_2 - \frac{\langle \hat{x}_2, \hat{v}_1 \rangle}{\langle \hat{v}_1, \hat{v}_1 \rangle} \hat{v}_1$$

is the residual between $\hat{x}_2$ and the best approximation that $\mathrm{span}\{\hat{v}_1\}$ can provide to $\hat{x}_2$.

Similarly, $\hat{x}_3 \notin \mathrm{span}\{\hat{x}_1, \hat{x}_2\} = \mathrm{span}\{\hat{v}_1, \hat{v}_2\}$ and

$$\frac{\langle \hat{x}_3, \hat{v}_1 \rangle}{\langle \hat{v}_1, \hat{v}_1 \rangle} \hat{v}_1 + \frac{\langle \hat{x}_3, \hat{v}_2 \rangle}{\langle \hat{v}_2, \hat{v}_2 \rangle} \hat{v}_2$$

is the best approximation that $\mathrm{span}\{\hat{v}_1, \hat{v}_2\}$ can provide to $\hat{x}_3$. The vector

$$\hat{v}_3 = \hat{x}_3 - \frac{\langle \hat{x}_3, \hat{v}_1 \rangle}{\langle \hat{v}_1, \hat{v}_1 \rangle} \hat{v}_1 - \frac{\langle \hat{x}_3, \hat{v}_2 \rangle}{\langle \hat{v}_2, \hat{v}_2 \rangle} \hat{v}_2$$

is the residual between $\hat{x}_2$ and the best approximation that $\mathrm{span}\{\hat{v}_1, \hat{v}_2\}$ can provide to $\hat{x}_3$.

Figure 6.2 gives a visual image of why $\hat{v}_3$ is orthogonal to any vector in $\mathrm{span}\{\hat{v}_1, \hat{v}_2\}$. The vector $\hat{x}_3$ is not in $\mathrm{span}\{\hat{v}_1, \hat{v}_2\}$ and $\hat{v}_3$ is the shortest distance from $\hat{x}_3$ to the plane spanned by $\{\hat{v}_1, \hat{v}_2\}$, which is the perpendicular distance from $\hat{x}_3$ to the plane spanned by $\{\hat{v}_1, \hat{v}_2\}$.

The next result highlights the importance of the Gram–Schmidt process.

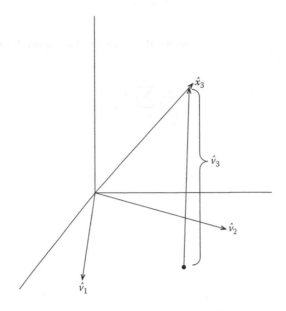

FIGURE 6.2

Theorem 5

Let V be a finite dimensional inner product space with basis $\{\hat{x}_1, \ldots, \hat{x}_n\}$ and let $\{\hat{v}_1, \ldots, \hat{v}_n\}$ be the vectors generated by the Gram–Schmidt process from $\{\hat{x}_1, \ldots, \hat{x}_n\}$. Then

(a) $\{\hat{v}_1, \ldots, \hat{v}_n\}$ is an orthogonal set.
(b) For any $k = 1, \ldots, n$, we have

$$\operatorname{span}\{\hat{v}_1, \ldots, \hat{v}_k\} = \operatorname{span}\{\hat{x}_1, \ldots, \hat{x}_k\}.$$

Proof

(a) We proceed by induction. We have

$$\langle \hat{v}_2, \hat{v}_1 \rangle = \left\langle \hat{x}_2 - \frac{\langle \hat{x}_2, \hat{v}_1 \rangle}{\langle \hat{v}_1, \hat{v}_1 \rangle} \hat{v}_1, \hat{v}_1 \right\rangle = \langle \hat{x}_2, \hat{v}_1 \rangle - \frac{\langle \hat{x}_2, \hat{v}_1 \rangle}{\langle \hat{v}_1, \hat{v}_1 \rangle} \langle \hat{v}_1, \hat{v}_1 \rangle = 0.$$

Suppose that $\{\hat{v}_1, \ldots, \hat{v}_{k-1}\}$ is an orthogonal set and

$$\hat{v}_k = \hat{x}_k - \frac{\langle \hat{x}_k, \hat{v}_1 \rangle}{\langle \hat{v}_1, \hat{v}_1 \rangle} \hat{v}_1 - \frac{\langle \hat{x}_k, \hat{v}_2 \rangle}{\langle \hat{v}_2, \hat{v}_2 \rangle} \hat{v}_2 - \cdots - \frac{\langle \hat{x}_k, \hat{v}_{k-1} \rangle}{\langle \hat{v}_{k-1}, \hat{v}_{k-1} \rangle} \hat{v}_{k-1}.$$

Then if $j < k$, we have

$$\langle \hat{v}_k, \hat{v}_j \rangle = \left\langle \hat{x}_k - \frac{\langle \hat{x}_k, \hat{v}_1 \rangle}{\langle \hat{v}_1, \hat{v}_1 \rangle} \hat{v}_1 - \frac{\langle \hat{x}_k, \hat{v}_2 \rangle}{\langle \hat{v}_2, \hat{v}_2 \rangle} \hat{v}_2 - \cdots - \frac{\langle \hat{x}_k, \hat{v}_{k-1} \rangle}{\langle \hat{v}_{k-1}, \hat{v}_{k-1} \rangle} \hat{v}_{k-1}, \hat{v}_j \right\rangle$$

$$= \langle \hat{x}_k, \hat{v}_j \rangle - \frac{\langle \hat{x}_k, \hat{v}_1 \rangle}{\langle \hat{v}_1, \hat{v}_1 \rangle} \langle \hat{v}_1, \hat{v}_j \rangle - \cdots - \frac{\langle \hat{x}_k, \hat{v}_{k-1} \rangle}{\langle \hat{v}_{k-1}, \hat{v}_{k-1} \rangle} \langle \hat{v}_{k-1}, \hat{v}_j \rangle$$

Since $\{\hat{v}_1, \ldots, \hat{v}_{k-1}\}$ is an orthogonal set, each term $\langle \hat{v}_i, \hat{v}_j \rangle = 0$ except $\langle \hat{v}_j, \hat{v}_j \rangle$. Thus,

$$\langle \hat{v}_k, \hat{v}_j \rangle = \langle \hat{x}_k, \hat{v}_j \rangle - \frac{\langle \hat{x}_k, \hat{v}_j \rangle}{\langle \hat{v}_j, \hat{v}_j \rangle} \langle \hat{v}_j, \hat{v}_j \rangle = 0$$

and the result is true by induction.

(b) By the construction, $\hat{v}_j$ is a linear combination of $\{\hat{x}_1, \ldots, \hat{x}_j\}$ and $\hat{x}_j$ is a linear combination of $\{\hat{v}_1, \ldots, \hat{v}_j\}$ for any $j = 1, \ldots, n$.

Since $\{\hat{v}_1, \ldots, \hat{v}_n\}$ is an orthogonal set, it can be normalized to give an orthonormal set. Thus, we have the following result.

Corollary

Every finite dimensional inner product space has an orthonormal basis.

Exercises

1. Apply the Gram–Schmidt process to the following sets of vectors to find an orthonormal basis:

(a) $\left\{ \begin{pmatrix} 1 \\ 0 \end{pmatrix}, \begin{pmatrix} 2 \\ 7 \end{pmatrix} \right\}$

(b) $\left\{ \begin{pmatrix} 3 \\ -5 \end{pmatrix}, \begin{pmatrix} 4 \\ 11 \end{pmatrix} \right\}$

(c) $\left\{ \begin{pmatrix} 2 \\ 7 \end{pmatrix}, \begin{pmatrix} -5 \\ -9 \end{pmatrix} \right\}$

(d) $\left\{ \begin{pmatrix} 1 \\ 0 \\ 1 \end{pmatrix}, \begin{pmatrix} 2 \\ -5 \\ 4 \end{pmatrix}, \begin{pmatrix} 1 \\ 1 \\ 1 \end{pmatrix} \right\}$

(e) $\left\{ \begin{pmatrix} -2 \\ 5 \\ -5 \end{pmatrix}, \begin{pmatrix} 1 \\ 3 \\ 1 \end{pmatrix}, \begin{pmatrix} 0 \\ 4 \\ 3 \end{pmatrix} \right\}$

(f) $\left\{ \begin{pmatrix} 3 \\ 1 \\ 0 \end{pmatrix}, \begin{pmatrix} 3 \\ 3 \\ 2 \end{pmatrix}, \begin{pmatrix} -1 \\ -2 \\ 5 \end{pmatrix} \right\}$

(g) $\left\{ \begin{pmatrix} 1 \\ 1 \\ 1 \end{pmatrix}, \begin{pmatrix} 0 \\ 2 \\ -9 \end{pmatrix}, \begin{pmatrix} 4 \\ 2 \\ 10 \end{pmatrix} \right\}$

(h) $\left\{ \begin{pmatrix} 1 \\ 0 \\ 1 \\ -2 \end{pmatrix}, \begin{pmatrix} 1 \\ 1 \\ -3 \\ 4 \end{pmatrix}, \begin{pmatrix} 3 \\ 4 \\ 0 \\ 1 \end{pmatrix}, \begin{pmatrix} 1 \\ 1 \\ 1 \\ 1 \end{pmatrix} \right\}$

(i) $\left\{ \begin{pmatrix} 2 \\ 3 \\ 0 \\ 0 \end{pmatrix}, \begin{pmatrix} 1 \\ -4 \\ 6 \\ 3 \end{pmatrix}, \begin{pmatrix} 5 \\ 1 \\ 1 \\ 2 \end{pmatrix}, \begin{pmatrix} 1 \\ 0 \\ 0 \\ 0 \end{pmatrix} \right\}$

2. Use the Gram–Schmidt process to find an orthonormal basis for $P_1(x)$ from $\{1-x, 4+3x\}$ where the inner product is

$$\langle f, g \rangle = \int_0^1 f(x)g(x)dx.$$

3. Use the Gram–Schmidt process to find an orthonormal basis for $P_2(x)$ from $\{3+x^2, 1+2x, 3x\}$ where the inner product is

$$\langle f, g \rangle = \int_{-1}^1 f(x)g(x)dx.$$

4. Find an orthonormal basis of $\mathbb{R}^2$ where the inner product is

$$\langle (x_1, y_1), (x_2, y_2) \rangle = x_1 x_2 + 2 y_1 y_2.$$

5. Use the Gram–Schmidt process to derive an orthonormal basis of $\mathbb{R}^2$ from $\{(-1,3), (2,8)\}$ where the inner product is

$$\langle (x_1, y_1), (x_2, y_2) \rangle = x_1 x_2 + 3 y_1 y_2.$$

6.5 Representation of a Linear Transformation on Inner Product Spaces (Optional)

For inner product spaces, finding the representation of a linear transformation with respect to given bases is easier if the basis of the image vector space is orthonormal. Recall the following facts.

Suppose that U and V are inner product spaces and

$$T : U \to V$$

is a linear transformation. We want to find the matrix representation of T with respect to the basis $B_1 = \{\hat{u}_1, \ldots, \hat{u}_m\}$ for U and the basis $B_2 = \{\hat{v}_1, \ldots, \hat{v}_n\}$ for V.

Recall that the ith column of the matrix we seek is $\left[T\left(\hat{u}_i\right)\right]_{B_2}$. That is, if

$$T\left(\hat{u}_i\right) = a_{1i}\hat{v}_1 + a_{2i}\hat{v}_2 + \cdots + a_{ni}\hat{v}_n$$

then the ith column of the matrix we seek is

$$\begin{pmatrix} a_{1i} \\ a_{2i} \\ \vdots \\ a_{ni} \end{pmatrix}.$$

Furthermore, recall that in an inner product space with *orthonormal* basis $\{\hat{v}_1, \ldots, \hat{v}_n\}$ we have

$$\hat{v} = a_1\hat{v}_1 + a_1\hat{v}_1 + \cdots + a_n\hat{v}_{n1} = \left\langle \hat{v}, \hat{v}_1 \right\rangle \hat{v}_1 + \left\langle \hat{v}, \hat{v}_2 \right\rangle \hat{v}_2 + \cdots + \left\langle \hat{v}, \hat{v}_n \right\rangle \hat{v}_n$$

so that

$$T\left(\hat{u}_i\right) = a_{1i}\hat{v}_1 + a_{2i}\hat{v}_2 + \cdots + a_{ni}\hat{v}_n = \left\langle T\left(\hat{u}_i\right), \hat{v}_1 \right\rangle \hat{v}_1 + \left\langle T\left(\hat{u}_i\right), \hat{v}_2 \right\rangle \hat{v}_2 + \cdots + \left\langle T\left(\hat{u}_i\right), \hat{v}_n \right\rangle \hat{v}_n.$$

So under these conditions, the matrix representation of T with respect to the basis $B_1 = \{\hat{u}_1, \ldots, \hat{u}_m\}$ for U and the basis $B_2 = \{\hat{v}_1, \ldots, \hat{v}_n\}$ for V is

$$\begin{pmatrix} \left\langle T(\hat{u}_1), \hat{v}_1 \right\rangle & \left\langle T(\hat{u}_2), \hat{v}_1 \right\rangle & \cdots & \left\langle T(\hat{u}_m), \hat{v}_1 \right\rangle \\ \left\langle T(\hat{u}_1), \hat{v}_2 \right\rangle & \left\langle T(\hat{u}_2), \hat{v}_2 \right\rangle & \cdots & \left\langle T(\hat{u}_m), \hat{v}_2 \right\rangle \\ \vdots & \vdots & \vdots & \vdots \\ \left\langle T(\hat{u}_1), \hat{v}_n \right\rangle & \left\langle T(\hat{u}_2), \hat{v}_n \right\rangle & \cdots & \left\langle T(\hat{u}_m), \hat{v}_n \right\rangle \end{pmatrix}.$$

Example

Let

$$T : \mathbb{R}^3 \to \mathbb{R}^2$$

have the matrix representation

$$\begin{pmatrix} 2 & -1 & 1 \\ 0 & 3 & 4 \end{pmatrix}$$

in the usual basis. Find the matrix representation of T with respect to the bases

$$\left\{ \begin{pmatrix} 1 \\ 0 \\ 0 \end{pmatrix}, \begin{pmatrix} 1 \\ 1 \\ 0 \end{pmatrix}, \begin{pmatrix} 1 \\ 1 \\ 1 \end{pmatrix} \right\} \quad \text{and} \quad \left\{ \begin{pmatrix} 1/\sqrt{2} \\ 1/\sqrt{2} \end{pmatrix} \begin{pmatrix} 1/\sqrt{2} \\ -1/\sqrt{2} \end{pmatrix} \right\}$$

Note that the basis for $\mathbb{R}^2$ is an orthonormal basis.
  We have

$$T \begin{pmatrix} 1 \\ 0 \\ 0 \end{pmatrix} = \begin{pmatrix} 2 \\ 0 \end{pmatrix}, \quad T \begin{pmatrix} 1 \\ 1 \\ 0 \end{pmatrix} = \begin{pmatrix} 1 \\ 3 \end{pmatrix}, \quad T \begin{pmatrix} 1 \\ 1 \\ 1 \end{pmatrix} = \begin{pmatrix} 2 \\ 7 \end{pmatrix}$$

$$\left\langle T \begin{pmatrix} 1 \\ 0 \\ 0 \end{pmatrix}, \begin{pmatrix} \frac{1}{\sqrt{2}} \\ \frac{1}{\sqrt{2}} \end{pmatrix} \right\rangle = \left\langle \begin{pmatrix} 2 \\ 0 \end{pmatrix}, \begin{pmatrix} \frac{1}{\sqrt{2}} \\ \frac{1}{\sqrt{2}} \end{pmatrix} \right\rangle = \frac{2}{\sqrt{2}} ;$$

$$\left\langle T \begin{pmatrix} 1 \\ 0 \\ 0 \end{pmatrix}, \begin{pmatrix} 1/\sqrt{2} \\ -1/\sqrt{2} \end{pmatrix} \right\rangle = \left\langle \begin{pmatrix} 2 \\ 0 \end{pmatrix}, \begin{pmatrix} 1/\sqrt{2} \\ -1/\sqrt{2} \end{pmatrix} \right\rangle = \frac{2}{\sqrt{2}}$$

$$\left\langle T \begin{pmatrix} 1 \\ 1 \\ 0 \end{pmatrix}, \begin{pmatrix} \frac{1}{\sqrt{2}} \\ \frac{1}{\sqrt{2}} \end{pmatrix} \right\rangle = \left\langle \begin{pmatrix} 1 \\ 3 \end{pmatrix}, \begin{pmatrix} \frac{1}{\sqrt{2}} \\ \frac{1}{\sqrt{2}} \end{pmatrix} \right\rangle = \frac{4}{\sqrt{2}} ;$$

$$\left\langle T \begin{pmatrix} 1 \\ 1 \\ 0 \end{pmatrix}, \begin{pmatrix} 1/\sqrt{2} \\ -1/\sqrt{2} \end{pmatrix} \right\rangle = \left\langle \begin{pmatrix} 1 \\ 3 \end{pmatrix}, \begin{pmatrix} 1/\sqrt{2} \\ -1/\sqrt{2} \end{pmatrix} \right\rangle = \frac{-2}{\sqrt{2}}$$

$$\left\langle T \begin{pmatrix} 1 \\ 1 \\ 1 \end{pmatrix}, \begin{pmatrix} \frac{1}{\sqrt{2}} \\ \frac{1}{\sqrt{2}} \end{pmatrix} \right\rangle = \left\langle \begin{pmatrix} 2 \\ 7 \end{pmatrix}, \begin{pmatrix} \frac{1}{\sqrt{2}} \\ \frac{1}{\sqrt{2}} \end{pmatrix} \right\rangle = \frac{9}{\sqrt{2}} ;$$

$$\left\langle T \begin{pmatrix} 1 \\ 1 \\ 1 \end{pmatrix}, \begin{pmatrix} 1/\sqrt{2} \\ -1/\sqrt{2} \end{pmatrix} \right\rangle = \left\langle \begin{pmatrix} 2 \\ 7 \end{pmatrix}, \begin{pmatrix} 1/\sqrt{2} \\ -1/\sqrt{2} \end{pmatrix} \right\rangle = \frac{-5}{\sqrt{2}}$$

so the matrix representation of T with respect to these bases is

$$\begin{pmatrix} \dfrac{2}{\sqrt{2}} & \dfrac{4}{\sqrt{2}} & \dfrac{9}{\sqrt{2}} \\ \dfrac{2}{\sqrt{2}} & \dfrac{-2}{\sqrt{2}} & \dfrac{-5}{\sqrt{2}} \end{pmatrix}.$$

If we use the earlier method to compute the matrix with the formula $P_{\mathcal{B}_2}^{-1} T P_{\mathcal{B}_1}$ with

$$P_{\mathcal{B}_1} = \begin{pmatrix} 1 & 1 & 1 \\ 0 & 1 & 1 \\ 0 & 0 & 1 \end{pmatrix}, \quad T = \begin{pmatrix} 2 & -1 & 1 \\ 0 & 3 & 4 \end{pmatrix}, \quad P_{\mathcal{B}_2} = \begin{pmatrix} 1/\sqrt{2} & 1/\sqrt{2} \\ 1/\sqrt{2} & -1/\sqrt{2} \end{pmatrix}$$

we of course get the same result.

Exercises

In Exercises 1 and 2, find the matrix representation of the linear transformation of T with the usual (standard) dot product.

1. Let

$$T : \mathbb{R}^2 \to \mathbb{R}^2$$

have the matrix representation

$$\begin{pmatrix} -2 & 0 \\ 4 & 3 \end{pmatrix}$$

in the usual basis. Find the matrix representation of T with respect to the bases

$$\left\{ \begin{pmatrix} 1 \\ -2 \end{pmatrix} \begin{pmatrix} -5 \\ 6 \end{pmatrix} \right\} \quad \text{and} \quad \left\{ \begin{pmatrix} 1/\sqrt{2} \\ 1/\sqrt{2} \end{pmatrix} \begin{pmatrix} 1/\sqrt{2} \\ -1/\sqrt{2} \end{pmatrix} \right\}$$

2. Let

$$T : \mathbb{R}^3 \to \mathbb{R}^2$$

have the matrix representation

$$\begin{pmatrix} 0 & 3 & -1 \\ 2 & -6 & 2 \end{pmatrix}$$

in the usual basis. Find the matrix representation of T with respect to the bases

$$\left\{\begin{pmatrix} -2 \\ 0 \\ 1 \end{pmatrix}, \begin{pmatrix} 4 \\ 5 \\ 0 \end{pmatrix}, \begin{pmatrix} 0 \\ 0 \\ 3 \end{pmatrix}\right\} \quad \text{and} \quad \left\{\begin{pmatrix} 1/\sqrt{2} \\ 1/\sqrt{2} \end{pmatrix}\begin{pmatrix} 1/\sqrt{2} \\ -1/\sqrt{2} \end{pmatrix}\right\}$$

6.6 Orthogonal Complement

Definition

Let V be an inner product space and let W be a nonempty subset of V. The orthogonal complement of W, denoted $W^\perp$, is the set of vectors

$$W^\perp = \left\{\hat{v} \in V \mid \langle \hat{v}, \hat{w} \rangle = 0 \text{ for every } \hat{w} \in W\right\}.$$

For any set W, the set $W^\perp$ is nonempty because $\hat{0} \in W^\perp$ for any set W.

The major goal of this section is to derive an algorithm to determine $W^\perp$ from any set W. This algorithm is demonstrated by an example at the end of the section. The next two theorems give the justification for the algorithm.

Theorem 6

If V is an inner product space and W is a nonempty subset of V, then $W^\perp$ is a subspace of V.

The proof is left as an exercise. Note that $W^\perp$ in the theorem is a subspace of V even if W is only a subset of V.

Theorem 7

Suppose that V is an inner product space and let W be a subspace of V. Then

(1) $W \cap W^\perp = \{\hat{0}\}$
(2) $W + W^\perp = V$

Proof

(1) Suppose that $\hat{v} \in W \cap W^\perp$. Then $\langle \hat{v}, \hat{v} \rangle = 0$, so $\hat{v} = \hat{0}$.
(2) Let $U = W + W^\perp$. We show that $U = V$. Suppose $U \neq V$. We have $U \subset V$ since $W \subset V$ and $W^\perp \subset V$. We suppose $U \neq V$ and will get a contradiction.

Form an orthonormal basis of U and extend this basis to be an orthonormal basis of V. Since $U \neq V$, there is a vector $\hat{v}$ in the orthonormal basis of V with $\hat{v} \notin U$. Since $\hat{v}$ is orthogonal to U, then $\hat{v}$ is orthogonal to W which means $\hat{v} \in W^\perp$ which is a contradiction.

Corollary

If V is an inner product space and W is a subspace of V then

(1) $V = W \oplus W^\perp$.
(2) Any vector $\hat{v} \in V$ is uniquely expressible as

$$\hat{v} = \hat{w}_1 + \hat{w}_2,$$

where $\hat{w}_1 \in W, \hat{w}_2 \in W^\perp$.
(3) $\dim(V) = \dim(W) + \dim(W^\perp)$

Example

Find the orthogonal complement of

$$W = \left\{ \begin{pmatrix} 1 \\ 2 \\ 0 \\ 1 \end{pmatrix} \begin{pmatrix} 1 \\ 1 \\ 3 \\ 3 \end{pmatrix} \right\}.$$

A vector

$$\hat{v} = \begin{pmatrix} v_1 \\ v_2 \\ v_3 \\ v_4 \end{pmatrix}$$

is in $W^\perp$ if and only if

$$\left\langle \begin{pmatrix} v_1 \\ v_2 \\ v_3 \\ v_4 \end{pmatrix}, \begin{pmatrix} 1 \\ 2 \\ 0 \\ 1 \end{pmatrix} \right\rangle = 0 \quad \text{and} \quad \left\langle \begin{pmatrix} v_1 \\ v_2 \\ v_3 \\ v_4 \end{pmatrix}, \begin{pmatrix} 1 \\ 1 \\ 3 \\ 3 \end{pmatrix} \right\rangle = 0.$$

This yields the system of equations

$$1v_1 + 2v_2 + 0v_3 + 1v_4 = 0$$
$$1v_1 + 1v_2 + 3v_3 + 3v_4 = 0.$$

The augmented matrix for this system is

$$\begin{pmatrix} 1 & 2 & 0 & 1 & 0 \\ 1 & 1 & 3 & 3 & 0 \end{pmatrix},$$

which, when row reduced, is

$$\begin{pmatrix} 1 & 0 & 6 & 5 & 0 \\ 0 & 1 & -3 & -2 & 0 \end{pmatrix}.$$

The free variables are v_3 and v_4.

$$v_1 = -6v_4 - 5v_4$$
$$v_2 = 3v_3 + 2v_4$$

so letting $v_3 = s$, $v_4 = t$ we have

$$\begin{pmatrix} v_1 \\ v_2 \\ v_3 \\ v_4 \end{pmatrix} = \begin{pmatrix} -6v_3 - 5v_4 \\ 3v_3 + 2v_4 \\ v_3 \\ v_4 \end{pmatrix} = \begin{pmatrix} -6s - 5t \\ 3s + 2t \\ s \\ t \end{pmatrix} = s\begin{pmatrix} -6 \\ 3 \\ 1 \\ 0 \end{pmatrix} + t\begin{pmatrix} -5 \\ 2 \\ 0 \\ 1 \end{pmatrix}.$$

Thus,

$$W^\perp = \text{span}\left\{ \begin{pmatrix} -6 \\ 3 \\ 1 \\ 0 \end{pmatrix}, \begin{pmatrix} -5 \\ 2 \\ 0 \\ 1 \end{pmatrix} \right\}.$$

Exercises

1. Find $W^\perp$ when W is the span of the following sets of vectors:

(a) $\left\{ \begin{pmatrix} 1 \\ 0 \\ -4 \end{pmatrix}, \begin{pmatrix} 2 \\ 2 \\ 1 \end{pmatrix} \right\}$

(b) $\left\{ \begin{pmatrix} 3 \\ -2 \\ 5 \end{pmatrix} \right\}$

(c) $\left\{ \begin{pmatrix} 2 \\ 2 \\ 1 \end{pmatrix}, \begin{pmatrix} 6 \\ 0 \\ 3 \end{pmatrix}, \begin{pmatrix} 10 \\ 4 \\ 5 \end{pmatrix} \right\}$

(d) $\left\{ \begin{pmatrix} 1 \\ 0 \\ 0 \\ 4 \end{pmatrix}, \begin{pmatrix} 2 \\ 1 \\ 3 \\ 5 \end{pmatrix}, \begin{pmatrix} 5 \\ 2 \\ 0 \\ -1 \end{pmatrix} \right\}$

$$(e) \left\{ \begin{pmatrix} 3 \\ 1 \\ 1 \\ 2 \end{pmatrix}, \begin{pmatrix} 0 \\ 1 \\ 4 \\ 0 \end{pmatrix} \right\}$$

2. Show that a vector $\hat{u}$ is in the orthogonal complement of the span of $\{\hat{v}_1 \ldots, \hat{v}_n\}$ if and only if $\langle u, \hat{v}_i \rangle = 0$, for $i = 1, \ldots, n$.

6.7 Four Subspaces Associated with a Matrix (Optional)

Let A be an $m \times n$ matrix. There are four fundamental subspaces associated with A:

1. The null space of A, $\mathcal{N}(A) = \{\hat{x} \in \mathbb{R}^n | A\hat{x} = \hat{0}\}$
2. The row space of A, which is the set of linear combinations of the rows of A.
3. The column space of A, which is the set of linear combinations of the columns of A, which is also the row space of A^T. Recall from Theorem 19, Chapter 3, the column space of A is the range of A, which we denote here as $\mathcal{R}(A)$.
4. The null space of A^T, $\mathcal{N}(A^T) = \{\hat{x} \in \mathbb{R}^m | A^T\hat{x} = \hat{0}\}$.

Theorem 8

If A is a linear transformation on a finite dimensional inner product space, then $\mathcal{N}(A) = \left(\mathcal{R}(A^T)\right)^{\perp}$.

Proof

We first give the idea of the proof when A is a 2×3 matrix. Let

$$A = \begin{pmatrix} a_{11} & a_{12} & a_{13} \\ a_{21} & a_{22} & a_{23} \end{pmatrix}, \quad \hat{x} \in \mathcal{N}(A), \quad \hat{y} \in \mathcal{R}(A^T).$$

Since $\hat{y} \in \mathcal{R}(A^T)$, there is a $\hat{z}$ with $\hat{y} = A^T\hat{z}$.
Then

$$A\hat{x} = \begin{pmatrix} a_{11} & a_{12} & a_{13} \\ a_{21} & a_{22} & a_{23} \end{pmatrix} \begin{pmatrix} x_1 \\ x_2 \\ x_3 \end{pmatrix} = \begin{pmatrix} a_{11}x_1 + a_{12}x_2 + a_{13}x_3 \\ a_{21}x_1 + a_{22}x_2 + a_{23}x_3 \end{pmatrix} = \begin{pmatrix} 0 \\ 0 \end{pmatrix}$$

So

$$\langle \hat{x}, \hat{y} \rangle = x_1 \left(a_{11}z_1 + a_{21}z_2 \right) + x_2 \left(a_{12}z_1 + a_{22}z_2 \right) + x_3 \left(a_{13}z_1 + a_{23}z_2 \right)$$

$$= z_1 \left(a_{11}x_1 + a_{12}x_2 + a_{13}x_3 \right) + z_2 \left(a_{21}x_1 + a_{22}x_2 + a_{23}x_3 \right) = z_1 (0) + z_2 (0) = 0.$$

This proof can be generalized to the case of an $m \times n$ matrix.

A more "elegant" proof is the following.
We have for $\hat{x} \in \mathcal{N}(A)$, $\hat{y} \in \mathcal{R}(A^T)$ with $\hat{y} = A^T \hat{z}$

$$\langle \hat{x}, \hat{y} \rangle = \langle \hat{x}, A^T \hat{z} \rangle = \langle A\hat{x}, \hat{z} \rangle = \langle \hat{0}, \hat{z} \rangle = 0.$$

Corollary

(a) $\left(\mathcal{N}(A) \right)^{\perp} = \mathcal{R}(A^T)$

(b) $\mathcal{N}(A^T) = \left(\mathcal{R}(A) \right)^{\perp}$

(c) $\mathcal{R}(A) = \left(\mathcal{N}(A^T) \right)^{\perp}$

Proof

Part (a) is true because in finite dimensions $S = (S^{\perp})^{\perp}$ for any subspace S. Note that this is not true if S is a subset that is not a subspace.

Parts (b) and (c) are true by replacing A with A^T.

6.8 Projections

Definition

Suppose that V is a finite dimensional vector space. A linear transformation

$$P: V \to V$$

is a projection if

$$P^2 = P.$$

The word projection suggests a shadow. If an object makes a shadow, then observing the shadow can tell us something about the object, but not everything. If V is a vector space with basis $\{\hat{v}_1, \ldots, \hat{v}_n\}$ and $\hat{v}$ is a vector in V, then $\hat{v}$ has a unique representation

$$\hat{v} = a_1 \hat{v}_1 + \cdots + a_n \hat{v}_n.$$

Knowing some of the a_i's tells us something about $\hat{v}$. A central topic of the section is, given a vector and a subspace, find "how much" of the vector is in the subspace.

Example

Suppose that V is a vector space with basis $\{\hat{v}_1, \ldots, \hat{v}_n\}$ and U is the subspace of V whose basis is $\{\hat{v}_1, \ldots, \hat{v}_k\}$. If

$$\hat{v} = a_1 \hat{v}_1 + \cdots + a_k \hat{v}_k + a_{k+1} \hat{v}_{k+1} + \cdots + a_n \hat{v}_n,$$

then the function

$$\text{Proj}_U \hat{v} = a_1 \hat{v}_1 + \cdots + a_k \hat{v}_k$$

is a projection of $\hat{v}$ onto U.

The proof of the following theorem is left as an exercise.

Theorem 9

If U and W are the range and null spaces respectively of a projection P, then

$$P\hat{x} = \hat{x} \text{ for all } \hat{x} \in U, \quad P\hat{y} = \hat{0} \text{ for all } \hat{y} \in W$$

and

$$V = U \oplus W.$$

Furthermore, $I - P$ is a projection onto W.

We will restrict our attention to inner product spaces for the remainder of this section.

Definition

In a real (complex) inner product space, the projection P is orthogonal if the matrix representation of P is symmetric (Hermitian) with respect to some basis.

Example

If $V = \mathbb{R}^3$ with the standard basis, and if U is the subspace generated by $\{\hat{e}_1, \hat{e}_2\}$, then

$$\text{Proj}_U(x,y,z) = (x,y,0).$$

The matrix for this projection with respect to the standard basis is

$$\begin{pmatrix} 1 & 0 & 0 \\ 0 & 1 & 0 \\ 0 & 0 & 0 \end{pmatrix},$$

which is symmetric.

Our next goal is, given an inner product space V and subspace W, to find the projection matrix P of V onto W. This will give us a method of finding the distance to a subspace from a vector not in the subspace. It also gives the best approximation for a vector *not* in a subspace by a vector *in* the subspace. Both of these questions have important applications. If one wishes to ignore the theory, the formula that you need is contained in Theorem 11. Here are the major facts that enable us to accomplish that theorem.

1. If A is a matrix, then by Corollary (c.) of Theorem 8

$$\mathcal{R}(A) = \left(\mathcal{N}(A^T) \right)^{\perp}$$

so that

$$V = \mathcal{R}(A) \oplus \left(\mathcal{N}\left(A^T\right)\right)$$

2. We have shown that if A is a matrix whose columns are linearly independent, then A is a one-to-one function as a linear transformation.
3. We show in Theorem 10 that if A is a matrix whose columns are linearly independent, then $A^T A$ is an invertible matrix.
4. We show in Theorem 11 that if W is a subspace of V with basis $\{\hat{a}_1, \ldots, \hat{a}_k\}$ then the projection of V onto W is given by the matrix

$$P = A\left(A^T A\right)^{-1} A^T,$$

where A is the matrix whose columns are $\hat{a}_1, \ldots, \hat{a}_k$.

Theorem 10

Suppose A is a matrix whose columns are linearly independent. Then $A^T A$ is invertible.

Proof

We have $A^T A$ as a square matrix. Suppose there is a nonzero vector $\hat{x}$ for which

$$\left(A^T A\right)\hat{x} = \hat{0}.$$

Then

$$0 = \left\langle \left(A^T A\right)\hat{x}, \hat{x}\right\rangle = \langle A\hat{x}, A\hat{x}\rangle = \|A\hat{x}\|^2$$

so

$$A\hat{x} = \hat{0}.$$

If

$$\hat{x} = \begin{pmatrix} x_1 \\ \vdots \\ x_n \end{pmatrix}$$

then

$$A\hat{x} = x_1\hat{a}_1 + \cdots + x_n\hat{a}_n = 0,$$

where $\hat{a}_i$ is the ith column of A. But since the columns of A are linearly independent, the only way this can occur is if every $x_i = 0$, that is, $\hat{x} = \hat{0}$.

Theorem 11

Let W be the subspace spanned by the linearly independent vectors $\{\hat{a}_1, \ldots, \hat{a}_k\}$ and let A be the matrix whose columns are the vectors $\hat{a}_1, \ldots, \hat{a}_k$.
   Then the projection matrix of V onto W is

$$A\left(A^T A\right)^{-1} A^T.$$

Proof

Let $\hat{b} \in V$ and let P be the projection onto the column space of A. Since $P\hat{b}$ is in the column space of A, we have $P\hat{b} = A\hat{x}$ for some $\hat{x} \in V$. Set

$$\hat{e} = \hat{b} - P\hat{b} = \hat{b} - A\hat{x}.$$

Now

$$P\hat{e} = P\left(\hat{b} - P\hat{b}\right) = P\hat{b} - P^2\hat{b} = P\hat{b} - P\hat{b} = \hat{0}$$

Therefore, $\hat{e}$ is in $\mathcal{N}\left(A\right) = \mathcal{R}(A)^{\perp} = \mathcal{N}\left(A^T\right)$.
   So

$$\hat{0} = A^T \hat{e} = A^T\left(\hat{b} - A\hat{x}\right)$$

which gives $A^T \hat{b} = A^T A\hat{x}$.
   Since $A^T A$ is invertible, we have

$$\hat{x} = \left(A^T A\right)^{-1} A^T \hat{b}.$$

Now $P\hat{b} = A\hat{x}$, so

$$P\hat{b} = A\left(A^T A\right)^{-1} A^T \hat{b}$$

and we have

$$P = A\left(A^T A\right)^{-1} A^T.$$

Example

Let

$$\hat{v}_1 = \begin{pmatrix} 1 \\ 2 \\ 1 \\ 3 \end{pmatrix}, \quad \hat{v}_2 = \begin{pmatrix} 1 \\ 1 \\ 1 \\ 1 \end{pmatrix}, \quad \text{and} \quad \text{so } A = \begin{pmatrix} 1 & 1 \\ 2 & 1 \\ 1 & 1 \\ 3 & 1 \end{pmatrix}.$$

Then

$$A^T A = \begin{pmatrix} 15 & 7 \\ 7 & 4 \end{pmatrix}$$

$$\left(A^T A\right)^{-1} = \begin{pmatrix} \dfrac{4}{11} & \dfrac{-7}{11} \\ \dfrac{-7}{11} & \dfrac{15}{11} \end{pmatrix}$$

$$A\left(A^T A\right)^{-1} A^T = \begin{pmatrix} \dfrac{5}{11} & \dfrac{2}{11} & \dfrac{5}{11} & -\dfrac{1}{11} \\ \dfrac{2}{11} & \dfrac{3}{11} & \dfrac{2}{11} & \dfrac{4}{11} \\ \dfrac{5}{11} & \dfrac{2}{11} & \dfrac{5}{11} & -\dfrac{1}{11} \\ -\dfrac{1}{11} & \dfrac{4}{11} & -\dfrac{1}{11} & \dfrac{9}{11} \end{pmatrix}.$$

Let

$$\hat{x} = \begin{pmatrix} 6 \\ 3 \\ 2 \\ 9 \end{pmatrix}.$$

Then

$$\left[A\left(A^T A\right)^{-1} A^T\right]\hat{x} = \begin{pmatrix} \dfrac{5}{11} & \dfrac{2}{11} & \dfrac{5}{11} & -\dfrac{1}{11} \\ \dfrac{2}{11} & \dfrac{3}{11} & \dfrac{2}{11} & \dfrac{4}{11} \\ \dfrac{5}{11} & \dfrac{2}{11} & \dfrac{5}{11} & -\dfrac{1}{11} \\ -\dfrac{1}{11} & \dfrac{4}{11} & -\dfrac{1}{11} & \dfrac{9}{11} \end{pmatrix}\begin{pmatrix} 6 \\ 3 \\ 2 \\ 9 \end{pmatrix} = \begin{pmatrix} \dfrac{37}{11} \\ \dfrac{61}{11} \\ \dfrac{37}{11} \\ \dfrac{85}{11} \end{pmatrix} = \dfrac{24}{11}\begin{pmatrix} 1 \\ 2 \\ 1 \\ 3 \end{pmatrix} + \dfrac{13}{11}\begin{pmatrix} 1 \\ 1 \\ 1 \\ 1 \end{pmatrix}.$$

Let V be a finite dimensional inner product space and let W be a subspace of V. Using the Gram–Schmidt process, we can find an orthonormal basis for W, say $\{\hat{u}_1,\ldots,\hat{u}_k\}$. For $\hat{v} \in V$, we have

$$\mathrm{proj}_W\hat{v} = \langle \hat{v},\hat{u}_1 \rangle \hat{u}_1 + \cdots + \langle \hat{v},\hat{u}_n \rangle \hat{u}_n.$$

This formulation will be used to show $\mathrm{proj}_W\hat{v}$ is the vector in W that is closest to the vector $\hat{v}$.

Theorem 12

Let V be a finite dimensional inner product space with W a subspace of V. Let $\{\hat{u}_1,\ldots,\hat{u}_k\}$ be an orthonormal basis for W. Let $\hat{v} \in V$. Then

$$\hat{w}_1 = \mathrm{proj}_W\hat{v}$$

is the vector in W that is closest to $\hat{v}$ in the sense that if $\hat{x} \in W$, $\hat{x} \neq \hat{w}_1$, then $\|\hat{v} - \hat{w}_1\| < \|\hat{v} - \hat{x}\|$.

Proof

Add the vectors $\hat{u}_{k+1},\ldots,\hat{u}_n$ to the set $\{\hat{u}_1,\ldots,\hat{u}_k\}$ so that $\{\hat{u}_1,\ldots,\hat{u}_k,\hat{u}_{k+1},\ldots,\hat{u}_n\}$ is an orthonormal basis of V. Let $\hat{v}\in V$. Then

$$\hat{v} = \langle\hat{v},\hat{u}_1\rangle\hat{u}_1 + \cdots + \langle\hat{v},\hat{u}_k\rangle\hat{u}_k + \langle\hat{v},\hat{u}_{k+1}\rangle\hat{u}_{k+1} + \cdots + \langle\hat{v},\hat{u}_n\rangle\hat{u}_n.$$

Let $\hat{x}\in W$, $\hat{x}\neq\hat{w}_1$. Then

$$\hat{x} = \langle\hat{x},\hat{u}_1\rangle\hat{u}_1 + \cdots + \langle\hat{x},\hat{u}_k\rangle\hat{u}_k.$$

Now

$$\|\hat{v}-\hat{x}\| = \left\|\left(\langle\hat{v},\hat{u}_1\rangle\hat{u}_1 + \cdots + \langle\hat{v},\hat{u}_k\rangle\hat{u}_k + \langle\hat{v},\hat{u}_{k+1}\rangle\hat{u}_{k+1} + \cdots + \langle\hat{v},\hat{u}_n\rangle\hat{u}_n\right) - \left(\langle\hat{x},\hat{u}_1\rangle\hat{u}_1 + \cdots + \langle\hat{x},\hat{u}_k\rangle\hat{u}_k\right)\right\|$$

$$= \left\|\left(\langle\hat{v},\hat{u}_1\rangle - \langle\hat{x},\hat{u}_1\rangle\right)\hat{u}_1 + \cdots + \left(\langle\hat{v},\hat{u}_k\rangle - \langle\hat{x},\hat{u}_k\rangle\right)\hat{u}_k\right\| + \left\|\langle\hat{v},\hat{u}_{k+1}\rangle\hat{u}_{k+1} + \cdots + \langle\hat{v},\hat{u}_n\rangle\hat{u}_n\right\|$$

and

$$\|\hat{v}-\hat{w}_1\| = \left\|\left(\langle\hat{v},\hat{u}_1\rangle\hat{u}_1 + \cdots + \langle\hat{v},\hat{u}_k\rangle\hat{u}_k + \langle\hat{v},\hat{u}_{k+1}\rangle\hat{u}_{k+1} + \cdots + \langle\hat{v},\hat{u}_n\rangle\hat{u}_n\right) - \left(\langle\hat{v},\hat{u}_1\rangle\hat{u}_1 + \cdots + \langle\hat{v},\hat{u}_k\rangle\hat{u}_k\right)\right\|$$

$$= \left\|\langle\hat{v},\hat{u}_{k+1}\rangle\hat{u}_{k+1} + \cdots + \langle\hat{v},\hat{u}_n\rangle\hat{u}_n\right\|.$$

Thus

$$\|\hat{v}-\hat{x}\| - \|\hat{v}-\hat{w}_1\| = \left\|\left(\langle\hat{v},\hat{u}_1\rangle - \langle\hat{x},\hat{u}_1\rangle\right)\hat{u}_1 + \cdots + \left(\langle\hat{v},\hat{u}_k\rangle - \langle\hat{x},\hat{u}_k\rangle\right)\hat{u}_k\right\| \geq 0$$

and

$$\left\|\left(\langle\hat{v},\hat{u}_1\rangle - \langle\hat{x},\hat{u}_1\rangle\right)\hat{u}_1 + \cdots + \left(\langle\hat{v},\hat{u}_k\rangle - \langle\hat{x},\hat{u}_k\rangle\right)\hat{u}_k\right\| = 0$$

if and only if

$$\langle\hat{v},\hat{u}_i\rangle - \langle\hat{x},\hat{u}_i\rangle = 0, \quad i = 1,\ldots,k$$

which is true if and only if $\hat{w}_1 = \hat{x}$.

Example

(a) Find the point in the plane

$$x + y - 2z = 0$$

that is nearest to the point

$$\begin{pmatrix} 5 \\ 6 \\ 7 \end{pmatrix}.$$

We have

$$x = -y + 2z$$

so y and z are free variables. We set

$$y = s, \quad z = t$$

so a vector in the plane is

$$\begin{pmatrix} -s + 2t \\ s \\ t \end{pmatrix} = \begin{pmatrix} -s \\ s \\ 0 \end{pmatrix} + \begin{pmatrix} 2t \\ 0 \\ t \end{pmatrix} = s\begin{pmatrix} -1 \\ 1 \\ 0 \end{pmatrix} + t\begin{pmatrix} 2 \\ 0 \\ 1 \end{pmatrix}$$

and a basis for the plane is

$$\left\{ \begin{pmatrix} -1 \\ 1 \\ 0 \end{pmatrix}, \begin{pmatrix} 2 \\ 0 \\ 1 \end{pmatrix} \right\}.$$

Thus, we take

$$A = \begin{pmatrix} -1 & 2 \\ 1 & 0 \\ 0 & 1 \end{pmatrix}$$

and

$$A\left(A^T A\right)^{-1} A^T = \begin{pmatrix} \dfrac{5}{6} & -\dfrac{1}{6} & \dfrac{1}{3} \\ -\dfrac{1}{6} & \dfrac{5}{6} & \dfrac{1}{3} \\ \dfrac{1}{3} & \dfrac{1}{3} & \dfrac{1}{3} \end{pmatrix}.$$

The projection of $\begin{pmatrix} 5 \\ 6 \\ 7 \end{pmatrix}$ onto the plane is

$$\begin{pmatrix} \dfrac{5}{6} & -\dfrac{1}{6} & \dfrac{1}{3} \\ -\dfrac{1}{6} & \dfrac{5}{6} & \dfrac{1}{3} \\ \dfrac{1}{3} & \dfrac{1}{3} & \dfrac{1}{3} \end{pmatrix} \begin{pmatrix} 5 \\ 6 \\ 7 \end{pmatrix} = \begin{pmatrix} \dfrac{11}{2} \\ \dfrac{13}{2} \\ 6 \end{pmatrix}.$$

(b) The distance between $\begin{pmatrix} 5 \\ 6 \\ 7 \end{pmatrix}$ and the plane is

$$\left\| \begin{pmatrix} 5 \\ 6 \\ 7 \end{pmatrix} - \begin{pmatrix} \dfrac{11}{2} \\ \dfrac{13}{2} \\ 6 \end{pmatrix} \right\| = \left\| \begin{pmatrix} -1/2 \\ -1/2 \\ 1 \end{pmatrix} \right\| = \sqrt{\dfrac{3}{2}}.$$

Exercises

1. Find the projection of the vector

$$\hat{v} = \begin{pmatrix} 3 \\ 2 \end{pmatrix}$$

onto the subspace $W = \text{span}\left\{ \begin{pmatrix} 4 \\ -5 \end{pmatrix} \right\}$ of $\mathbb{R}^2$.

In Exercises 2 through 7, W is the subspace spanned by the set of vectors S.
(a) Find the projection of the vector $\hat{y}$ onto W.
(b) Find the distance from the point to the subspace.

2. $S = \left\{ \begin{pmatrix} 1 \\ 2 \\ 2 \end{pmatrix}, \begin{pmatrix} -1 \\ 0 \\ 2 \end{pmatrix} \right\}$ $\hat{y} = \begin{pmatrix} 3 \\ 2 \\ -4 \end{pmatrix}$.

3. $S = \left\{ \begin{pmatrix} 1 \\ -1 \\ 1 \end{pmatrix}, \begin{pmatrix} 1 \\ 0 \\ 1 \end{pmatrix} \right\}$ $\hat{y} = \begin{pmatrix} 0 \\ 6 \\ 5 \end{pmatrix}$.

4. $S = \left\{ \begin{pmatrix} 0 \\ 1 \\ 1 \end{pmatrix}, \begin{pmatrix} 1 \\ 0 \\ 1 \end{pmatrix} \right\}$ $\hat{y} = \begin{pmatrix} 8 \\ 1 \\ 3 \end{pmatrix}$.

5. $S = \left\{ \begin{pmatrix} 1 \\ 1 \\ 1 \\ 1 \end{pmatrix}, \begin{pmatrix} 4 \\ -1 \\ 4 \\ 1 \end{pmatrix} \right\}$ $\hat{y} = \begin{pmatrix} 2 \\ -3 \\ 0 \\ 0 \end{pmatrix}$.

6. $S = \left\{ \begin{pmatrix} 1 \\ 0 \\ 0 \\ 1 \end{pmatrix}, \begin{pmatrix} 0 \\ 1 \\ 1 \\ 0 \end{pmatrix} \right\}$ $\hat{y} = \begin{pmatrix} 4 \\ 2 \\ 0 \\ 7 \end{pmatrix}$.

7. $S = \left\{ \begin{pmatrix} -1 \\ 1 \\ -1 \\ 1 \end{pmatrix}, \begin{pmatrix} 1 \\ 3 \\ 1 \\ -1 \end{pmatrix} \right\}$ $\hat{y} = \begin{pmatrix} -3 \\ 4 \\ 4 \\ 1 \end{pmatrix}$.

8. Show that if P is the projection in problem 6, then $P^2 = P$ and $P^T = P$.

9. Show that if P is any projection, then $I - P$ is a projection.

10. (a) Find the projection of the point $(4, -1, 2)$ onto the plane

$$3x - 2y + z = 0.$$

(b) Find the distance between the point $(4, -1, 2)$ and the plane

$$3x - 2y + z = 0.$$

6.9 Least-Squares Estimates in Statistics (Optional)

One of the most widely used and versatile methods in statistics is linear regression. The general form of the regression model is

$$Y = \beta_0 + \beta_1 X_1 + \beta_2 X_2 + \ldots + \beta_n X_n + \varepsilon.$$

Y is called the response variable, the X_i's are explanatory or predictor variables, and ε is a random error term. The versatility comes from the possibility that some X's can be functions of others; the model is linear in the β_i's but not necessarily in the X's. Some examples of linear regression models are

$$Y = \beta_0 + \beta_1 X_1 + \beta_2 X_2 + \varepsilon$$
$$Y = \beta_0 + \beta_1 X_1 + \beta_2 X_2 + \beta_3 X_1 X_2 + \varepsilon$$
$$Y = \beta_0 + \beta_1 X_1 + \beta_2 X_2 + \beta_3 X_2^2 + \varepsilon.$$

In application, data is obtained by experiment or observation and used to produce estimates of the coefficients. A linear regression model may be used to predict Y for given values of the explanatory variables, or to describe the relationship of the explanatory variables to Y. Analysis may show which variables are useful in the model and which ones can be dropped without much loss of information. There are many excellent books about linear regression models, including Mendenhall and Sincich (2012).

The purpose of this section is to show how linear algebra provides the estimates of the coefficients in a linear regression model.

To make the notation simple, suppose we have two predictor variables, X_1 and X_2, and we have four vectors of observed data (x_{i1}, x_{i2}, y_i), $i = 1, 2, 3, 4$. Each of the data vectors exemplifies the model

$$y_i = \beta_0 + \beta_1 x_{i1} + \beta_2 x_{i2} + \varepsilon_i,$$

where ε_i is not known.

These four equations may be expressed in matrix form as

$$\hat{Y} = X\hat{\beta} + \hat{\varepsilon},$$

where

$$\hat{Y} = \begin{bmatrix} y_1 \\ y_2 \\ y_3 \\ y_4 \end{bmatrix}, \quad X = \begin{bmatrix} 1 & x_{11} & x_{12} \\ 1 & x_{21} & x_{22} \\ 1 & x_{31} & x_{32} \\ 1 & x_{41} & x_{42} \end{bmatrix}, \quad \hat{\beta} = \begin{bmatrix} \beta_0 \\ \beta_1 \\ \beta_2 \end{bmatrix}, \quad \text{and} \quad \hat{\varepsilon} = \begin{bmatrix} \varepsilon_1 \\ \varepsilon_2 \\ \varepsilon_3 \\ \varepsilon_4 \end{bmatrix}.$$

We want to find a vector value of $\hat{\beta}$ that makes the vector $\hat{\varepsilon}$ as small as possible.

We begin by ignoring $\hat{\varepsilon}$ temporarily and working on the rest of the equation, $\hat{Y} = X\hat{\beta}$, which in all likelihood does not have an exact solution. Although the matrix X can't be inverted, $X^t X$ can be, so we left-multiply both sides of the equation by $(X^t X)^{-1} X^t$:

$$\left(X^t X\right)^{-1} X^t \hat{Y} = \left(X^t X\right)^{-1} X^t X \hat{\beta} = \hat{\beta}.$$

In statistics, it is customary to denote this solution by $\hat{\beta}$, but since our vectors already have hats, we will resort to using an asterisk to denote an estimate obtained from data:

$$\hat{\beta}^* = \left(X^t X\right)^{-1} X^t \hat{Y}.$$

Using the estimated coefficients, the predicted value (also called the fitted value) of $\hat{Y}$, which we denote with an asterisk, is

$$\hat{Y}^* = X\hat{\beta}^* = X\left(X^t X\right)^{-1} X^t \hat{Y}.$$

It is now apparent that the predicted value of $\hat{Y}$ is the projection of $\hat{Y}$ into the column space of the data matrix X. Furthermore, the difference

$$\hat{E} = \hat{Y} - \hat{Y}^* = \hat{Y} - X\hat{\beta}^*$$

is the difference between the actual, observed value of $\hat{Y}$ and the value $\hat{Y}^*$ predicted by the linear model; since $\hat{\beta}^*$ is an estimate of $\hat{\beta}$, $\hat{E}$ is an estimate of $\hat{\varepsilon}$. The components of $\hat{E}$ are called residuals and $\hat{E}$ is the residual vector. The theory of projections guarantees that $\hat{E}$ is in the orthogonal complement of the column space of X, and that $\hat{E}$ is the smallest possible residual among all choices for $\hat{\beta}$.

Note that $\left|\hat{E}\right|^2 = \sum_{i=1}^{n} \left(y_i - y_i^*\right)^2$ is the sum of the squares of the residuals.

This technique for estimating $\hat{\beta}$ is called *the method of least squares*, $\hat{\beta}^*$ is called the least-squares estimate of $\hat{\beta}$, and the resulting equation $\hat{Y}^* = X\hat{\beta}^*$ is called the least-squares regression equation.

Example

Suppose our four data points are $(1, 2, 4)$, $(0, 4, -3)$, $(2, 0, 10)$, and $(3, 3, 9)$.

$$\text{Then } Y = \begin{bmatrix} 4 \\ -3 \\ 10 \\ 9 \end{bmatrix} \text{ and } X = \begin{bmatrix} 1 & 1 & 2 \\ 1 & 0 & 4 \\ 1 & 2 & 0 \\ 1 & 3 & 3 \end{bmatrix}.$$

$$\text{So } X^t X = \begin{bmatrix} 4 & 6 & 9 \\ 6 & 14 & 11 \\ 9 & 11 & 29 \end{bmatrix} \text{ and } \left(X^t X\right)^{-1} = \frac{1}{30} \begin{bmatrix} 57 & -15 & -12 \\ -15 & 7 & 2 \\ -12 & 2 & 4 \end{bmatrix}.$$

It follows that

$$\hat{\beta}^* = \left(X^tX\right)^{-1}X^t\hat{Y} = \begin{bmatrix} 33/10 \\ 103/30 \\ -23/15 \end{bmatrix} \quad \text{and} \quad \widehat{Y^*} = X\hat{\beta}^* = \begin{bmatrix} 11/3 \\ -17/6 \\ 61/6 \\ 9 \end{bmatrix} \quad \hat{E} = \hat{Y} - \widehat{Y^*} = \begin{bmatrix} 1/3 \\ -1/6 \\ -1/6 \\ 0 \end{bmatrix}.$$

The least-squares regression equation is

$$\widehat{Y^*} = (33/10) + (103/30)X_1 + (-23/15)X_2.$$

The sum of squared residuals is $\left(\frac{1}{3}\right)^2 + \left(-\frac{1}{6}\right)^2 + \left(-\frac{1}{6}\right)^2 + 0^2 = \frac{1}{6}$.
Check that $X^t\hat{E} = 0$.

Summary

The general form of the least-squares regression has k explanatory variables $X_1, X_2, \ldots, X_k$, and the response variable Y. The linear model is

$$Y = \beta_0 + \beta_1X_1 + \ldots + \beta_kX_k + \epsilon.$$

Data for estimating the coefficients of the model consists of n vectors in $\mathbb{R}^{k+1}$ of the form

$$\left(x_{i1}, x_{i2}, \ldots, x_{ik}, y_i\right), \quad 1 \le i \le n, \text{ with } n \ge k+1.$$

We construct the vector $\hat{Y} = \begin{bmatrix} y_1 \\ y_2 \\ \vdots \\ y_n \end{bmatrix}$ and the matrix $X = \begin{bmatrix} 1 & x_{11} & x_{12} & \ldots & x_{1k} \\ 1 & x_{21} & x_{22} & \ldots & x_{2k} \\ \vdots & \vdots & \vdots & \ddots & \vdots \\ 1 & x_{n1} & x_{n2} & \ldots & x_{nk} \end{bmatrix}.$

Provided that the rank of X is $k+1$, so that X^tX is invertible, the least-squares estimate of $\hat{\beta}$ is

$$\hat{\beta}^* = \left(X^tX\right)^{-1}X^t\hat{Y}.$$

Then $\widehat{Y^*} = X\hat{\beta}^*$ is the least-squares regression equation.
The residual vector is $\hat{E} = \hat{Y} - \widehat{Y^*} = \hat{Y} - X\hat{\beta}^*$
Because $\widehat{Y^*}$ is the projection of $\hat{Y}$ into the column space of X, the least-squares estimate minimizes the sum of the squared residuals.

Exercise

1. The following table gives percent body fat measured by a water immersion method, with weight (in pounds), height (in inches), and neck circumference (in centimeters) for 8 men. (Excerpted from a larger data set in Lock, Lock, Lock, Lock, Lock, *Statistics: Unlocking the Power of Data*, page 562)

Body Fat	Weight	Height	Neck
32.3	247.25	73.5	42.1
22.5	177.25	71.5	36.2
22	156.25	69	35.5
12.3	154.25	67.75	36.2
20.5	177	70	37.2
22.6	198	72	39.9
28.7	200.5	71.5	37.9
21.3	163	70.25	35.3

   a. Taking Y = Body fat, X_1 = Weight, X_2 = Height, and X_3 = Neck circumference, set up the matrix X for the linear model $\hat{Y} = X\hat{\beta} + \hat{\varepsilon}$. Using matrix operations, calculate the least-squares estimate of $\hat{\beta}$ and the vector of fitted values, $\widehat{Y^*}$.

   b. Calculate the vector $\hat{E}$ of residuals and find $\left|\hat{E}\right|^2$.

   c. Let $\hat{b}$ = [6, 0.3, 1, –2]. Compare $\hat{b}$ with the least-squares estimate of $\hat{\beta}$. Use $\hat{b}$ to calculate $\hat{e} = \hat{Y} - X\hat{b}$. Will $\left|\hat{e}\right|^2$ be larger or smaller than $\left|\hat{E}\right|^2$ from part b above? Verify this by calculation.

6.10 Weighted Inner Products (Optional)

While the Euclidean inner product is the most often used inner product, there are applications where other inner products are useful. Among these are weighted inner products that we now consider. The role of the weights is to distinguish between components if some components have more impact than others.

On $\mathbb{R}^n$, for

$$\hat{x} = \begin{pmatrix} x_1 \\ \vdots \\ x_n \end{pmatrix}, \quad \hat{y} = \begin{pmatrix} y_1 \\ \vdots \\ y_n \end{pmatrix}, \quad \hat{z} = \begin{pmatrix} z_1 \\ \vdots \\ z_n \end{pmatrix}$$

we show that the function

$$\hat{x} \cdot \hat{y} = c_1 x_1 y_1 + \cdots + c_n x_n y_n; \quad c_i > 0, \quad i = 1, \ldots, n$$

is a dot product. The "weights" are the numbers c_i. We verify that the axioms are satisfied.

1. $\hat{x} \cdot \hat{x} = c_1 x_1 x_1 + \cdots + c_n x_n x_n \geq 0$ and $\hat{x} \cdot \hat{x} = 0$ if and only if $\hat{x} = \hat{0}$.

2. $\hat{x} \cdot \hat{y} = c_1 x_1 y_1 + \cdots + c_n x_n y_n = c_1 y_1 x_1 + \cdots + c_n y_n x_n = \hat{y} \cdot \hat{x}$.

3. $\hat{x} \cdot (\hat{y} + \hat{z}) = c_1 x_1 (y_1 + z_1) \cdots + c_n x_n (y_n + z_n) = c_1 x_1 y_1 + c_1 x_1 z_1 + \cdots + c_n x_n y_n + c_n x_n z_n$

$$= c_1 x_1 y_1 + \cdots + c_n x_n y_n + c_1 x_1 z_1 + \cdots + c_n x_n z_n = \hat{x} \cdot \hat{y} + \hat{x} \cdot \hat{z}.$$

4. $a(\hat{x} \cdot \hat{y}) = a(c_1 x_1 y_1 + \cdots + c_n x_n y_n) = \left[c_1 (a x_1) \right] y_1 + \cdots + \left[c_n (a x_n) \right] y_n = (a\hat{x} \cdot \hat{y})$.

Furthermore,

$$a(\hat{x} \cdot \hat{y}) = a(c_1 x_1 y_1 + \cdots + c_n x_n y_n) = c_1 x_1 (a y_1) + \cdots + c_n x_n (a y_n) = (\hat{x} \cdot a\hat{y}).$$

Reference

Mendenhall, W. and Sincich, T.T. *A Second Course in Statistics: Regression Analysis*, 7th edn., Pearson, Upper Saddle River, NJ, 2012.

7

Linear Functionals, Dual Spaces, and Adjoint Operators*

7.1 Linear Functionals

In previous chapters, we have studied linear transformations between vector spaces. We denoted the class of linear transformations from the vector space V to the vector space W by $L(V,W)$. Since the scalar field $\mathcal{F}$ is a vector space (the simplest of all nontrivial vector spaces), one class of linear transformations from V is $L(V, \mathcal{F})$. This class of linear transformations is the subject of this section.

Definition

If V is a vector space with scalar field $\mathcal{F}$, then a linear transformation

$$T : V \to \mathcal{F}$$

is called a linear functional on V.

We denote the set of linear functionals on V by V'. Thus, in our earlier notation we are now describing $L(V, \mathcal{F})$ as V'. These linear transformations take vectors from V as inputs and give scalars as outputs.

Examples

1. Let $T : \mathbb{R}^2 \to \mathbb{R}$ be defined by $T(x, y) = 4x - 3y$.
2. Let $T : C[0,1] \to \mathbb{R}$ be defined

$$T(f) = \int_0^1 f(x)\, dx.$$

3. If V is an inner product space we can define a linear functional on V as follows: Fix a vector $\hat{w} \in V$. Define $T : V \to \mathcal{F}$ by

$$T(\hat{v}) = \langle \hat{v}, \hat{w} \rangle.$$

---

* *Note*: In this chapter, there are some subsections that are marked as optional. The material in these sections is beyond what is normally done in a first linear algebra course. Omitting them will not compromise later material.

Note that if the scalar field is the complex numbers, then

$$T(\hat{v}) = \langle \hat{w}, \hat{v} \rangle$$

would not be a linear functional but it would be if the scalar field is the real numbers. (Why?)

For $v_1', v_2' \in V'$, and $\alpha \in \mathcal{F}$, we define

$$v_1' + v_2' \in V' \quad \text{and} \quad \alpha v_1' \in V'$$

by

$$\left(v_1' + v_2'\right)(\hat{x}) = v_1'(\hat{x}) + v_2'(\hat{x}) \quad \text{for } \hat{x} \in V$$

$$\left(\alpha v_1'\right)(\hat{x}) = \alpha\left(v_1'(\hat{x})\right) \quad \text{for } \hat{x} \in V.$$

We know V' is a vector space because $L(V, W)$ is a vector space for any vector space W.

Definition

With these operations, V' is a vector space, called the dual of the vector space V.

There are actually two duals of a vector space, the topological dual and the algebraic dual. The algebraic dual is what we have just defined. The topological dual consists of continuous linear functionals (a term that we will not define). For finite dimensional vector spaces these coincide, but for infinite dimensional vector spaces the topological dual space is a proper subset of the algebraic dual space. Since we are dealing only with finite dimensions, we will simply use the term dual space.

Theorem 1

Suppose that V is a finite dimensional vector space with basis $\mathcal{B} = \{\hat{v}_1, \ldots, \hat{v}_n\}$. Let

$$v_i' : V \to \mathcal{F}; \quad i = 1, \ldots, n$$

be the linear functional defined by

$$v_i'(\hat{v}_j) = \delta_{ij} = \begin{cases} 1 & \text{if } i = j \\ \\ 0 & \text{if } i \neq j \end{cases}.$$

Then $\mathcal{B}' = \{v_1', \ldots, v_n'\}$ is a basis for V' called the basis of V' dual to the basis $\mathcal{B}$.

Proof

We first show that $\mathcal{B}'$ is a linearly independent set.

Let $0'$ denote the zero linear functional on V, that is,

$$0'(v) = 0 \quad \text{for all } v \in V$$

and suppose that

$$0' = \alpha_1 v_1' + \cdots + \alpha_n v_n'.$$

Choose $\hat{v}_j$ from the basis. Then

$$0 = 0'\left(\hat{v}_j\right) = \left(\alpha_1 v_1' + \cdots + \alpha_n v_n'\right)\hat{v}_j = \alpha_1 v_1'\left(\hat{v}_j\right) + \cdots + \alpha_j v_j'\left(\hat{v}_j\right) + \cdots + \alpha_n v_n'\left(\hat{v}_j\right)$$

$$= \alpha_1 0 + \cdots + \alpha_j v_j'\left(\hat{v}_j\right) + \cdots + \alpha_n 0 = \alpha_j.$$

Thus, $\alpha_j = 0$. Since j was arbitrary, $\mathcal{B}'$ is linearly independent.

We next show that $\mathcal{B}'$ spans V'. Choose $v' \in V'$. Now v' is determined by its effect on the elements of $\mathcal{B}$. Suppose that $v'\left(\hat{v}_j\right) = \alpha_j, j = 1, \ldots, n$. Then we claim

$$v' = \sum_{i=1}^{n} \alpha_i v_i'.$$

We demonstrate the claim by showing each has the same effect on a basis element. We have

$$\left(\sum_{i=1}^{n} \alpha_i v_i'\right)\left(\hat{v}_j\right) = \sum_{i=1}^{n} \alpha_i v_i'\left(\hat{v}_j\right) = \alpha_j$$

and $v'\left(\hat{v}_j\right) = \alpha_j$, so the claim follows.

Corollary

If the dimension of the vector space V is n, then the dimension of V' is n.

The vector spaces in this chapter will be finite dimensional inner product spaces. In that setting, the most important result is the next theorem.

Theorem 2 (Riesz Representation Theorem)

If U is a finite dimensional inner product space, then for y', a linear functional on U, there is a unique vector $\hat{y} \in U$ for which

$$y'\left(\hat{x}\right) = \langle \hat{x}, \hat{y} \rangle \quad \text{for all } \hat{x} \in U.$$

Proof

Let $y' \in U'$ and let $\{\hat{u}_1, \ldots, \hat{u}_n\}$ be an orthonormal basis of U (which is possible by the Gram–Schmidt process). Suppose that $y'\left(\hat{u}_i\right) = \alpha_i$. Let $\hat{y} = \overline{\alpha_1}\hat{u}_1 + \cdots + \overline{\alpha_n}\hat{u}_n$. If $\hat{x} = \beta_1\hat{u}_1 + \cdots + \beta_n\hat{u}_n$, then

$$y'\left(\hat{x}\right) = y'\left(\beta_1\hat{u}_1 + \cdots + \beta_n\hat{u}_n\right) = \beta_1 y'\left(\hat{u}_1\right) + \cdots + \beta_n y'\left(\hat{u}_n\right) = \beta_1\alpha_1 + \cdots + \beta_n\alpha_n.$$

But

$$\langle \hat{x}, \hat{y} \rangle = \langle \beta_1 \hat{u}_1 + \cdots + \beta_n \hat{u}_n, \; \overline{\alpha_1} \hat{u}_1 + \cdots + \overline{\alpha_n} \hat{u}_n \rangle,$$

which, by the orthonormality of $\{\hat{u}_1, \ldots, \hat{u}_n\}$, is equal to

$$\langle \beta_1 \hat{u}_1, \overline{\alpha_1} \hat{u}_1 \rangle + \cdots + \langle \beta_1 \hat{u}_1, \overline{\alpha_n} \hat{u}_n \rangle + \langle \beta_2 \hat{u}_2, \overline{\alpha_1} \hat{u}_1 \rangle + \cdots + \langle \beta_2 \hat{u}_2, \overline{\alpha_n} \hat{u}_n \rangle + \cdots$$
$$+ \langle \beta_n \hat{u}_n, \overline{\alpha_1} \hat{u}_1 \rangle + \cdots + \langle \beta_n \hat{u}_n, \overline{\alpha_n} \hat{u}_n \rangle = \beta_1 \alpha_1 + \cdots + \beta_n \alpha_n.$$

To show uniqueness, suppose that there are $\hat{y}_1, \hat{y}_2 \in U$ for which

$$y'(\hat{x}) = \langle \hat{x}, \hat{y}_1 \rangle = \langle \hat{x}, \hat{y}_2 \rangle$$

for all $\hat{x} \in U$.
   Then

$$0 = \langle \hat{x}, \hat{y}_1 \rangle - \langle \hat{x}, \hat{y}_2 \rangle = \langle \hat{x}, \hat{y}_1 - \hat{y}_2 \rangle$$

for all $\hat{x} \in U$. Taking $\hat{x} = \hat{y}_1 - \hat{y}_2$, we have

$$0 = \langle \hat{x}, \hat{y}_1 - \hat{y}_2 \rangle = \langle \hat{y}_1 - \hat{y}_2, \hat{y}_1 - \hat{y}_2 \rangle = \| \hat{y}_1 - \hat{y}_2 \|^2$$

so $\hat{y}_1 = \hat{y}_2$.
   In the proof of the theorem, it was demonstrated how to find the vector $\hat{y}$ from the linear functional y'. The corollary that follows highlights this connection.

Corollary

Let U be a vector space with orthonormal basis $\{\hat{u}_1, \ldots, \hat{u}_n\}$. Let $y' \in U'$ with

$$y'(\hat{u}_i) = \alpha_i.$$

Then the vector

$$\hat{y} = \overline{\alpha_1} \hat{u}_1 + \cdots + \overline{\alpha_n} \hat{u}_n$$

satisfies

$$y'(\hat{x}) = \langle \hat{x}, \hat{y} \rangle \quad \text{for all } \hat{x} \in U.$$

Example

Let $T : \mathbb{C}^3 \to \mathbb{C}$ be defined by

$$T(x_1, x_2, x_3) = (2 + 3i)x_1 + 4x_2 + 2ix_3.$$

So T is y' in the theorem.

We want to find the vector $\hat{y}$ for which

$$T(x_1, x_2, x_3) = \langle \hat{x}, \hat{y} \rangle = x_1 \overline{y_1} + x_2 \overline{y_2} + x_3 \overline{y_3}.$$

This will be true if

$$\overline{y_1} = (2 + 3i), \quad \overline{y_2} = 4, \quad \overline{y_3} = 2i$$

so

$$y_1 = 2 - 3i, \quad y_2 = 4, \quad y_3 = -2i$$

or

$$\hat{y} = \begin{pmatrix} 2 - 3i \\ 4 \\ -2i \end{pmatrix}.$$

What occurs in this example highlights what the proof of Theorem 2 says. Namely, the components of $\hat{y}$ are the complex conjugates of the coefficients of x_i in the definition of $T(\hat{x})$.

7.1.1 The Second Dual of a Vector Space (Optional)

Since V' is a vector space, it has a dual space that is denoted V''. The elements of V'' are linear functionals on V'. In this section, we describe the relationship between V and V''. We can use the elements of V to build the elements of V'', as we now describe.

Let V be a vector space. Fix $x_0 \in V$. For each $y \in V'$, we have $y(x_0) \in \mathcal{F}$. Also for each $z \in V''$ and each $y \in V'$, we have $z(y) \in \mathcal{F}$. We demonstrate that if V is a finite dimensional vector space, then there is a special relationship between V and V''. To that end, for the fixed $x_0 \in V$ we seek a unique $z_0 \in V''$ for which $y(x_0) = z_0(y)$ for every $y \in V'$.

Let

$\{e_1, \dots, e_n\}$ be a basis for V

$\{e_1', \dots, e_n'\}$ be the basis for V', where $e_i'(e_j) = \delta_{ij}$

and

$\{e_1'', \dots, e_n''\}$ be the basis for V'', where $e_i''(e_j') = \delta_{ij}$

Let the fixed $x_0 \in V$ that we know be given by

$$x_0 = a_1 e_1 + \cdots + a_n e_n$$

and the $z_0 \in V''$ to be determined be given by

$$z_0 = c_1 e_1'' + \cdots + c_n e_n''.$$

An element $y \in V'$ can be represented as

$$y = b_1 e_1' + \cdots + b_n e_n'.$$

Thus,

$$y(x_0) = a_1b_1 + \cdots + a_nb_n$$

and

$$z_0(y) = b_1c_1 + \cdots + b_nc_n$$

so the condition that

$$y(x_0) = z_0(y) \quad \text{for every } y \in V'$$

can be expressed as

$$a_1b_1 + \cdots + a_nb_n = b_1c_1 + \cdots + b_nc_n$$

for every choice of b_i, $i = 1, \ldots, n$.
  This means that $a_i = c_i$, $i = 1, \ldots, n$.
  The function

$$\Phi : V \to V''$$

defined by

$$\Phi(a_1e_1 + \cdots + a_ne_n) = a_1e_1'' + \cdots + a_ne_n''$$

is one-to-one and onto and satisfies

$$\Phi(a_1x_1 + a_2x_2) = a_1\Phi(x_1) + a_2\Phi(x_2).$$

This says that V and V'' are (naturally) isomorphic.

Exercises

1. For each of the linear functionals $T:V \to \mathcal{F}$ given here, find the vector $\hat{w}$ for which

$$T(\hat{v}) = \langle \hat{v}, \hat{w} \rangle.$$

(a) $T:\mathbb{R}^2 \to \mathbb{R}$; $T(x,y) = 4x - 2y$
(b) $T:\mathbb{R}^3 \to \mathbb{R}$; $T(x,y,z) = 3x - 7z$
(c) $T:\mathbb{R}^3 \to \mathbb{R}$; $T(x,y,z) = x$
(d) $T:\mathbb{C}^2 \to \mathbb{C}$; $T(x,y) = (2-4i)x - 8y$
(e) $T:\mathbb{C}^3 \to \mathbb{C}$; $T(x,y,z) = ix + (3-2i)y + (5+4i)z$
(f) $T:\mathbb{C}^3 \to \mathbb{C}$; $T(x,y,z) = (3-i)x + (7+5i)y + (1-2i)z.$

2. Suppose $T: \mathbb{R}^2 \to \mathbb{R}$; $T(x, y) = 5x + 3y$. Find the vector $\hat{w}$ for which

$$T(\hat{v}) = \langle \hat{v}, \hat{w} \rangle.$$

(a) In the standard basis.

(b) In the basis

$$\mathcal{B} = \left\{ \begin{pmatrix} 3 \\ 1 \end{pmatrix} \begin{pmatrix} -2 \\ 5 \end{pmatrix} \right\}.$$

(c) Find $T \begin{pmatrix} 6 \\ 1 \end{pmatrix}$ in the standar basis.

(d) Find $T \begin{pmatrix} 6 \\ 1 \end{pmatrix}$ in the basis $\mathcal{B}$.

3. Suppose $T: \mathbb{R}^3 \to \mathbb{R}$; $T(x, y, z) = 5x + 3y - 4z$. Find the vector $\hat{w}$ for which

$$T(\hat{v}) = \langle \hat{v}, \hat{w} \rangle.$$

(a) In the basis $\mathcal{B} = \left\{ \begin{pmatrix} 1 \\ 1 \\ 1 \end{pmatrix} \begin{pmatrix} 1 \\ 1 \\ 0 \end{pmatrix} \begin{pmatrix} 2 \\ 0 \\ 0 \end{pmatrix} \right\}.$

(b) Find $T \begin{pmatrix} a \\ b \\ c \end{pmatrix}$ in the basis $\mathcal{B}$.

7.2 The Adjoint of a Linear Operator

The adjoint of a linear operator has classically been defined for all vector spaces (see Halmos, *Finite Dimensional Vector Spaces*), but the most common applications occur with inner product spaces. It has become common in the literature to either differentiate between the adjugate of a linear transformation on a vector space and reserve the term adjoint for inner product spaces, or to only deal with inner product spaces. We will only consider inner product spaces.

7.2.1 The Adjoint Operator

In this section, V and W are finite dimensional inner product spaces and $T: V \to W$ is a linear transformation. We want to find a linear transformation

$$T^*: W \to V$$

that satisfies

$$\langle T\hat{v}, \hat{w} \rangle_W = \langle \hat{v}, T^*\hat{w} \rangle_V$$

for every $\hat{v} \in V$, $\hat{w} \in W$.

The notation $\langle\,,\,\rangle_W$ and $\langle\,,\,\rangle_V$ is used to emphasize that V and W may have different inner products.

The major tool in this construction will be the Riesz representation theorem.

To establish some intuition, we consider an example that has the features of the abstract theory.

Let

$$T : \mathbb{R}^3 \to \mathbb{R}^2$$

be defined by

$$T\begin{pmatrix} x \\ y \\ z \end{pmatrix} = \begin{pmatrix} 3x + 2y - z \\ -4x + 3y \end{pmatrix}.$$

For a given $\hat{w} \in \mathbb{R}^2$, we can define a linear functional

$$T_{\hat{w}} : \mathbb{R}^3 \to \mathbb{R}$$

by

$$T_{\hat{w}}\begin{pmatrix} x \\ y \\ z \end{pmatrix} = \langle T\hat{v}, \hat{w}\rangle_{\mathbb{R}^2}.$$

If we take $\hat{w} = \begin{pmatrix} -1 \\ 4 \end{pmatrix}$ in our example, we get

$$\langle T\hat{v}, \hat{w}\rangle_{\mathbb{R}^2} = \left\langle T\begin{pmatrix} x \\ y \\ z \end{pmatrix}, \begin{pmatrix} -1 \\ 4 \end{pmatrix}\right\rangle_{\mathbb{R}^2} = \left\langle \begin{pmatrix} 3x + 2y - z \\ -4x + 3y \end{pmatrix}, \begin{pmatrix} -1 \\ 4 \end{pmatrix}\right\rangle_{\mathbb{R}^2}$$

$$= -3x - 2y + z - 16x + 12y = -19x + 10y + z = T_{\hat{w}}\begin{pmatrix} x \\ y \\ z \end{pmatrix}.$$

Thus,

$$T_{\hat{w}} : \mathbb{R}^3 \to \mathbb{R}.$$

By the Riesz representation theorem, associated with $T_{\hat{w}}$ is a unique vector in $\mathbb{R}^3$ that we denote $T^*(\hat{w})$ for which

$$T_{\hat{w}}(\hat{v}) = \langle \hat{v}, T^*(\hat{w})\rangle_{\mathbb{R}^3}.$$

Thus,

$$\langle T(\hat{v}), \hat{w}\rangle_{\mathbb{R}^2} = \langle \hat{v}, T^*(\hat{w})\rangle_{\mathbb{R}^3}$$

Definition

The function $T^*: W \to V$ defined earlier is called the adjoint of T.
  We now

  (i)   Show T^* is a linear transformation
  (ii) Show how to find the representation of T^* given the representation of T

Theorem 3

Let V and W be finite dimensional inner product spaces and

$$T : V \to W$$

be a linear transformation. The function $T^*: W \to V$ defined by

$$\left\langle T(\hat{v}), \hat{w} \right\rangle_W = \left\langle \hat{v}, T^*(\hat{w}) \right\rangle_V$$

is a linear transformation.

Proof

Let $\hat{w}_1, \hat{w}_2 \in W$, and $\alpha \in \mathcal{F}$. We show

$$T^*\left(\hat{w}_1 + \hat{w}_2\right) = T^*\left(\hat{w}_1\right) + T^*\left(\hat{w}_2\right).$$

For $\hat{v} \in V$, we have

$$\left\langle \hat{v}, T^*\left(\hat{w}_1 + \hat{w}_2\right) \right\rangle_V = \left\langle T\hat{v}, \hat{w}_1 + \hat{w}_2 \right\rangle_W = \left\langle T\hat{v}, \hat{w}_1 \right\rangle_W + \left\langle T\hat{v}, \hat{w}_2 \right\rangle_W$$
$$= \left\langle \hat{v}, T^*\left(\hat{w}_1\right) \right\rangle_V + \left\langle \hat{v}, T^*\left(\hat{w}_2\right) \right\rangle_V$$

and so by the uniqueness of the Riesz representation, it follows that

$$T^*\left(\hat{w}_1 + \hat{w}_2\right) = T^*\left(\hat{w}_1\right) + T^*\left(\hat{w}_2\right).$$

We next show $T^*\left(\alpha\hat{w}_1\right) = \alpha T^*\left(\hat{w}_1\right)$.
  We have

$$\left\langle \hat{v}, T^*\left(\alpha\hat{w}_1\right) \right\rangle_V = \left\langle T\hat{v}, \alpha\hat{w}_1 \right\rangle_W = \bar{\alpha}\left\langle T\hat{v}, \hat{w}_1 \right\rangle_W = \bar{\alpha}\left\langle \hat{v}, T^*\left(\hat{w}_1\right) \right\rangle_V = \left\langle \hat{v}, \alpha T^*\left(\hat{w}_1\right) \right\rangle_V$$

and again by the uniqueness of the Riesz representation, it follows that

$$T^*\left(\alpha\hat{w}_1\right) = \alpha T^*\left(\hat{w}_1\right).$$

We want to determine a formula for T^* in terms of T. The following example shows how to do this. In our subsequent discussion, we assume that the inner product is the usual inner product for the different vector spaces.

Example

Suppose $T:\mathbb{R}^3 \to \mathbb{R}^2$ is defined by

$$T(x,y,z) = (2x - 5y + 2z, 3x + 4y - 5z)$$

so that the matrix representation of T is

$$\begin{pmatrix} 2 & -5 & 2 \\ 3 & 4 & -5 \end{pmatrix}$$

We find $T^*\hat{w}$ by solving $\langle T\hat{v}, \hat{w}\rangle = \langle \hat{v}, T^*\hat{w}\rangle$.

Let $\hat{v} = (x,y,z)$ and $\hat{w} = (a,b)$.
We have

$$\langle T\hat{v}, \hat{w}\rangle = \langle (2x - 5y + 2z, 3x + 4y - 5z), (a,b)\rangle = a(2x - 5y + 2z) + b(3x + 4y - 5z).$$

We want to express

$$a(2x - 5y + 2z) + b(3x + 4y - 5z)$$

as

$$xf(a,b) + yg(a,b) + zh(a,b).$$

Since $T^*\hat{w} \in V$ this will allow us to express our answer in the proper form, which is

$$T^*\hat{w} = (f(a,b), g(a,b), h(a,b)).$$

Now

$$a(2x - 5y + 2z) + b(3x + 4y - 5z) = x(2a + 3b) + y(-5a + 4b) + z(2a - 5b)$$

so

$$T^*\hat{w} = T^*(a,b) = (2a + 3b, -5a + 4b, 2a - 5b).$$

Notice that the matrix representation of T^* is

$$\begin{pmatrix} 2 & 3 \\ -5 & 4 \\ 2 & -5 \end{pmatrix},$$

which is the transpose of the matrix T.

We show later that if the scalar field is $\mathbb{R}$ then $T^* = T^T$ and if the scalar field is $\mathbb{C}$ then the representation of the adjoint is the matrix whose entries are the complex conjugate of the entries of the transpose of T.

We next formally address the question of finding the matrix representation of T^* with respect to a particular basis. We reiterate a point we have made several times earlier that while a linear transformation does not change with a change in basis, the matrix representation of the transformation does. The following "equation" might be helpful in keeping this straight:

$$\text{Linear transformation} + \text{basis} \Rightarrow \text{matrix representation}.$$

We first relate the matrix representation of T^* to that of T if we have the usual basis for both vector spaces.

Theorem 4

Let U and V be finite dimensional inner product spaces over $\mathbb{C}$ with the usual inner product. Let

$$A: U \to V$$

be a linear transformation with matrix

$$A = \begin{pmatrix} a_{11} & a_{12} & \cdots & a_{1n} \\ a_{21} & a_{22} & \cdots & a_{2n} \\ \vdots & \vdots & & \vdots \\ a_{m1} & a_{m2} & \cdots & a_{mn} \end{pmatrix}.$$

If the i,j entry of A is a_{ij}, then the i,j entry of the adjoint of A is $\overline{a_{ji}}$.

Proof

Let

$$A = \begin{pmatrix} a_{11} & a_{12} & \cdots & a_{1n} \\ a_{21} & a_{22} & \cdots & a_{2n} \\ \vdots & \vdots & & \vdots \\ a_{m1} & a_{m2} & \cdots & a_{mn} \end{pmatrix} \quad A^* = \begin{pmatrix} b_{11} & b_{12} & \cdots & b_{1m} \\ b_{21} & b_{22} & \cdots & b_{2m} \\ \vdots & \vdots & & \vdots \\ b_{n1} & b_{n2} & \cdots & b_{nm} \end{pmatrix}$$

$$\hat{x} = \begin{pmatrix} x_1 \\ \vdots \\ x_n \end{pmatrix} \quad \text{and} \quad \hat{y} = \begin{pmatrix} y_1 \\ \vdots \\ y_m \end{pmatrix}.$$

Then

$$A\hat{x} = \begin{pmatrix} a_{11}x_1 + \cdots + a_{1n}x_n \\ \vdots \\ a_{m1}x_1 + \cdots + a_{mn}x_n \end{pmatrix}$$

and

$$\langle A\hat{x}, \hat{y} \rangle = \left(a_{11}x_1 + \cdots + a_{1n}x_n \right)\overline{y_1} + \cdots + \left(a_{m1}x_1 + \cdots + a_{mn}x_n \right)\overline{y_m}.$$

Also

$$A^* \hat{y} = \begin{pmatrix} b_{11}y_1 + \cdots + b_{1m}y_m \\ \vdots \\ b_{1n}y_1 + \cdots + b_{nm}y_m \end{pmatrix}$$

and

$$\langle \hat{x}, A^* \hat{y} \rangle = x_1 \left(\overline{b_{11}y_1 + \cdots + b_{1m}y_m} \right) + \cdots + x_n \left(\overline{b_{n1}y_1 + \cdots + b_{nm}y_m} \right)$$

$$= \left(x_1 \overline{b_{11}} + x_2 \overline{b_{21}} + \cdots + x_n \overline{b_{n1}} \right) \overline{y_1} + \cdots + \left(x_1 \overline{b_{1m}} + \cdots + x_n \overline{b_{nm}} \right) \overline{y_m}$$

for all $x_i, y_j; i = 1, \ldots, n, j = 1, \ldots, m$. Thus,

$$b_{ij} = \overline{a}_{ji}; \quad i, j = 1, \ldots, n.$$

Notation

The matrix A^* is the conjugate transpose of the matrix A, and it is obtained by first taking the transpose of A and then taking the complex conjugate of each entry of the resulting matrix.

Example

Suppose $T: \mathbb{C}^3 \to \mathbb{C}^2$ is defined by

$$T(x, y, z) = \left(3x + (2 + 5i)y + iz, (2 - 4i)x + (1 - 3i)y + 9z \right).$$

The matrix of T in the standard basis is

$$\begin{pmatrix} 3 & 2 + 5i & i \\ 2 - 4i & 1 - 3i & 9 \end{pmatrix}$$

so the matrix representation of T^* in the standard basis is

$$A^* = \begin{pmatrix} 3 & 2 + 4i \\ 2 - 5i & 1 + 3i \\ -i & 9 \end{pmatrix}.$$

Not surprisingly, the situation when other than the usual bases are used is not as simple. However, if orthonormal bases are used, it is still manageable. This is because if $\{\hat{v}_1, \ldots, \hat{v}_n\}$ is an orthonormal basis for V and $\hat{v} \in V$ then

$$\hat{v} = \langle \hat{v}, \hat{v}_1 \rangle \hat{v}_1 + \cdots + \langle \hat{v}, \hat{v}_n \rangle \hat{v}_n$$

so that the representation of $\hat{v}$ as a column vector in the basis $\{\hat{v}_1, \ldots, \hat{v}_n\}$ is

$$\begin{pmatrix} \langle \hat{v}, \hat{v}_1 \rangle \\ \vdots \\ \langle \hat{v}, \hat{v}_n \rangle \end{pmatrix}.$$

This fact, together with the work we did on representing a linear transformation in other than the usual bases in Section 6.5, gives the following result.

Theorem 5

If V and W are finite dimensional inner product spaces with orthonormal bases $\mathcal{B} = \{\hat{v}_1, \ldots, \hat{v}_n\}$ and $\mathcal{C} = \{\hat{w}_1, \ldots, \hat{w}_m\}$, respectively and

$$T: V \to W$$

is a linear transformation, then the matrix representation of T with respect to these bases is

$$A = P_{\mathcal{C}}^{-1} T P_{\mathcal{B}}.$$

where $P_{\mathcal{B}}$ and $P_{\mathcal{C}}$ are the transition matrices for $\mathcal{B}$ and $\mathcal{C}$, respectively.

With an argument identical to that of Theorem 3, we get the following.

Corollary

The matrix for T^* under the hypotheses of Theorem 4 is the conjugate transpose of the matrix A. That is, if

$$T^*: W \to V$$

with orthonormal bases $\mathcal{B}$ of V and $\mathcal{C}$ of W, then the matrix of T^* with respect to these bases is A^* where

$$\left(A^*\right)_{ij} = \overline{A}_{ji}.$$

7.2.2 Adjoint on Weighted Inner Product Spaces (Optional)

The adjoint of a linear operator depends on the inner product as well as the operator. Consider the following example.

Define an inner product on $\mathbb{R}^2$ by

$$\langle \hat{x} \cdot \hat{y} \rangle = 2x_1 y_1 + x_2 y_2$$

and let

$$A = \begin{pmatrix} a_{11} & a_{12} \\ a_{21} & a_{22} \end{pmatrix}.$$

So

$$\langle A\hat{x} \cdot \hat{y} \rangle = \left\langle \begin{pmatrix} a_{11}x_1 + a_{12}x_2 \\ a_{21}x_1 + a_{22}x_2 \end{pmatrix}, \left(\begin{pmatrix} y_1 \\ y_2 \end{pmatrix} \right) \right\rangle$$
$$= 2\left(a_{11}x_1 + a_{12}x_2\right)y_1 + \left(a_{21}x_1 + a_{22}x_2\right)y_2.$$

Suppose that

$$B = \begin{pmatrix} b_{11} & b_{12} \\ b_{21} & b_{22} \end{pmatrix}$$

is the adjoint of A. Then

$$\langle \hat{x}, B\hat{y} \rangle = \left\langle \begin{pmatrix} x_1 \\ x_2 \end{pmatrix}, \begin{pmatrix} b_{11}y_1 + b_{12}y_2 \\ b_{21}y_1 + b_{22}y_2 \end{pmatrix} \right\rangle$$
$$= 2x_1 \left(b_{11}y_1 + b_{12}y_2 \right) + x_2 \left(b_{21}y_1 + b_{22}y_2 \right).$$

This gives

$$2a_{11}x_1y_1 = 2b_{11}x_1y_1 \quad \text{so } a_{11} = b_{11}$$
$$2a_{12}x_2y_1 = b_{21}x_2y_1 \quad \text{so } 2a_{12} = b_{21}$$
$$a_{21}x_1y_2 = 2b_{12}x_1y_2 \quad \text{so } a_{21} = 2b_{12} \quad \text{or} \quad b_{12} = \frac{a_{21}}{2}$$
$$a_{22}x_2y_2 = b_{22}x_2y_2 \quad \text{so } a_{22} = b_{22}$$

Thus, the adjoint of A is

$$A' = \begin{pmatrix} b_{11} & b_{12} \\ b_{21} & b_{22} \end{pmatrix} = \begin{pmatrix} a_{11} & \dfrac{a_{21}}{2} \\ 2a_{12} & a_{22} \end{pmatrix}.$$

Exercises

1. For the linear transformations given below, find (i) $T^*\hat{w}$, (ii) the matrix representation of T in the usual basis, and (iii) the matrix representation of T^* in the usual basis.

   (a) $T: \mathbb{R}^3 \rightarrow \mathbb{R}^2$ is defined by

   $$T(x,y,z) = (x + 2y - 3z, 3x - 5z).$$

   (b) $T: \mathbb{R}^2 \rightarrow \mathbb{R}^4$ is defined by

   $$T(x,y) = (4y, 3x + 2z, x + 2y - 6z, 2x).$$

   (c) $T: \mathbb{R} \rightarrow \mathbb{R}^2$ is defined by

   $$T(x) = (4x, -3x).$$

   (d) $T: \mathbb{C}^3 \rightarrow \mathbb{C}^2$ is defined by

   $$T(x,y,z) = \left((2-3i)x + (2-4i)y + (1-3i)y + iz, 4x - 3iy + z \right).$$

(e) $T:\mathbb{C}^2 \rightarrow \mathbb{C}^3$ is defined by

$$T(x,y) = \left(3x + (4+5i)y, 6ix - 2iy, (6-2i)x + (1-i)y\right).$$

(f) $T:\mathbb{C}^3 \rightarrow \mathbb{C}^3$ is defined by

$$T(x,y,z) = \left(ix + y, 4iy - 2z, x + (3+i)z\right).$$

2. (a) If
$T:\mathbb{R}^2 \rightarrow \mathbb{R}^3$ is defined by

$$T(x,y) = (x - 2y, 2x + 4y, 3x)$$

find $T_{\hat{w}}$ for
   (i) $\hat{w} = (1, -2, 0)$
   (ii) $\hat{w} = (-3, 5, -2)$

(b) If
$T:\mathbb{R}^3 \rightarrow \mathbb{R}^2$ is defined by

$$T(x,y,z) = (2x - 2y + 3z, x + 4y + z)$$

find $T_{\hat{w}}$ for
   (i) $\hat{w} = (-4, -2)$
   (ii) $\hat{w} = (1,5)$

3. Show if $T:U \rightarrow V$ is a linear transformation then
   (a) $(T^*)^* = T$
   (b) $(aT)^* = \bar{a}(T^*)$

4. Recall that if $T:U \rightarrow V$ is a linear transformation then

$$\mathcal{N}(T) = \left\{\hat{u} \in U \mid T\hat{u} = \hat{0}\right\}, \quad \mathcal{R}(T) = \left\{\hat{v} \in V \mid \hat{v} = T\hat{u} \text{ for some } \hat{u} \in U\right\}$$

Show that
   (a) $\mathcal{N}(T^*) = \left(\mathcal{R}(T)\right)^{\perp}$
   (b) $\mathcal{R}(T^*) = \left(\mathcal{N}(T)\right)^{\perp}$
   (c) $\mathcal{N}(T) = \left(\mathcal{R}(T)^*\right)^{\perp}$
   (d) $\mathcal{R}(T) = \left(\mathcal{N}(T^*)\right)^{\perp}$
   Compare with Theorem 9, Chapter 6.

5. Let $T:U \rightarrow V$ be a linear transformation. Show
   (a) T is one-to-one if and only if T^* is onto.
   (b) T is onto if and only if T^* is one-to-one.

7.3 The Spectral Theorem

In some areas of science, including mathematical physics, eigenvectors on inner product spaces are of central importance.

Of particular interest are symmetric matrices when the vector space is $\mathbb{R}^n$ and Hermitian matrices when the vector space is $\mathbb{C}^n$.

We repeat a definition.

Definition

A symmetric matrix A is an $n \times n$ matrix for which $A^T = A$, that is, $A_{ij} = A_{ji}$ $i, j = 1, \ldots, n$.

A Hermitian matrix A is an $n \times n$ matrix for which $\bar{A}_{ji} = A_{ij}$ $i, j = 1, \ldots, n$.

Note that if the scalar field is $\mathbb{R}$, then a symmetric matrix has the property that $A = A^*$, and if the scalar field is $\mathbb{C}$, then a Hermitian matrix has the property that $A = A^*$.

Also, note that the main diagonal entries of a Hermitian matrix are real.

Definition

A linear operator A on the inner product space V is self-adjoint if $A = A^*$, that is, if

$$\langle A\hat{v}, \hat{w} \rangle = \langle \hat{w}, A\hat{v} \rangle \quad \text{for all } \hat{v}, \hat{w} \in V.$$

Theorem 6

Let $V = \mathcal{F}^n$ be an inner product space and let A be a symmetric matrix. Then

1. The eigenvalues of A are real.
2. Eigenvectors corresponding to different eigenvalues are orthogonal.

Proof

1. Suppose that λ is an eigenvalue of A with eigenvector $\hat{v}$. Then

$$\lambda \|\hat{v}\|^2 = \lambda \langle \hat{v}, \hat{v} \rangle = \langle \lambda \hat{v}, \hat{v} \rangle = \langle A\hat{v}, \hat{v} \rangle = \langle \hat{v}, A^T \hat{v} \rangle = \langle \hat{v}, A\hat{v} \rangle = \langle \hat{v}, \lambda \hat{v} \rangle = \bar{\lambda} \langle \hat{v}, \hat{v} \rangle = \bar{\lambda} \|\hat{v}\|^2$$

and since $\|\hat{v}\|^2 \neq 0$, we have $\lambda = \bar{\lambda}$.

2. Suppose that λ is an eigenvalue of A with eigenvector $\hat{v}$ and μ is an eigenvalue of A with eigenvector $\hat{u}$. (So λ and μ are real.) Then

$$\lambda \langle \hat{v}, \hat{u} \rangle = \langle \lambda \hat{v}, \hat{u} \rangle = \langle A\hat{v}, \hat{u} \rangle = \langle \hat{v}, A\hat{u} \rangle = \langle \hat{v}, \mu \hat{u} \rangle = \mu \langle \hat{v}, \hat{u} \rangle.$$

Thus,

$$\lambda \langle \hat{v}, \hat{u} \rangle - \mu \langle \hat{v}, \hat{u} \rangle = (\lambda - \mu) \langle \hat{v}, \hat{u} \rangle = 0$$

but

$$(\lambda - \mu) \neq 0 \quad \text{so } \langle \hat{v}, \hat{u} \rangle = 0.$$

Corollary

If A is a self-adjoint matrix, then

1. The eigenvalues of A are real.
2. Eigenvectors corresponding to different eigenvalues are orthogonal.

The spectral theorem, which we now study, is arguably the most important theorem in linear algebra. The proof of the spectral theorem in the real case is (perhaps surprisingly) more difficult than in the complex case. The reason is that the proof requires that we know that a symmetric matrix has an eigenvalue. We know this is true in the complex case because, by the fundamental theorem of algebra, the characteristic polynomial factors into linear factors over the complex numbers. To prove the spectral theorem in the real case, we need the next two theorems.

Theorem 7

Suppose that A is a symmetric matrix. If α and β are real numbers with $\alpha^2 < 4\beta$, then $A^2 + \alpha A + \beta I$ is invertible.

Proof

Let $\hat{v}$ be a nonzero vector. Now

$$\left\langle \left(A^2 + \alpha A + \beta I\right)\hat{v}, \hat{v}\right\rangle = \left\langle A^2 v, \hat{v}\right\rangle + \alpha\left\langle A\hat{v}, \hat{v}\right\rangle + \beta\|\hat{v}\|^2 = \left\langle A\hat{v}, A^T\hat{v}\right\rangle + \alpha\left\langle A\hat{v}, \hat{v}\right\rangle + \beta\|\hat{v}\|^2$$

$$= \left\langle A\hat{v}, A\hat{v}\right\rangle + \alpha\left\langle A\hat{v}, \hat{v}\right\rangle + \beta\|\hat{v}\|^2 = \|A\hat{v}\|^2 + \alpha\left\langle A\hat{v}, \hat{v}\right\rangle + \beta\|\hat{v}\|^2$$

since A is symmetric.

By the Cauchy–Schwartz inequality

$$|\alpha|\|A\hat{v}\|\|\hat{v}\| \geq \alpha\left\langle A\hat{v}, \hat{v}\right\rangle$$

so

$$\|A\hat{v}\|^2 + \alpha\left\langle A\hat{v}, \hat{v}\right\rangle + \beta\|\hat{v}\|^2 \geq \|A\hat{v}\|^2 - |\alpha|\|A\hat{v}\|\|\hat{v}\| + \beta\|\hat{v}\|^2.$$

Now

$$\|A\hat{v}\|^2 - |\alpha|\|A\hat{v}\|\|\hat{v}\| + \beta\|\hat{v}\|^2 = \left(\|A\hat{v}\|^2 - |\alpha|\|A\hat{v}\|\|\hat{v}\| + \frac{\alpha^2}{4}\|\hat{v}\|^2\right) + \left(\beta\|\hat{v}\|^2 - \frac{\alpha^2}{4}\|\hat{v}\|^2\right)$$

$$= \left(\|A\hat{v}\| - \frac{|\alpha|}{2}\|\hat{v}\|\right)^2 + \left(\beta - \frac{\alpha^2}{4}\right)\|\hat{v}\|^2$$

and

$$\left(\|A\hat{v}\| - \frac{|\alpha|}{2}\|\hat{v}\|\right)^2 + \left(\beta - \frac{\alpha^2}{4}\right)\|\hat{v}\|^2 > 0$$

if $\alpha^2 < 4\beta$.

Thus, $(A^2 + \alpha A + \beta I)\hat{v} \neq \hat{0}$, so $\mathcal{N}(A^2 + \alpha A + \beta I) = \{\hat{0}\}$ and so $(A^2 + \alpha A + \beta I)$ is invertible.

Theorem 8

If A is a real $n \times n$ symmetric matrix, then A has a real eigenvalue.

Proof

Suppose that $\hat{v} \neq \hat{0}$. The set $\{\hat{v}, A\hat{v}, A^2\hat{v}, \ldots, A^n\hat{v}\}$ has $n+1$ vectors and so is linearly dependent. Thus, there are real numbers $c_0, c_1, \ldots, c_n$ not all zero for which

$$c_0\hat{v} + c_1 A\hat{v} + \cdots + c_n A^n\hat{v} = \hat{0}.$$

Consider the polynomial

$$p(x) = c_0 + c_1 x + c_2 x^2 + \cdots + c_n x^n.$$

This is a polynomial with real coefficients and may be factored over the real numbers so that each factor is a linear factor or an irreducible quadratic factor. Thus, we have

$$p(x) = \gamma(x - \lambda_1)\cdots(x - \lambda_k)\left(x^2 + \alpha_1 x + \beta_1\right)\cdots\left(x^2 + \alpha_j x + \beta_j\right).$$

According to the quadratic formula, there are no real roots to $x^2 + \alpha x + \beta = 0$ exactly when $\alpha^2 < 4b$. According to Theorem 7, this means

$$A^2 + \alpha A + \beta I$$

is invertible.

Now consider

$$p(A) = c_0 I + c_1 A + c_2 A^2 + \cdots + c_n A^n = \gamma(A - \lambda_1 I)\cdots(A - \lambda_k I)\left(A^2 + \alpha_1 A + \beta_1\right)\cdots\left(A^2 + \alpha_j A + \beta_j\right).$$

Now

$$\gamma(A - \lambda_1 I)\cdots(A - \lambda_k I)\left(A^2 + \alpha_1 A + \beta_1\right)\cdots\left(A^2 + \alpha_j A + \beta_j\right)\hat{v} = \hat{0}$$

and each $(A^2 + \alpha_i A + \beta_i)$ is invertible, so

$$(A - \lambda_1 I)\cdots(A - \lambda_k I)\hat{v} = \hat{0}.$$

Thus, there is a k for which $(A - \lambda_k I)\hat{v} = \hat{0}$ and so λ_k is an eigenvalue for A.

There are different ways to state the spectral theorem. One of the more intuitive ways is given here.

Theorem 9 (Spectral Theorem)

Let V be an inner product space. A linear operator T on V is self-adjoint if and only if there is an orthonormal basis of V consisting of eigenvectors of T.

Proof

Suppose that T is a self-adjoint operator on V. We show there is an orthonormal basis of eigenvectors.

We prove the result by induction on the dimension of V. Let n denote the dimension of V.

For $n=1$, let $\{\hat{u}\}$ be an orthonormal basis for V. Then $T\hat{u} = c\hat{u}$ for some $c \in \mathcal{F}$. Since T is self-adjoint, c is real.

Assume the result holds for $n=k$.

Suppose that $n=k+1$. If the field is the complex numbers, then the characteristic polynomial factors into linear factors, and there is an eigenvalue. If the field is the real numbers, there is an eigenvalue by Theorem 8. Since T is self-adjoint, the eigenvalues are real. Let $\hat{u}$ be a normalized eigenvector of T with eigenvalue λ. Let W be the subspace of T spanned by $\hat{u}$.

Now $T(W) \subset W$. We show

$$T(W^{\perp}) \subset W^{\perp}.$$

Let $c\hat{u} \in U$ and $\hat{y} \in W^{\perp}$. We show that $T(W^{\perp}) \subset W^{\perp}$ by showing

$$\langle c\hat{u}, T\hat{y} \rangle = 0.$$

We have

$$\langle c\hat{u}, T\hat{y} \rangle = c\langle \hat{u}, T\hat{y} \rangle = c\langle T\hat{u}, \hat{y} \rangle = c\langle \lambda\hat{u}, \hat{y} \rangle = c\lambda\langle \hat{u}, \hat{y} \rangle = 0.$$

Consider the matrix for T that we have constructed thus far.

Let the basis to be constructed be $\mathcal{B} = \{\hat{u}, *, *, \ldots, *\}$ where the vectors other than $\hat{u}$ are not yet determined.

Let A be the matrix representation of T in the basis $\mathcal{B}$. Because $T\hat{u} = \lambda\hat{u}$, the first column of A is

$$\begin{pmatrix} \lambda \\ 0 \\ \vdots \\ 0 \end{pmatrix}$$

and because $T(W^{\perp}) \subset W^{\perp}$ the first row of A is

$$\begin{pmatrix} \lambda & 0 & \cdots & 0 \end{pmatrix}.$$

Thus, the matrix for T to this point is

$$\begin{pmatrix} \lambda & 0 & \cdots & 0 \\ 0 & * & \cdots & * \\ \vdots & \vdots & & \vdots \\ 0 & * & \cdots & * \end{pmatrix}.$$

Thus,

$$V = W \oplus W^\perp$$

and the dimension of $W^\perp$ is k.

There is a somewhat subtle point that needs to be addressed before we can claim to use the induction hypothesis. Namely, that T restricted to $W^\perp$ is a self-adjoint operator on $W^\perp$. We denote T restricted to $W^\perp$ by $T_{W^\perp}$.

Suppose that $\hat{u}, \hat{v} \in W^\perp$. We have

$$\left\langle \left(T_{W^\perp} \hat{u} \right), \hat{v} \right\rangle = \left\langle T\hat{u}, \hat{v} \right\rangle$$

because if $\hat{u} \in W^\perp \subset V$ then $T_{W^\perp}(\hat{u}) = T(\hat{u})$.

Also,

$$\left\langle T\hat{u}, \hat{v} \right\rangle = \left\langle \hat{u}, T\hat{v} \right\rangle$$

because T is self-adjoint on V and

$$\left\langle \hat{u}, T\hat{v} \right\rangle = \left\langle \hat{u}, T_{W^\perp}(\hat{v}) \right\rangle$$

because $T(W^\perp) \subset W^\perp$ so if $\hat{v} \in W^\perp$ then $T_{W^\perp}(\hat{v}) = T(\hat{v})$.

Thus, we may apply the induction principle to conclude that the result holds for all n.

Conversely, suppose there is an orthonormal basis of eigenvectors of T. We show that T is self-adjoint.

In the orthonormal basis, T is diagonal so $T = T^T$ or $T = T^*$

We collect the results of the spectral theorem.

If A is a self-adjoint matrix, then

(i)   The eigenvalues of A are real.
(ii) The algebraic dimension of each eigenvalue is equal to the geometric dimesion.
(iii) Eigenvectors corresponding to different eigenvalues are orthogonal.
(iv) The matrix P whose columns are the normalized eigenvectors of A has the properties

$$P = P^T = P^{-1}$$

and

$$A = PDP^{-1},$$

where D is a diagonal matrix whose diagonal elements are the eigenvalues of A.

Example

We construct a spectral diagonalization of the matrix

$$A = \begin{pmatrix} 4 & 1 & 1 \\ 1 & 4 & 1 \\ 1 & 1 & 4 \end{pmatrix}.$$

The characteristic polynomial of A is

$$-\lambda^3 + 12\lambda^2 - 45\lambda + 54 = -(\lambda - 3)^2(\lambda - 6).$$

For $\lambda = 3$, we have

$$A - 3I = \begin{pmatrix} 1 & 1 & 1 \\ 1 & 1 & 1 \\ 1 & 1 & 1 \end{pmatrix}$$

which row reduces to

$$\begin{pmatrix} 1 & 1 & 1 \\ 0 & 0 & 0 \\ 0 & 0 & 0 \end{pmatrix}.$$

Two linearly independent eigenvectors are

$$\hat{v}_1 = \begin{pmatrix} -1 \\ 1 \\ 0 \end{pmatrix} \quad \text{and} \quad \hat{v}_2 = \begin{pmatrix} -1 \\ 0 \\ 1 \end{pmatrix}.$$

For $\lambda = 6$, We have

$$A - 6I = \begin{pmatrix} -2 & 1 & 1 \\ 1 & -2 & 1 \\ 1 & 1 & -2 \end{pmatrix}$$

which row reduces to

$$\begin{pmatrix} 1 & 0 & -1 \\ 0 & 1 & -1 \\ 0 & 0 & 0 \end{pmatrix} \quad \text{and} \quad \hat{v}_3 = \begin{pmatrix} 1 \\ 1 \\ 1 \end{pmatrix}$$

is an eigenvector.

When we normalize the eigenvectors, we have

$$\frac{\hat{v}_1}{\|\hat{v}_1\|} = \hat{w}_1 = \begin{pmatrix} -\dfrac{1}{\sqrt{2}} \\ \dfrac{1}{\sqrt{2}} \\ 0 \end{pmatrix}, \quad \frac{\hat{v}_2}{\|\hat{v}_2\|} = \hat{w}_2 = \begin{pmatrix} -\dfrac{1}{\sqrt{2}} \\ 0 \\ \dfrac{1}{\sqrt{2}} \end{pmatrix}, \quad \frac{\hat{v}_3}{\|\hat{v}_3\|} = \hat{w}_3 = \begin{pmatrix} \dfrac{1}{\sqrt{3}} \\ \dfrac{1}{\sqrt{3}} \\ \dfrac{1}{\sqrt{3}} \end{pmatrix}.$$

One choice of P is

$$P = \begin{pmatrix} -\dfrac{1}{\sqrt{2}} & -\dfrac{1}{\sqrt{2}} & \dfrac{1}{\sqrt{3}} \\ \dfrac{1}{\sqrt{2}} & 0 & \dfrac{1}{\sqrt{3}} \\ 0 & \dfrac{1}{\sqrt{2}} & \dfrac{1}{\sqrt{3}} \end{pmatrix}$$

and for this choice of P we have

$$D = \begin{pmatrix} 3 & 0 & 0 \\ 0 & 3 & 0 \\ 0 & 0 & 6 \end{pmatrix}.$$

Exercises

Give a spectral diagonalization for the following matrices.

1. $A = \begin{pmatrix} 2 & 1 \\ 1 & 2 \end{pmatrix}$

2. $A = \begin{pmatrix} 0 & 3 \\ 3 & 0 \end{pmatrix}$

3. $A = \begin{pmatrix} 3 & 2 \\ 2 & 3 \end{pmatrix}$

4. $A = \begin{pmatrix} 1 & 2 & 2 \\ 2 & 1 & 2 \\ 2 & 2 & 1 \end{pmatrix}$

5. $A = \begin{pmatrix} 3 & 1 & -1 \\ 1 & 3 & -1 \\ -1 & -1 & 5 \end{pmatrix}$

6. $A = \begin{pmatrix} 0 & 1 & 1 \\ 1 & 0 & 1 \\ 1 & 1 & 0 \end{pmatrix}$

Appendix A: A Brief Guide to MATLAB®

MATLAB (MATrix LABoratory) is a numerical computation tool that is particularly good at matrix computations and linear algebra algorithms. It does many other things, and we will use only a small part of its capabilities.

It is easy to get started. Here is a view of the screen when a MATLAB® session begins:

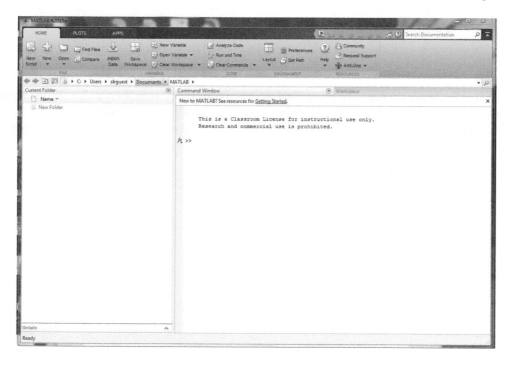

Matrices are assigned, and operations are carried out in the Command window.

A.1 Creating Matrices and Vectors

To enter the matrix $\begin{bmatrix} 3 & -2 & 9 \\ 4 & 0 & 8 \\ -1 & 5 & 4 \end{bmatrix}$ and assign it the name A, type the following after the >> prompt in the Command window:

$$A = \begin{bmatrix} 3 & -2 & 9; & 4 & 0 & 8; & -1 & 5 & 4 \end{bmatrix}$$

and press Enter. Notice that in the assignment statement, one row of the matrix is separated from the next by a semicolon. Successive entries can be separated by one or more spaces or by commas.

```
>> A = [3     -2     9;    4     0     8;    -1     5     4]

A =

       3     -2     9
       4      0     8
      -1      5     4
```

MATLAB names are case-sensitive, that is, *A* is a different name than *a*. A variable name must start with a letter, which can be followed by letters, digits, or underscores.

At this point, if we enter *A*, then *A* is displayed, but if we enter "*a*," MATLAB reminds us that *a* hasn't been defined yet.

```
>> A

A =

       3     -2     9
       4      0     8
      -1      5     4

>> a
Undefined function or variable 'a'.

Did you mean:
>> A
```

An efficient way to create a diagonal matrix is the *diag* function, as shown here.

```
>>     diag([2, 5, -1, 4])

ans =

       2      0      0      0
       0      5      0      0
       0      0     -1      0
       0      0      0      4
```

If a matrix or other result from a computation is not assigned to a name, its default name is *ans*.

We could create a 3-by-3 identity matrix using *diag*. Another way is with the MATLAB command *eye*.

```
>>  I=eye(3)

I    =

      1        0        0
      0        1        0
      0        0        1
```

Matrices can be combined to create larger matrices. For example, try out

```
>> D = [I, A]
```

A *vector* is a matrix consisting of either one column or one row. Not surprisingly, MATLAB distinguishes between an n-by-1 matrix and a 1-by-n matrix.

A single entry in a matrix can be assigned a value by referencing its row and column. Here is an example:

```
I =

      1        0        0
      0        1        0
      0        0        1
>> I(3,1)=6

I =

      1        0        0
      0        1        0
      6        0        1
```

The notation A(3, :) refers to all of row 3 of matrix A. Similarly, A(:, 2) refers to the second column of matrix A. This is a nice way to make separate vectors of the rows or columns of A.

```
A =
       3       -2        9
       4        0        8
      -1        5        4

>> c1=A( :,  1)

c1 =
       3
       4
      -1
```

```
>> c2=A( :  , 2)

c2 =
    -2
     0
     5

>> r3=A(3,  :)

r3 =
    -1     5     4
```

A.2 Correcting Typing Errors

Pressing the up-arrow on the keyboard will scroll back through the commands in a current session. This is handy if you need to correct a typing error. Rather than retype the full line, you can up-arrow to bring the line back, correct the error, and reenter. This is illustrated below.

```
>> B=[1  3  5 , 2,5,7 ; 4  6  8]

Error using vertcat
Dimensions of matrices being concatenated are not consistent.
```

Use up-arrow to recall the last command.
Correct the error and press enter.

```
 >> B=[1  3  5 , 2,5,7 ; 4  6  8]
>> B=[1  3  5 ; 2,5,7 ; 4  6  8]
B =

     1     3     5
     2     5     7
     4     6     8
```

A.3 Operations

Matrix addition in MATLAB is straightforward. The sum of matrices A and B is denoted A+B.

In MATLAB, the symbol for matrix multiplication is *. To try it out, create the matrix

$$B = \begin{bmatrix} 3 & 1 & 4 \\ 2 & -1 & 1 \end{bmatrix}$$

MATLAB calculates the product B*A, but not the product A*B.

```
>> B

B =

     3     1     4
     2    -1     1

>> C = B*A

C =

     9    14    51
     1     1    14

>> A*B
Error using  *
Inner matrix dimensions must agree.
```

The * symbol also serves for scalar multiplication, for example, 3 times matrix A is 3*A.

The transpose of a matrix A is denoted in MATLAB by A' where ' is the apostrophe on the keyboard.

```
C =

     2     3     4
     1    -1     6
     9     1     2

>> C'

ans =

     2     1     9
     3    -1     1
     4     6     2
```

There are at least two ways to calculate the dot product of two vectors in MATLAB. One way is to use the transpose operator, and another way is to use the function dot().

Let x = [3, 4, 1, 6] and y = [−1, 0, 2, 4], both row vectors. Then their dot product is x*y' or dot(x,y).

```
>> x*y'

ans =

    23

>> dot(x, y)

ans =

    23
```

One way to find the inverse of a square matrix is the function inv() as shown in the following.

```
>> F=inv(A)
F =
   -0.3704     0.4907    -0.1481
   -0.2222     0.1944     0.1111
    0.1852    -0.1204     0.0741
>> A*F
ans =
    1.0000          0          0
         0     1.0000          0
   -0.0000     0.0000     1.0000
```

The minus sign in the last line of output results from roundoff in the computations. You can change the precision of calculations in MATLAB—see *digits* in the documentation.

Another way to get a matrix inverse is with the exponentiation operator, ^.

```
>> A^(-1)

ans =

   -0.3704     0.4907    -0.1481
   -0.2222     0.1944     0.1111
    0.1852    -0.1204     0.0741
```

The determinant of a square matrix is calculated by the function det(). What is the relationship between the matrices K and L in the following?

```
K =
     2      1
     3     -1

>> L=(1/det(K))*[-1,-1; -3 2]
```

```
L =
    0.2000      0.2000
    0.6000     -0.4000
```

A.4 Help, Workspace, Saving

MATLAB has thorough documentation that a user can search via the Help button in the upper right corner of the screen, or via the F1 key.

Clicking the Workspace tab on the upper right opens a window that shows what objects are currently in working memory. You can SAVE the workspace contents to have them available later. To do this and to see other Workspace actions, click on the symbol for the drop-down menu on the Workspace tab.

A.5 Creating Elementary Matrices

We can reference particular rows of a matrix by specifying the row numbers in square brackets and using a colon to include all columns, as illustrated in the following:

```
G =
     1       2       3       4
     5       6       7       8
     9      10      11      12
    13      14      15      16
 >> G([1 3],:)

ans =

     1       2       3       4
     9      10      11      12
```

We can use this to switch the first and third rows of G, as shown.

```
>> G([1 3],:)=G([3 1],:)

G =

     9      10      11      12
     5       6       7       8
     1       2       3       4
    13      14      15      16
```

To multiply a row by a constant, replace the row with the multiple.

```
>> G(4,:)= (1/13)*G(4,:)
```

```
G =

      9.0000     10.0000     11.0000     12.0000
      5.0000      6.0000      7.0000      8.0000
      1.0000      2.0000      3.0000      4.0000
      1.0000      1.0769      1.1538      1.2308
```

Similarly, we can add to one row a scalar multiple of another row. For example

```
>> G(2,:)=G(2,:)  +  (-5)*G(3,:)

G =

      9.0000     10.0000     11.0000     12.0000
           0     -4.0000     -8.0000    -12.0000
      1.0000      2.0000      3.0000      4.0000
      1.0000      1.0769      1.1538      1.2308
```

Now any desired elementary matrix can be created by starting with the right size identity matrix and applying row operations.

A.6 Row Reduced Form

The MATLAB function for producing the row reduced form of matrix A is `rref(A)`.

```
>> A = [3 4 -2 5; 1 -2 1 0; 7 4 -3 3]
A =
      3      4     -2      5
      1     -2      1      0
      7      4     -3      3
>> B = rref(A)
B =
      1.0000           0           0      1.0000
           0      1.0000           0      3.5000
           0           0      1.0000      6.0000
```

A.7 Complex Numbers

MATLAB uses the lowercase letter i for the square root of -1, unless i has been specifically defined otherwise (which is not recommended.) MATLAB carries out complex number calculations and will return complex number results of calculations. Examples are

```
>> (2+3i)*(-6+i)
ans =
 -15.0000 -16.0000i
```

```
>> (-4)^.5
ans =
   0.0000 + 2.0000i
```

A.8 Orthonormal Basis

The function `orth()` takes a matrix as input and returns a matrix whose columns are an orthonormal basis for the range of the matrix.

```
M =
     1     2     1
     0    -5     1
     1     4     1

>> N=orth(M)
N =
   -0.3170   -0.5788   -0.7514
    0.7267   -0.6573    0.1997
   -0.6094   -0.4827    0.6290

>>N'*N
ans =
    1.0000   -0.0000   -0.0000
   -0.0000    1.0000   -0.0000
   -0.0000   -0.0000    1.0000
```

A.9 Characteristic Polynomial, Eigenvalues, and Eigenvectors

To find the characteristic polynomial of a matrix, first declare the variable x as symbolic:

```
>> syms x
```

Now you can have MATLAB evaluate $\det(A - xI)$ using the `eye()` function for the identity.

```
A =
     2     0     0
    -3     0     1
     0     1     0

>> det(A-x*eye(3))
 ans =
- x^3 + 2*x^2 + x - 2
```

To calculate eigenvalues of A, MATLAB provides the function `eig(A)`, which returns a column vector containing the eigenvalues of A.

A more informative option is described in MATLAB Help as follows:

> `[V,D]` = `eig(A)` returns diagonal matrix D of eigenvalues and matrix V whose columns are the corresponding right eigenvectors, so that `A*V` = `V*D`.

```
A =
     2      0      0
    -3      0      1
     0      1      0


>> eig(A)
ans =
     1
    -1
     2


>> [V,D]=eig(A)
V =
         0           0      0.4082
    0.7071      0.7071     -0.8165
    0.7071     -0.7071     -0.4082


D =
     1      0      0
     0     -1      0
     0      0      2
>> A*V
ans =
         0           0      0.8165
    0.7071     -0.7071     -1.6330
    0.7071      0.7071     -0.8165


>> V*D
ans =
         0           0      0.8165
    0.7071     -0.7071     -1.6330
    0.7071      0.7071     -0.8165
```

A.10 The Norm Function

The MATLAB function for finding the length of a vector, that is, $\sqrt{x \cdot x'}$, is norm().

```
r3 =
    -1      5      4
>> norm(r3)
ans =
    6.4807
>> norm(r3)^2
ans =
    42
>> x=[2+i, 3, 4-i]
```

```
x =
   2.0000 + 1.0000i    3.0000 + 0.0000i    4.0000 - 1.0000i
>> norm(x)
ans =
   5.5678
```

A.11 Matrix Decompositions and Factorizations

A.11.1 LU Factorization

The LU factorization gives a square matrix A as the product LU where L is a permutation of a lower triangular matrix with ones on its diagonal, and U is an upper triangular matrix. The MATLAB function is lu().

```
A =
      3     -2      9
      4      0      8
     -1      5      4

>> [L,U]=lu(A)

L =
    0.7500    -0.4000     1.0000
    1.0000          0          0
   -0.2500     1.0000          0

U =
    4.0000          0     8.0000
         0     5.0000     6.0000
         0          0     5.4000

>> L*U

ans =
      3     -2      9
      4      0      8
     -1      5      4
```

A.11.2 Cholesky Factorization

The Cholesky factorization expresses a positive definite, symmetric matrix as the product of a triangular matrix and its transpose

$$A = R'R,$$

where R is an upper triangular matrix. The MATLAB function for this factorization is chol().

```
B =
     10      2      3
      2     16      5
      3      5      6

>> R=chol(B)
```

```
R =
    3.1623        0.6325        0.9487
         0        3.9497        1.1140
         0             0        1.9644

>> R'*R

ans =
   10.0000        2.0000        3.0000
    2.0000       16.0000        5.0000
    3.0000        5.0000        6.0000
```

A.11.3 Singular Value Decomposition

Here is the MATLAB documentation for the singular value decomposition.

```
s = svd(X)
[U,S,V] = svd(X)
[U,S,V] = svd(X,0)
[U,S,V] = svd(X,'econ')
```

The svd command computes the matrix singular value decomposition.

   s = svd(X) returns a vector of singular values.

   [U,S,V] = svd(X) produces a diagonal matrix S of the same dimension as X, with nonnegative diagonal elements in decreasing order, and unitary matrices U and V so that X = U*S*V'.

   [U,S,V] = svd(X,0) produces the "economy size" decomposition. If X is m-by-n with m > n, then svd computes only the first n columns of U and S is n-by-n.

   [U,S,V] = svd(X,'econ') also produces the "economy size" decomposition. If X is m-by-n with m >= n, it is equivalent to svd(X,0). For m < n, only the first m columns of V are computed and S is m-by-m.

For the matrix

```
X =
     1     2
     3     4
     5     6
     7     8
```

the statement

```
[U,S,V] = svd(X)
```

produces

```
U =
   -0.1525       -0.8226       -0.3945       -0.3800
   -0.3499       -0.4214        0.2428        0.8007
   -0.5474       -0.0201        0.6979       -0.4614
   -0.7448        0.3812       -0.5462        0.0407

S =
   14.2691             0
         0        0.6268
         0             0
         0             0
```

```
V =
    -0.6414      0.7672
    -0.7672     -0.6414
```

The economy size decomposition generated by

```
[U,S,V] = svd(X,0)
```

produces

```
U =
    -0.1525     -0.8226
    -0.3499     -0.4214
    -0.5474     -0.0201
    -0.7448      0.3812
S =
    14.2691           0
         0      0.6268
V =
    -0.6414      0.7672
    -0.7672     -0.6414
```

A.11.4 QR Factorization

Finally, there is the QR factorization. MATLAB documentation says: "The orthogonal, or QR, factorization expresses any rectangular matrix as the product of an orthogonal or unitary matrix and an upper triangular matrix. A column permutation might also be involved:

$$A = QR$$

or

$$AP = QR,$$

where Q is orthogonal or unitary, R is upper triangular, and P is a permutation."
    The MATLAB function is qr(). For example

```
D =

     1      2      3      4
     5      2      4      3
    -2      4      1      0

>> [Q,R]= qr(D)

Q =
    -0.1826      0.3853     -0.9045
    -0.9129      0.2752      0.3015
     0.3651      0.8808      0.3015

R =
    -5.4772     -0.7303     -3.8341     -3.4689
          0      4.8442      3.1377      2.3671
          0           0     -1.2060     -2.7136

>> Q*R
```

```
ans =
    1.0000    2.0000    3.0000    4.0000
    5.0000    2.0000    4.0000    3.0000
   -2.0000    4.0000    1.0000    0.0000

>> Q'*Q
ans = 1.0000     0.0000    -0.0000
       0.0000     1.0000    -0.0000
      -0.0000    -0.0000     1.0000
```

Appendix B: An Introduction to R

B.1 Getting Started with R

R is a computer language for statistical computing that is widely used by statisticians and statistics students. Not only is it flexible and powerful, it is free! Because matrices and matrix operations and decompositions are useful in statistics, R is also a powerful tool for doing linear algebra computations. This little manual is a starting point for learning the language; it is intended to give a minimal yet sufficient set of R commands for an introductory linear algebra course. If you are interested in more details, there are many textbooks and references for R, as well as online help. Web headquarters for the R Project is http://www.r-project.org/; from this website you can download R onto your own computer.

This manual is designed to accompany a linear algebra textbook; it only describes *how* to do a computation, not *why* or under what conditions.

- You will probably want to save your inputs and results in a personal folder. Set up a folder now on your flash drive.
- Now double click on the icon for R on the desktop.
- The first thing you should do at every session with R is to change into the directory where you are keeping your work. Click on the File menu on the upper left corner of the screen, and select "Change dir...." Select your folder.

B.2 Calculations in R

We communicate commands to R by typing them at the ">" prompt and pressing the Enter key. Try these out:

```
>2/6*12
>3^2
```

You may add comments if they are preceded by a # symbol. This is a good way to make notes about what you are doing. Try out:

```
>100/10^2                    # exponentiation takes precedence over division.
```

Try:

```
>3 + 7*
```

Instead of the usual prompt, R returns a + prompt to indicate that the command is incomplete; R waits for the rest of the command:

```
> 3 + 7*
+ 5
[1] 38
```

NOTE: In this guide, R commands and output are given in a different font to make them easier to distinguish from the rest of the text.

B.3 Assigning Values to Variables

In R, acceptable variable names use letters, numbers, dot, or underscore. A variable name must start with a letter or a dot. R is case-sensitive, so, for example, A and a are not the same variable. Some examples of acceptable variable names are x3, yr12, yr.birth, birth_month, and R2D2.

Variables are assigned values using an equal sign on the right of the variable. Try out the following commands in R:

```
>x=12

> x^2

>year = 2011

>year=year+1

>year

>e               # the letter e is not a constant in R.
>e=exp(1)        # this assigns the "usual" value to the variable e.

>pi              # "pi" cannot be used as a variable name, since it
                   represents a defined constant in R.

>3=x             # this will generate an error message.

>wt<-12.63       # the combination <- is another assignment operator
>wt
```

B.4 Vectors and Matrices in R

We can create a vector using the c () function, as follows:

```
>b = c(9, 8, 5, 12, 4, 17)
>b
[1]  9  8  5 12  4 17
```

The index [1] denotes the first entry of the vector. If the vector were long enough to continue onto the next line, there would be an index at the beginning of the next line to indicate the position in the vector for the next data value. We can inspect any position in a vector using its index. To demonstrate, enter

```
>b[4]
```

You can add components to a vector in a variety of ways. Try these:

```
>b[7]=9

>b=c(b,c(5,8,11))
>b
```

  Note 1: Since the character c is a defined function in R, it cannot be used as a variable name.
  Note 2: Changing the position of the cursor on a line *cannot* be accomplished in R with a mouse click; the left and right arrow keys must be used. If you want to move left to correct a typing error, you must either backspace or use the left arrow key; move back to the end of the line using the right arrow key or the "End" key.
  We can apply mathematical operations and functions to vectors. Investigate the following; inspect the value of variable *b* after each operation:

```
>b -9      # This operation does not alter the entries of the object b.

>1/b

>sort(b)

>x=c(1,5,3,5)

>b + x

>sum(b)

>length(b)
```

The sequence of integers from n to m is denoted by n:m in R. For example,

```
> 2:9
[1] 2 3 4 5 6 7 8 9
> s=1:12
> s
 [1]  1  2  3  4  5  6  7  8  9 10 11 12
```

An R workspace is a portion of computer memory where objects such as matrices and vectors are stored while R is running. It is frequently helpful to remind yourself what objects you have currently available in your workspace. To see the contents of your workspace, use ls() or objects(). To delete an object, use the "remove" operator, rm(). For example, try

```
>rm(x)
>ls()
```

If you discover that you have mistyped a command, of course you can retype it and reenter it. But it may be more efficient to backtrack using arrow keys and then edit. The up-arrow and down-arrow keys let you scroll through previous commands. Try this out.

You can edit the current line and re-execute it. For example, scroll up to

```
>sort(b).
```

Use the left arrow to move to the > prompt and change the line to

```
>y=sort(b)
```

and enter.

Now the variables y and b contain the same entries but in different orders.

To enter the matrix $\begin{bmatrix} 3 & -2 & 9 \\ 4 & 0 & 8 \\ -1 & 5 & 4 \end{bmatrix}$ and assign it the name A, type the following after the

> prompt:

```
> A=matrix(c(3,4,-1,-2,0,5,9,8,4),ncol=3)
> A
     [,1] [,2] [,3]
[1,]   3   -2    9
[2,]   4    0    8
[3,]  -1    5    4
>
```

Notice that the entries are listed column-by-column in the assignment statement. If you would prefer to type the entries row-by-row, use

```
> B=matrix(c(3,-2,9,4,0,8,-1,5,4), ncol=3, byrow = T)
> B
     [,1] [,2] [,3]
[1,]   3   -2    9
[2,]   4    0    8
[3,]  -1    5    4
>
```

There is a shortcut for creating a diagonal matrix:

```
> D=diag(c(4,6,3,2))
> D
     [,1] [,2] [,3] [,4]
[1,]   4    0    0    0
[2,]   0    6    0    0
[3,]   0    0    3    0
[4,]   0    0    0    2
>
```

Try out

```
>diag(1,4)
```

A single entry of a matrix can be edited by referencing its row and column. For example,

```
> D[2,1]=7
> D
     [,1] [,2] [,3] [,4]
[1,]   2    0    0    0
[2,]   7    5    0    0
[3,]   0    0   -1    0
[4,]   0    0    0    4
```

Parts of a matrix can be extracted by specifying rows and columns in square brackets. Study the following examples:

```
> G=matrix(1:28,ncol=7,byrow=T)
> G
     [,1] [,2] [,3] [,4] [,5] [,6] [,7]
[1,]    1    2    3    4    5    6    7
[2,]    8    9   10   11   12   13   14
[3,]   15   16   17   18   19   20   21
[4,]   22   23   24   25   26   27   28
> G[3,]                              # the third row of G
[1] 15 16 17 18 19 20 21
> G[,6]                              # the sixth column of G
[1]  6 13 20 27
> G[c(1,4),]                   # rows 1 and 4 of G
     [,1] [,2] [,3] [,4] [,5] [,6] [,7]
[1,]    1    2    3    4    5    6    7
[2,]   22   23   24   25   26   27   28
>
```

Matrices can be "pasted together" using the functions cbind() and rbind(). Test out the following examples:

```
> A
     [,1] [,2] [,3]
[1,]    3   -2    9
[2,]    4    0    8
[3,]   -1    5    4
> B=matrix(c(5,7,2,0,1,-3),ncol=2,byrow=T)
> B
     [,1] [,2]
[1,]    5    7
[2,]    2    0
[3,]    1   -3
> C=cbind(A,B)
> C
     [,1] [,2] [,3] [,4] [,5]
[1,]    3   -2    9    5    7
[2,]    4    0    8    2    0
[3,]   -1    5    4    1   -3
> D=rbind(A,t(B))   # t(B) is the transpose of B.
```

```
> D
     [,1] [,2] [,3]
[1,]    3   -2    9
[2,]    4    0    8
[3,]   -1    5    4
[4,]    5    2    1
[5,]    7    0   -3
```

B.5 Matrix Operations

The R notation for matrix multiplication is `%*%`. For example

```
> M=matrix(c(1,2,1,3,1,3,2,1,2), ncol=3)
> M
     [,1] [,2] [,3]
[1,]    1    3    2
[2,]    2    1    1
[3,]    1    3    2
> M%*%A
     [,1] [,2] [,3]
[1,]   13    8   41
[2,]    9    1   30
[3,]   13    8   41
```

R declines to multiply matrices if their dimensions are not conformable:

```
> M%*%D
Error in M %*% D: nonconformable arguments
```

However, R will treat a vector as either a row or a column, in order to make it conform to multiplication. For example

```
> V
[1] 1 5 4
> A%*%V
     [,1]
[1,]   29
[2,]   36
[3,]   40
> V%*%A
     [,1] [,2] [,3]
[1,]   19   18   65
```

Observe the crucial distinction between the two operations, `*` and `%*%`, when applied to vectors:

```
> V1=c(1,4,-1,2)
> V2=c(3,0,-2,5)
> V1*V2
[1]  3  0  2 10
> V1%*%V2
     [,1]
[1,]   15
```

A straightforward way to find the length of a vector is to use `sqrt()` and `%*%`.

```
> V1
[1]  1   4  -1   2
> sqrt(V1%*%V1)
   [,1]
[1,] 4.690416
```

The symbols for matrix addition and subtraction are the usual + and – signs.

```
> A+M
    [,1] [,2] [,3]
[1,]  4    1   11
[2,]  6    1    9
[3,]  0    8    6
> A-M
    [,1] [,2] [,3]
[1,]  2   -5    7
[2,]  2   -1    7
[3,] -2    2    2
```

The inverse of a matrix is produced by the `solve()` function in R, as illustrated:

```
> B=solve(A)
> B
    [,1]       [,2]         [,3]
[1,] -0.3703704  0.4907407 -0.14814815
[2,] -0.2222222  0.1944444  0.11111111
[3,]  0.1851852 -0.1203704  0.07407407
> A%*%B
       [,1]        [,2]   [,3]
[1,]  1.000000e+00 2.081668e-16    0
[2,]  0.000000e+00 1.000000e+00    0
[3,] -1.110223e-16 5.551115e-17    1
```

Because of inevitable tiny roundoff errors, the product of A with its inverse is very nearly but not exactly the identity matrix.

We can round off the last product by applying the `round()` function:

```
> round(A%*%B, digits = 5)
    [,1] [,2] [,3]
[1,]  1    0    0
[2,]  0    1    0
[3,]  0    0    1
```

The solve function will produce the solution of the matrix equation $Ay = x$, as in the following example:

```
> A
    [,1] [,2] [,3]
[1,]  3   -2    9
[2,]  4    0    8
[3,] -1    5    4

> x = c(4, -2,5)
> y = solve(A,x)
```

```
> y
[1]  -3.2037037  -0.7222222   1.3518519

> A%*%y
    [,1]
[1,]   4
[2,]  -2
[3,]   5
```

The determinant function is denoted by det().

```
> det(A)
[1]  108
> det(B)
[1]  0.009259259
> 1/108                                # as expected, det(A⁻¹) = 1/(det A).
[1]  0.009259259
```

The symbol for scalar multiplication is *.

```
> 5*A
     [,1]  [,2]  [,3]
[1,]  15   -10    45
[2,]  20     0    40
[3,]  -5    25    20
```

The transpose of matrix A is t(A).
```
> t(A)
     [,1]  [,2]  [,3]
[1,]   3     4    -1
[2,]  -2     0     5
[3,]   9     8     4
```

B.6 Elementary Row Operations

To multiply a row by a constant, replace the row with the multiple. In the following, the second row of matrix G is replaced by 1/5 times the second row.

```
> G
     [,1]  [,2]  [,3]  [,4]
[1,]   1     2     3     4
[2,]   5     6     7     8
[3,]   9    10    11    12
[4,]  13    14    15    16
> G[2,]=(1/5)*G[2,]
> G
     [,1]  [,2]  [,3]  [,4]
[1,]   1   2.0   3.0   4.0
[2,]   1   1.2   1.4   1.6
[3,]   9  10.0  11.0  12.0
[4,]  13  14.0  15.0  16.0
```

To interchange two rows, follow this example:

```
> G
     [,1] [,2] [,3] [,4] [,5] [,6] [,7]
[1,]  1    2    3    4    5    6    7
[2,]  8    9   10   11   12   13   14
[3,] 15   16   17   18   19   20   21
[4,] 22   23   24   25   26   27   28
> G[c(1,3),]=G[c(3,1),]          # interchange the first and third rows.
> G
     [,1] [,2] [,3] [,4] [,5] [,6] [,7]
[1,] 15   16   17   18   19   20   21
[2,]  8    9   10   11   12   13   14
[3,]  1    2    3    4    5    6    7
[4,] 22   23   24   25   26   27   28
>
```

To add to a row a scalar multiple of another row:

```
> G[4,]=G[4,]+(-22)*G[3,]                  # add to row 4, -22 times row 3.
> G
     [,1] [,2] [,3] [,4] [,5] [,6]  [,7]
[1,] 15   16   17   18   19   20    21
[2,]  8    9   10   11   12   13    14
[3,]  1    2    3    4    5    6     7
[4,]  0  -21  -42  -63  -84 -105  -126
```

B.7 Row Reduced Form

To obtain the row reduced form of a matrix, we need to install a package called *pracma*. The command for this is

```
> install.packages("pracma")
```

You will be prompted to choose a CRAN mirror. Installation will be faster from a mirror that is nearer to your geographic location. During the installation, respond "yes" to allow the code to be stored in a library on your computer.

Once the download is complete, you need to add the package to the working library for the current session. Use the command

```
> library("pracma")
```

This command must be used again whenever you start up a new session in R. Once the package is accessible, the row echelon form of a matrix is produced using the `rref()` function.

```
> K
     [,1] [,2] [,3] [,4] [,5] [,6]
[1,]  1    2    3    4    5    6
[2,]  7    8    9   10   11   12
[3,] 13   14   15   16   17   18
[4,] 19   20   21   22   23   24
```

```
> rref(K)
     [,1] [,2] [,3] [,4] [,5] [,6]
[1,]   1    0   -1   -2   -3   -4
[2,]   0    1    2    3    4    5
[3,]   0    0    0    0    0    0
[4,]   0    0    0    0    0    0
```

B.8 Complex Numbers in R

The lowercase i is not automatically recognized as the complex i, but it can be added to your workspace by defining it, as follows:

```
> i=complex(real=0,im=1)
> i
[1] 0+1i
```

R does not automatically give complex numbers as output when inputs are real but does give complex number results when the inputs are complex. For example

```
> sqrt(-1)
[1] NaN                                # NaN means Not a Number.
Warning message:
In sqrt(-1) : NaNs produced
```

However

```
> sqrt(-1+0i)
[1] 0+1i
```

You can convert a real number to the real part of a complex number using as.complex().

```
  > as.complex(15)
[1] 15+0i

> sqrt(-3)
[1] NaN
Warning message:
In sqrt(-3) : NaNs produced

> sqrt(as.complex(-3))
[1] 0+1.732051i
```

B.9 Finding Eigenvalues and Eigenvectors

The `eigen()` function takes a matrix as input and produces the eigenvalues and eigenvectors in an object called a list. This particular list consists of two components, $values and $vectors. Consider the next example:

```
> K=matrix(1:16,ncol=4,byrow = T)
> K
     [,1] [,2] [,3] [,4]
[1,]   1    2    3    4
[2,]   5    6    7    8
[3,]   9   10   11   12
[4,]  13   14   15   16
> KE=eigen(K)
> KE                                # the whole list is displayed.
$values
[1]   3.620937e+01 -2.209373e+00 -1.941536e-15 -3.467987e-16


$vectors
           [,1]      [,2]       [,3]        [,4]
[1,] -0.1511543  0.7270500 -0.3051507  0.05761073
[2,] -0.3492373  0.2832088  0.7458883  0.32916941
[3,] -0.5473203 -0.1606324 -0.5763245 -0.83117101
[4,] -0.7454033 -0.6044736  0.1355869  0.44439087
```

The vector of eigenvalues and the matrix of eigenvectors can be referenced separately:

```
> KE$values
[1] 3.620937e+01 -2.209373e+00 -1.941536e-15 -3.467987e-16
> KE$vectors
           [,1]      [,2]       [,3]        [,4]
[1,] -0.1511543  0.7270500 -0.3051507  0.05761073
[2,] -0.3492373  0.2832088  0.7458883  0.32916941
[3,] -0.5473203 -0.1606324 -0.5763245 -0.83117101
[4,] -0.7454033 -0.6044736  0.1355869  0.44439087
```

B.10 Orthonormal Basis

Suppose we want to find an orthonormal basis for the column space of the matrix

$$M = \begin{bmatrix} 1 & 2 & 1 & 5 \\ 0 & -5 & 1 & 1 \\ 1 & 4 & 1 & 0 \end{bmatrix}.$$

The R function svd(M) will return the singular value decomposition of M as a list. The list consists of matrices u and v and a vector d with the properties:

- The columns of u are an orthonormal basis for the column space of M.
- The columns of v are an orthonormal basis for the column space of the transpose of M (i.e., the row space of M.)
- If D is a diagonal matrix with diagonal d, then uDv = M.
- The nonzero entries of d are the singular values of M.

All of this is demonstrated in the following script:

```
> M
    [,1] [,2] [,3] [,4]
[1,]   1    2    1    5
[2,]   0   -5    1    1
[3,]   1    4    1    0
> SVDM=svd(M)
> SVDM              #display the list produced by the svd function.
$d
[1]  6.893554 5.154879 1.380627


$u
          [,1]       [,2]       [,3]
[1,]   0.5061532 -0.8400709 -0.1951661
[2,]  -0.6369880 -0.5166982  0.5720745
[3,]   0.5814251  0.1652389  0.7966435
$v
          [,1]       [,2]        [,3]
[1,] 0.1577674 -0.1309113  0.43565527
[2,] 0.9462386  0.3034609 -0.04645035
[3,] 0.0653640 -0.2311461  0.85001375
[4,] 0.2747172 -0.9150657 -0.29244398
> U=SVDM$u                  # for convenience, assign the elements of the
                             list to shorter names.
> V=SVDM$v
> D=diag(SVDM$d)
> U%*%t(U)              # check that the columns of U are orthonormal.
          [,1]        [,2]        [,3]
[1,] 1.000000e+00 -6.289728e-17 2.032879e-19
[2,] -6.289728e-17  1.000000e+00 6.732895e-17
[3,] 2.032879e-19  6.732895e-17 1.000000e+00
> t(V)%*%V              # check that the rows of V are orthonormal.
          [,1]        [,2]        [,3]
[1,] 1.000000e+00 1.196146e-16 1.980702e-17
[2,] 1.196146e-16  1.000000e+00 1.246832e-17
[3,] 1.980702e-17 1.246832e-17 1.000000e+00


> U%*%D%*%t(V)          #check that M = UDVt.
          [,1] [,2] [,3]      [,4]
[1,]   1.000000e+00    2    1 5.000000e+00
[2,]  -5.811324e-17   -5    1 1.000000e+00
[3,]   1.000000e+00    4    1 3.760555e-16
```

B.11 Characteristic Polynomial

The "pracma" package includes a function `charpoly()` to calculate the coefficients of the characteristic function. Here is an example:

```
> A
     [,1] [,2] [,3]
[1,]   3   -2    9
[2,]   4    0    8
[3,]  -1    5    4
> charpoly(A)
[1]   1   -7  -11 -108        # The characteristic polynomial of A is
                                 x³ - 7x² - 11x - 108.
```

B.12 Matrix Decompositions and Factorizations

B.12.1 LU Factorization

The "pracma" package provides a function `lu()` that accepts a positive definite square matrix A as input. The output is a pair of matrices L and U where L is lower triangular, U is upper triangular, and A=LU. This is called the *LU factorization* of A.

```
> A
     [,1] [,2] [,3]
[1,]   3   -2    9
[2,]   4    0    8
[3,]  -1    5    4
> LUA=lu(A)
> LUA
$L
           [,1]    [,2] [,3]
[1,]  1.0000000 0.000    0
[2,]  1.3333333 1.000    0
[3,] -0.3333333 1.625    1


$U
     [,1]      [,2] [,3]
[1,]   3 -2.000000  9.0
[2,]   0  2.666667 -4.0
[3,]   0  0.000000 13.5


> LUA$L%*%LUA$U        #  Check that LU = A.
     [,1] [,2] [,3]
[1,]   3   -2    9
[2,]   4    0    8
[3,]  -1    5    4
```

B.12.2 Cholesky Factorization

The *Cholesky factorization* expresses a positive definite symmetric matrix as the product of an upper triangular matrix and its transpose. The function chol() returns the upper triangular matrix, as shown in the following example. Note that this function is part of base R; no added package is needed to obtain it.

```
> B
    [,1]  [,2]  [,3]
[1,]  10    2    3
[2,]   2   16    5
[3,]   3    5    6
> R=chol(B)
>R
       [,1]      [,2]      [,3]
[1,] 3.162278 0.6324555 0.9486833
[2,] 0.000000 3.9496835 1.1140133
[3,] 0.000000 0.0000000 1.9644272
> t(R)%*%R              # check that B is the product of R by its
                          transpose.
    [,1]  [,2]  [,3]
[1,]  10    2    3
[2,]   2   16    5
[3,]   3    5    6
```

B.12.3 Singular Value Decomposition

The *singular value decomposition* of a matrix A consists of a diagonal matrix S of the same dimension as A, with nonnegative diagonal elements in decreasing order, and unitary matrices U and V such that $A = U*S*V^t$. The R function svd() takes A as input and returns a list containing U, V, and the diagonal of S.

Example:

```
> A
    [,1]  [,2]  [,3]
[1,]   3   -2    9
[2,]   4    0    8
[3,]  -1    5    4
> K=svd(A)

> K
$d
[1]  13.438450   5.786093   1.388958

$u
        [,1]       [,2]       [,3]
[1,] -0.7075215 -0.28897604 -0.6449079
[2,] -0.6606568 -0.05351371  0.7487783
[3,] -0.2508904  0.95583949 -0.1530519
```

```
$v
          [,1]      [,2]        [,3]
[1,] -0.33592425 -0.3520204   0.8736341
[2,]  0.01195011  0.9258665   0.3776618
[3,] -0.94181319  0.1373058  -0.3068143

> KU=K$u                         #rename the components of the list K, for
                                  convenience.
> KV=K$v
> D=diag(K$d)
> KU %*%t(KU)          #check that KU is unitary.
          [,1]      [,2]        [,3]
[1,]  1.000000e+00 -2.298509e-17  9.147278e-17
[2,] -2.298509e-17  1.000000e+00 -2.254802e-16
[3,]  9.147278e-17 -2.254802e-16  1.000000e+00
> KV%*%t(KV)  #Check that KV is unitary.
          [,1]      [,2]        [,3]
[1,]  1.000000e+00 -1.661540e-17 -4.816568e-17
[2,] -1.661540e-17  1.000000e+00  9.845911e-17
[3,] -4.816568e-17  9.845911e-17  1.000000e+00
> KU%*%D%*%t(KV)       #Check that A is the product.
    [,1]       [,2]   [,3]
[1,]   3 -2.000000e+00      9
[2,]   4 -8.637839e-16      8
[3,]  -1  5.000000e+00      4
```

B.13 Saving and Retrieving the Workspace

Before you close an R session, you should carefully save the workspace so that you can retrieve it later. Click on the floppy disk icon in the toolbar above the R console. Locate the folder where you want to save the workspace, and assign a filename to the workspace. Use the extension .RData. For example, save the workspace as "Chapter1Data.RData".

It is a good practice to save the workspace periodically even if you aren't exiting the program, since in the event of a system crash or power outage, what hasn't been saved is lost.

Using separate workspaces for different projects can help reduce clutter and improve organization.

When you're ready to end your session with R, enter the quit command: q().

When you want to access the workspace later, open R, change to the appropriate directory, click on the "load workspace" icon, and click on the workspace file. You may have to wait while the workspace loads and it may seem slow. If it takes very long, click on the Stop sign in the toolbar.

You may want to bookmark or print off this resource:

Quick reference list for r syntax and commands. http://cran.r-project.org/doc/contrib/Short-refcard.pdf.

Appendix C: Downloading R to Your Computer

The main page for the R Project is http://www.r-project.org/.

Locate the box titled "Getting Started" and click on the link to Cran Mirrors. In general, the download will be faster if you choose a location closer to you.

From the box titled "Download and Install R," click on the version (Linux, MacOS X, or Windows) that you need. If you are using Windows, the next step is to click on the link to the "base" subdirectories. "Base" is the essential part of the R software; many additional packages are available for specialized computations and functions. One or two more clicks will download the software. Once the download is complete, install R by double-clicking on the icon of the downloaded file.

Answers to Selected Exercises

Section 1.1

1. (a) $\begin{pmatrix} 4 & -14 \\ 1 & 9 \end{pmatrix}$

   (b) $\begin{pmatrix} -7 & -5 & 10 \\ -9 & -15 & 6 \\ -11 & 5 & 26 \end{pmatrix}$

   (c) Undefined as a single matrix

   (d) Undefined as a single matrix

   (e) $\begin{pmatrix} 4 & 2 \\ 0 & 2 \\ 14 & 4 \end{pmatrix}$

3. (a) $\begin{pmatrix} 4 & -2 & 0 \\ 0 & 0 & 3 \\ 0 & 0 & 0 \end{pmatrix}$

   (b) $\begin{pmatrix} 0 & 0 & 1 \\ 0 & 1 & 0 \\ 1 & 0 & 0 \end{pmatrix}$

   (c) $\begin{pmatrix} 1 & 1 & 1 \\ 2 & 2 & 2 \\ -1 & -1 & -1 \end{pmatrix}$

11. $\begin{pmatrix} 1/2 & -9 & 7/2 \\ 5/2 & -5/2 & 45/2 \end{pmatrix}$

Section 1.2

3. $\begin{pmatrix} a & 0 \\ c & a \end{pmatrix}$

5. $A - 3B = \begin{pmatrix} -5 & -3 \\ 7 & -15 \end{pmatrix}$

   $BA^T = \begin{pmatrix} 2 & 8 \\ -16 & -4 \end{pmatrix}$

$$AB^T = \begin{pmatrix} 2 & -16 \\ 8 & -4 \end{pmatrix}$$

$$\left(AB^T\right)^T = \begin{pmatrix} 2 & 8 \\ -16 & -4 \end{pmatrix}$$

(b) $\begin{pmatrix} -1 & 0 \\ c & 1 \end{pmatrix}$

7. A is a square matrix

9. (a) The second column of AB is all zeroes

11. (a) $(A+B)^2 = \begin{pmatrix} 116 & 144 \\ 180 & 224 \end{pmatrix}$ $A^2 + 2AB + B^2 = \begin{pmatrix} 112 & 132 \\ 192 & 228 \end{pmatrix}$

15. (a) $AB = \begin{pmatrix} a_{11}b_{11} & 0 & 0 \\ * & a_{22}b_{22} & 0 \\ ** & *** & a_{33}b_{33} \end{pmatrix}$

    (b) $(AB)_{23} = a_{21}b_{13} + a_{22}b_{23} + a_{23}b_{33} = a_{21}0 + a_{22}0 + 0b_{33} = 0$

19. (a) $det \begin{pmatrix} a & b & c \\ a & b & c \\ d & e & f \end{pmatrix} = 0$

23. (a) $a = -2, b = -1$

    (b) $a = -n, b = -1$

25. (b) A 2×2 matrix whose trace is 0 is of the form $\begin{pmatrix} a & b \\ c & -a \end{pmatrix}$

27. (a) $\begin{pmatrix} 1 & 0 & 0 \\ 0 & 1/3 & 0 \\ 0 & 0 & 1 \end{pmatrix}$

    (c) $\begin{pmatrix} 0 & 1 & 0 \\ 1 & 0 & 0 \\ 0 & 0 & 1 \end{pmatrix}$

29. (a) $\begin{pmatrix} 1 & 0 & -2 \\ 3 & 4 & -1 \\ 1 & 1 & 2 \end{pmatrix}$

    (b) The action done on the rows when multiplied by the elementary matrix on the left is instead action done on the columns when multiplied by the elementary on the right.

30. In this problem, while the inverse is unique, it can be achieved by different orderings of the elementary matrices

    (a) $\begin{pmatrix} 1 & -2 \\ 0 & 1 \end{pmatrix}\begin{pmatrix} 1 & 0 \\ -1 & 1 \end{pmatrix}\begin{pmatrix} 1/2 & 0 \\ 0 & 1 \end{pmatrix}\begin{pmatrix} 2 & 4 \\ 1 & 3 \end{pmatrix} = \begin{pmatrix} 1 & 0 \\ 0 & 1 \end{pmatrix}$

    $\begin{pmatrix} 2 & 4 \\ 1 & 3 \end{pmatrix}^{-1} = \begin{pmatrix} 3/2 & -2 \\ -1/2 & 1 \end{pmatrix}$

(c) $\begin{pmatrix} 1 & -4 \\ 0 & 1 \end{pmatrix}\begin{pmatrix} 1 & 0 \\ 0 & -\dfrac{1}{3} \end{pmatrix}\begin{pmatrix} 1 & 0 \\ -1 & 1 \end{pmatrix}\begin{pmatrix} 1 & 0 \\ 0 & \dfrac{1}{5} \end{pmatrix}\begin{pmatrix} 1 & 4 \\ 5 & 5 \end{pmatrix} = \begin{pmatrix} 1 & 0 \\ 0 & 1 \end{pmatrix}$

$$\begin{pmatrix} 1 & 4 \\ 5 & 5 \end{pmatrix}^{-1} = \begin{pmatrix} -1/3 & 4/15 \\ 1/3 & -1/15 \end{pmatrix}$$

(e) $\begin{pmatrix} 1 & 0 & -4 \\ 0 & 1 & 0 \\ 0 & 0 & 1 \end{pmatrix}\begin{pmatrix} 1 & 0 & 0 \\ 0 & 1 & -11 \\ 0 & 0 & 1 \end{pmatrix}\begin{pmatrix} 1 & 0 & 0 \\ 0 & 1 & 0 \\ 0 & 0 & -1/21 \end{pmatrix}\begin{pmatrix} 1 & 0 & 0 \\ 0 & 1 & 0 \\ 0 & -2 & 1 \end{pmatrix}$

$\begin{pmatrix} 1 & 0 & 0 \\ 0 & 1 & 0 \\ -1 & 0 & 1 \end{pmatrix}\begin{pmatrix} 1 & 0 & 0 \\ 2 & 1 & 0 \\ 0 & 0 & 1 \end{pmatrix}\begin{pmatrix} 1 & 0 & 4 \\ -2 & 1 & 3 \\ 1 & 2 & 5 \end{pmatrix} = \begin{pmatrix} 1 & 0 & 0 \\ 0 & 1 & 0 \\ 0 & 0 & 1 \end{pmatrix}$

$$\begin{pmatrix} 1 & 0 & 4 \\ -2 & 1 & 3 \\ 1 & 2 & 5 \end{pmatrix}^{-1} = \begin{pmatrix} 1/21 & -8/21 & 4/21 \\ -13/21 & -1/21 & 11/21 \\ 5/21 & 2/21 & -1/21 \end{pmatrix}$$

Section 1.3

1. $L = \begin{pmatrix} 1 & 0 & 0 \\ 2 & 1 & 0 \\ 3 & -1 & 1 \end{pmatrix}$ $U = \begin{pmatrix} 1 & 3 & 6 \\ 0 & 3 & -14 \\ 0 & 0 & -27 \end{pmatrix}$

3. $L = \begin{pmatrix} 1 & 0 & 0 \\ 2 & 1 & 0 \\ 0 & -8 & 1 \end{pmatrix}$ $U = \begin{pmatrix} 3 & 1 & 4 \\ 0 & -1 & 0 \\ 0 & 0 & 5 \end{pmatrix}$

5. $L = \begin{pmatrix} 1 & 0 & 0 \\ 2 & 1 & 0 \\ 4 & 2/9 & 1 \end{pmatrix}$ $U = \begin{pmatrix} 1 & 2 & 4 \\ 0 & -9 & 16 \\ 0 & 0 & 193/9 \end{pmatrix}$

7. $L = \begin{pmatrix} 2 & 0 & 0 & 0 \\ 4 & 1 & 0 & 0 \\ 1 & -1/2 & 1 & 0 \\ -3 & -2 & 2/3 & 1 \end{pmatrix}$ $U = \begin{pmatrix} 1 & 1 & 0 & 1/2 \\ 0 & -4 & -3 & 3 \\ 0 & 0 & 3/2 & 7 \\ 0 & 0 & 0 & 17/6 \end{pmatrix}$

Section 2.1

1. Solution is valid

3. Solution is not valid

Section 2.2

1. $\begin{pmatrix} 3 & 5 & 9 \\ 1 & -2 & -7 \\ 0 & 1 & 5 \end{pmatrix}$

3. $\begin{pmatrix} 1 & 1 & 1 & 0 \\ 1 & 1 & 1 & 3 \\ 1 & 2 & 1 & 5 \end{pmatrix}$

5. $8x_1 + 3x_1 = 6$
   $-4x_1 + 2x_2 + x_3 = 2$

Section 2.4

1. Free variables x_2, x_4, x_5, leading variables x_1, x_3
   $-2x_2 + x_2 - 3x_2 + 4, \quad x_3 = -2x_4 - 2; \quad x_2 = -2x_2 + x_2 - 3x_2 + 4$

2. $x_2 = r, \quad x_4 = s, \quad x_5 = t;$
   $x_1 = -2r + s - 3t + 4, \quad x_3 = -2s - 2$

3. No solution

5. $\begin{pmatrix} 2 & -3 & 6 \\ 1 & 5 & 9 \end{pmatrix} \rightarrow \begin{pmatrix} 1 & 0 & \dfrac{57}{13} \\ 0 & 1 & \dfrac{12}{13} \end{pmatrix}$ $x = \dfrac{57}{13},\, y = \dfrac{12}{13}$

7. $\begin{pmatrix} 3 & -5 & 1 & -2 & 0 \\ 6 & -10 & 2 & -4 & 6 \end{pmatrix} \rightarrow \begin{pmatrix} 1 & -\dfrac{5}{3} & \dfrac{1}{3} & -\dfrac{2}{3} & 0 \\ 0 & 0 & 0 & 0 & 1 \end{pmatrix}$ no solution

9. $\begin{pmatrix} 4 & 3 & -2 & 14 \\ -2 & 1 & -4 & -2 \\ 2 & 4 & -6 & 12 \end{pmatrix} \rightarrow \begin{pmatrix} 1 & 0 & 1 & 2 \\ 0 & 1 & -2 & 2 \\ 0 & 0 & 0 & 0 \end{pmatrix}$ z is the free variable;

   x and y are leading variables; $x = -z + 2, y = 2z + 2$
   $z = t; x = -t + 2, y = 2t + 2$

13. There is a unique solution unless $h = -8/3$.

15. $\left(-\dfrac{8}{11}t + \dfrac{43}{11}, -\dfrac{31}{22}t - \dfrac{3}{22}, t \right)$

17. (a) $49x + 15y + 52z = 116$
    (c) $41x - 6y - 5z = 80$

19. (a) $\left(\dfrac{170}{23}, \dfrac{-31}{23}, \dfrac{-55}{23} \right)$

Section 2.5

1. (a) (i) $x_1 \begin{pmatrix} 2 \\ -4 \\ 1 \end{pmatrix} + x_2 \begin{pmatrix} -5 \\ 0 \\ 6 \end{pmatrix} + x_3 \begin{pmatrix} 1 \\ 1 \\ -4 \end{pmatrix} = \begin{pmatrix} -2 \\ 7 \\ 0 \end{pmatrix}$

   (ii) $\begin{pmatrix} 2 & -5 & 1 \\ -4 & 0 & 1 \\ 1 & 6 & -4 \end{pmatrix} \begin{pmatrix} x_1 \\ x_2 \\ x_3 \end{pmatrix} = \begin{pmatrix} -2 \\ 7 \\ 0 \end{pmatrix}$

   (b) (i) $x_1 \begin{pmatrix} 1 \\ 0 \end{pmatrix} + x_2 \begin{pmatrix} -1 \\ 0 \end{pmatrix} + x_3 \begin{pmatrix} 1 \\ 1 \end{pmatrix} + x_4 \begin{pmatrix} 1 \\ 0 \end{pmatrix} = \begin{pmatrix} 0 \\ 9 \end{pmatrix}$

   (ii) $\begin{pmatrix} 1 & -1 & 1 & 1 \\ 0 & 0 & 1 & 0 \end{pmatrix} \begin{pmatrix} x_1 \\ x_2 \\ x_3 \\ x_4 \end{pmatrix} = \begin{pmatrix} 0 \\ 9 \end{pmatrix}$

3. (a) (i) $-3x + 5y + 2z = -5$

$$2w + 4x - 2y + 3z = 7$$

$$w + x + 2y + 6z = 0$$

   (ii) $w \begin{pmatrix} 1 \\ 2 \\ 1 \end{pmatrix} + x \begin{pmatrix} -3 \\ 4 \\ 1 \end{pmatrix} + y \begin{pmatrix} 5 \\ -2 \\ 2 \end{pmatrix} + z \begin{pmatrix} 2 \\ 3 \\ -6 \end{pmatrix} = \begin{pmatrix} -5 \\ 7 \\ 0 \end{pmatrix}$

5. No solution

9. (a) Yes

   (b) Yes

Section 3.1

1. $\hat{u} \cdot \hat{v} = 6$, $\|\hat{u}\| = \sqrt{13}$, $\|\hat{v}\| = \sqrt{40}$, $\|\hat{u} + \hat{v}\| = \sqrt{65}$

3. (a) $\|\hat{u}\| = \sqrt{17}$

   (b) $\dfrac{\hat{u}}{\|\hat{u}\|} = \left(\dfrac{1}{\sqrt{17}}, 0, \dfrac{-4}{\sqrt{17}} \right)$

   (c) $\left\| \dfrac{\hat{u}}{\|\hat{u}\|} \right\| = 1$

5. 12.03

7. $\left(\dfrac{-a}{\sqrt{a^2+b^2}}, \dfrac{b}{\sqrt{a^2+b^2}}\right), \left(\dfrac{a}{\sqrt{a^2+b^2}}, \dfrac{-b}{\sqrt{a^2+b^2}}\right)$

9. Analytically, this is because the equations $2x + 4y = 6$ and $3x + 6y = 6$ are inconsistent. Geometrically, this is because the vectors $\begin{pmatrix} 2 \\ 4 \end{pmatrix}$ and $\begin{pmatrix} 3 \\ 6 \end{pmatrix}$ are parallel and the vector $\begin{pmatrix} 6 \\ 6 \end{pmatrix}$ is not parallel to them.

13. (b) $\left(1, x^3\right) = \displaystyle\int_{-1}^{1} 1 \cdot x^3 dx = \dfrac{x^4}{4}\Big|_{-1}^{1} = \dfrac{1^4}{4} - \dfrac{(-1)^4}{4} = 0$

Section 3.2

1. Is a vector space.

3. Is a vector space.

5. Not a vector space. Not closed under vector addition or scalar multiplication. Does not have the zero vector.

7. Is a vector space.

9. Not a vector space. $\begin{pmatrix} -2 \\ -5 \end{pmatrix} + \begin{pmatrix} 5 \\ 2 \end{pmatrix} = \begin{pmatrix} 3 \\ -3 \end{pmatrix}$

11. Is a vector space.

Section 3.3

1. (a) Subspace
   (b) Subspace
   (c) Not a subspace

3. (a) Subspace
   (b) Subspace
   (c) Not a subspace

7. (a) It is linear combinations of $\begin{pmatrix} -1/2 \\ 1 \\ 0 \end{pmatrix}$ and $\begin{pmatrix} -1/2 \\ 0 \\ 1 \end{pmatrix}$

Section 3.4

1. Independent

3. Dependent

5. Independent

11. (a) Not in span

(c) Not in span

13. (a) $c = 0$, a and b any values

(c) a any value, $b = 2a$, $c = 3a$

19. $\left\{ \begin{pmatrix} 1 \\ 3 \\ 6 \end{pmatrix}, \begin{pmatrix} 2 \\ 4 \\ 0 \end{pmatrix} \right\} \begin{pmatrix} 1 \\ 0 \\ 2 \end{pmatrix}$

Section 3.5

1. $\left\{ \begin{pmatrix} 1 \\ -2 \\ 4 \end{pmatrix}, \begin{pmatrix} 11 \\ 14 \\ 8 \end{pmatrix} \right\}$ is a maximal independent set but not a basis.

$\left\{ \begin{pmatrix} 1 \\ -2 \\ 4 \end{pmatrix}, \begin{pmatrix} 11 \\ 14 \\ 8 \end{pmatrix}, \begin{pmatrix} 1 \\ 0 \\ 0 \end{pmatrix} \right\}$ is a basis.

3. $\left\{ \begin{pmatrix} 5 \\ 6 \end{pmatrix} \right\}$ is not a basis

$\left\{ \begin{pmatrix} 5 \\ 6 \end{pmatrix}, \begin{pmatrix} 1 \\ 0 \end{pmatrix} \right\}$ is a basis

5. The set given is a basis.

7. $\left\{ \begin{pmatrix} 1 \\ 2 \\ 0 \\ 5 \end{pmatrix}, \begin{pmatrix} 1 \\ 0 \\ 0 \\ 0 \end{pmatrix}, \begin{pmatrix} 0 \\ 1 \\ 0 \\ 0 \end{pmatrix}, \begin{pmatrix} 0 \\ 0 \\ 1 \\ 0 \end{pmatrix} \right\}$

9. $\left\{ \begin{pmatrix} 1 \\ 2 \\ 1 \\ 0 \end{pmatrix} \right\}$ is a basis for $U \cap W$;

$\left\{ \begin{pmatrix} 1 \\ 0 \\ 2 \\ 1 \end{pmatrix}, \begin{pmatrix} 1 \\ 1 \\ 0 \\ 0 \end{pmatrix}, \begin{pmatrix} 6 \\ -1 \\ 2 \\ 3 \end{pmatrix} \right\}$ is a basis for $U + W$.

11. $h = 38/3$

Section 3.6

1. $[\hat{u}]_B = \begin{pmatrix} 6 \\ -\dfrac{22}{3} \end{pmatrix}$

3. $[\hat{u}]_B = \begin{pmatrix} \dfrac{3}{13} \\ \dfrac{11}{13} \\ \dfrac{6}{13} \end{pmatrix}$

5. $[\hat{u}]_B = \begin{pmatrix} 71/2 \\ -15/2 \\ -11/2 \\ 59/2 \end{pmatrix}$

7. $[\hat{u}]_B = \begin{pmatrix} -\dfrac{12}{5} \\ \dfrac{4}{5} \\ \dfrac{9}{5} \end{pmatrix} = \left(-\dfrac{12}{5}\right)(1+2x) + \left(\dfrac{4}{5}\right)(x-x^2) + \left(\dfrac{9}{5}\right)(3+x^2)$

9. $[\hat{u}]_B = \begin{pmatrix} 5 \\ -4 \\ 2 \\ 1 \end{pmatrix} = 5\begin{pmatrix} 1 & 1 \\ 1 & 1 \end{pmatrix} - 4\begin{pmatrix} 1 & 1 \\ 1 & 0 \end{pmatrix} + 2\begin{pmatrix} 1 & 0 \\ 0 & 1 \end{pmatrix} + \begin{pmatrix} 0 & 2 \\ 1 & 1 \end{pmatrix} = \begin{pmatrix} 3 & 3 \\ 2 & 8 \end{pmatrix}$

13. $P_{B_1}[\hat{x}]_{B_1} = P_{B_2}[\hat{x}]_{B_2} \quad [\hat{v}]_{B_1} = \begin{pmatrix} 2 \\ 5 \end{pmatrix} \; [\hat{v}]_{B_2} = \begin{pmatrix} 6 \\ -2 \end{pmatrix}$

$P_{B_1} = \begin{pmatrix} 0 & 1 \\ 3 & 2 \end{pmatrix} \; P_{B_2} = \begin{pmatrix} 4 & 1 \\ -1 & 1 \end{pmatrix}$

$[\hat{v}]_{B_2} = P_{B_2}{}^{-1}P_{B_1}[\hat{v}]_{B_1} = \begin{pmatrix} \dfrac{-11}{5} \\ \dfrac{69}{5} \end{pmatrix}$

$[\hat{v}]_{B_1} = P_{B_1}{}^{-1}P_{B_2}[\hat{v}]_{B_2} = \begin{pmatrix} \dfrac{-52}{3} \\ 22 \end{pmatrix}$

15. $P_{B_1} = \begin{pmatrix} 1 & -3 & 4 \\ 5 & 0 & 1 \\ -2 & 2 & 4 \end{pmatrix}$ $P_{B_2} = \begin{pmatrix} 2 & 1 & 0 \\ 2 & 6 & 1 \\ -2 & 3 & 0 \end{pmatrix}$

$$\left[\hat{v}\right]_{B_1} = \begin{pmatrix} 2 \\ -3 \\ 1 \end{pmatrix} \quad \left[\hat{v}\right]_{B_2} = \begin{pmatrix} 2 \\ 0 \\ 5 \end{pmatrix}$$

$$\left[\hat{v}\right]_{B_2} = P_{B_2}^{-1}P_{B_1}\left[\hat{v}\right]_{B_1} = \begin{pmatrix} 51/8 \\ 9/4 \\ -61/4 \end{pmatrix} \quad \left[\hat{v}\right]_{B_1} = P_{B_1}^{-1}P_{B_2}\left[\hat{v}\right]_{B_2} = \begin{pmatrix} 23/13 \\ -7/13 \\ 2/13 \end{pmatrix}$$

Section 3.7

3. (a) Row reduced form of the matrix $\begin{pmatrix} 1 & 3 \\ 0 & 0 \end{pmatrix}$ basis for the row space $\{(1\ 3)\}$; basis for the column space $\left\{\begin{pmatrix} 1 \\ 2 \end{pmatrix}\right\}$; basis for the null space $\left\{\begin{pmatrix} -3 \\ 1 \end{pmatrix}\right\}$

(b) Row reducd form of the matrix $\begin{pmatrix} 1 & 0 & -1/3 \\ 0 & 1 & 2 \end{pmatrix}$ basis for the row space $\{(1\ \ 0\ \ -1/3),(0\ \ 1\ \ 2)\}$; basis for the column space $\left\{\begin{pmatrix} 3 \\ 3 \end{pmatrix},\begin{pmatrix} 0 \\ 1 \end{pmatrix}\right\}$; basis for the null space $\left\{\begin{pmatrix} 1/3 \\ -2 \\ 1 \end{pmatrix}\right\}$

Section 3.8

1. $\left\{\begin{pmatrix} 0 \\ 0 \\ 1 \\ 0 \end{pmatrix}\begin{pmatrix} 0 \\ 0 \\ 0 \\ 1 \end{pmatrix}\right\}$

3. $\begin{pmatrix} a & c \\ c & b \end{pmatrix} + \begin{pmatrix} d & e \\ e & f \end{pmatrix} = \begin{pmatrix} a+d & c+e \\ c+e & b+f \end{pmatrix}$

so the sum of symmetric matrices is symmetric

$$\alpha\begin{pmatrix} a & c \\ c & b \end{pmatrix} = \begin{pmatrix} \alpha a & \alpha c \\ \alpha c & \alpha b \end{pmatrix}$$

so a scalar multiple of a symmetric matrix is symmetric.

Thus, S is a subspace of $M_{2\times2}(\mathbb{R})$. Likewise, T is a subspace of $M_{2\times2}(\mathbb{R})$.

$\left\{ \begin{pmatrix} 1 & 0 \\ 0 & 0 \end{pmatrix}, \begin{pmatrix} 0 & 0 \\ 0 & 1 \end{pmatrix}, \begin{pmatrix} 0 & 1 \\ 1 & 0 \end{pmatrix} \right\}$ is a basis for S so the dimension of S is 3 and

$\left\{ \begin{pmatrix} 0 & 1 \\ -1 & 0 \end{pmatrix} \right\}$ is a basis for T so the dimension of T is 1.

The problem can be finished by showing $S \cap T = \emptyset$.

Section 4.1

1. (a) Linear
   (b) Not linear
   (c) Not linear
3. (a) Not linear
   (b) Linear

Section 4.2

1. Matrix of $T = \begin{pmatrix} 4 & -3 \\ 1 & 0 \\ 0 & 5 \end{pmatrix}$;

3. (a) Matrix of $T = (3)$.

5. $T\begin{pmatrix} x \\ y \end{pmatrix} = \begin{pmatrix} x - 3y \\ 7x \\ 5x + 9y \end{pmatrix}$

7. $T(x) = (2x)$

9. (a) $T\begin{pmatrix} 1 \\ 0 \end{pmatrix} = \begin{pmatrix} -3/4 \\ -1/2 \\ -1 \end{pmatrix}$, $T\begin{pmatrix} 0 \\ 1 \end{pmatrix} = \begin{pmatrix} -5/4 \\ -5/2 \\ -1 \end{pmatrix}$

   (b) $T\begin{pmatrix} 5 \\ -6 \end{pmatrix} = \begin{pmatrix} 15/4 \\ 25/2 \\ 1 \end{pmatrix}$

   (c) $T\begin{pmatrix} x \\ y \end{pmatrix} = \begin{pmatrix} \dfrac{-3}{4}x - \dfrac{5}{4}y \\ \dfrac{-1}{2}x - \dfrac{5}{2}y \\ -x - y \end{pmatrix}$

11. $m = 2/3$

15. (a) $\begin{pmatrix} 0 & 1 & 0 \\ 1 & 0 & 0 \\ 0 & 0 & 1 \end{pmatrix}$

(b) $\left\{ \begin{pmatrix} 0 \\ 0 \\ 0 \end{pmatrix} \right\}$

(c) Range $= \mathbb{R}^3$

17. (a) $\begin{pmatrix} 0 & 1 & 0 & 0 & \cdots & 0 \\ 0 & 0 & 2 & 0 & & 0 \\ \vdots & 0 & 0 & 3 & & \vdots \\ & \vdots & 0 & 0 & & \\ & & \vdots & \vdots & & 0 \\ 0 & 0 & 0 & 0 & & n \end{pmatrix}$

(b) $\{1\}$

(c) $\{1, x, x^2, \ldots, x^{n-1}\}$

18. (a) $\begin{pmatrix} 0 & 1 & 0 \\ 1 & 0 & 1 \\ 1 & 1 & 0 \end{pmatrix}$

(b) Matrix is invertible so T is $1 - 1$

19. (a) $\begin{pmatrix} 7 & -2 \\ 2 & 7 \end{pmatrix}$

(c) $\left(\dfrac{7}{53} \right) e^{3x} \sin x - \left(\dfrac{2}{53} \right) e^{3x} \cos x$

Section 4.3

1. (a) Matrix of $T = \begin{pmatrix} \dfrac{76}{7} & \dfrac{240}{7} \\ -\dfrac{18}{7} & -\dfrac{62}{7} \end{pmatrix}$.

(b) Null space of $T = \hat{0}$

(c) Basis for the range of $T = \left\{ \begin{pmatrix} \dfrac{76}{7} \\ -\dfrac{18}{7} \end{pmatrix}, \begin{pmatrix} \dfrac{240}{7} \\ -\dfrac{62}{7} \end{pmatrix} \right\}$

2. $P_C = \begin{pmatrix} 1 & 0 & 0 \\ 1 & 1 & 1 \\ 0 & 0 & 1 \end{pmatrix}$ $P_B = \begin{pmatrix} 1 & -1 & 3 \\ 0 & 1 & 2 \\ 0 & 2 & 0 \end{pmatrix}$ $T = \begin{pmatrix} 1 & 0 & 4 \\ 4 & 0 & 0 \\ 1 & 1 & 1 \end{pmatrix}$

$P_C^{-1}TP_B = \begin{pmatrix} 1 & 7 & 3 \\ 2 & -13 & 4 \\ 1 & 2 & 5 \end{pmatrix}$

4. $P_C = \begin{pmatrix} 1 & 0 & 0 \\ 1 & 1 & 1 \\ 0 & 0 & 1 \end{pmatrix}$ $P_B = \begin{pmatrix} 1 & -1 & 1 \\ 0 & 1 & 2 \\ 1 & 0 & 0 \end{pmatrix}$ $T = \begin{pmatrix} 3 & 0 & -4 \\ 4 & 0 & 0 \\ 1 & 1 & 1 \end{pmatrix}$

$P_C^{-1}TP_B = \begin{pmatrix} -1 & -3 & 3 \\ 3 & -1 & -2 \\ 2 & 0 & 3 \end{pmatrix}$

Section 4.4

1. (a) $S \circ T = T \circ S = \text{Identity}$

(b) Matrix for $T = \begin{pmatrix} 1 & -1 & 1 \\ 0 & 1 & -2 \\ 0 & 0 & 1 \end{pmatrix}$, matrix for $S = \begin{pmatrix} 1 & 1 & 1 \\ 0 & 1 & 2 \\ 0 & 0 & 1 \end{pmatrix}$

(c) $AB = BA = \begin{pmatrix} 1 & 0 & 0 \\ 0 & 1 & 0 \\ 0 & 0 & 1 \end{pmatrix}$

5. (a) $f(x) = 5x^3 + 3x^2$; $D(5x^3 + 3x^2) = 15x^2 + 6x$;

$$In\left(D\left(5x^3 + 3x^2\right)\right) = \int_0^x \left(15x^2 + 6x\right)dx = 5x^3 + 3x^2$$

$$In\left(5x^3 + 3x^2\right) = \int_0^x \left(5x^3 + 3x^2\right)dx = 5\frac{x^4}{4} + x^3; \ D\left(In\left(5x^3 + 3x^2\right)\right) = 5x^3 + 3x^2$$

(b) $f(x) = 3x + 4; D(3x + 4) = 3$;

$$In\left(D(3x + 4)\right) = \int_0^x (3)dx = 3x; \ In(3x + 4) = \int_0^x (3x + 4)dx = \frac{3}{2}x^2 + 4x$$

$$D\left(In(3x + 4)\right) = 3x + 4$$

(c) $f(x) = e^x; D(e^x) = e^x; ln(D(e^x)) = \int_0^x (e^x) dx = e^x - 1$

$$ln(e^x) = \int_0^x (e^x) dx = e^x - 1; D(ln(e^x)) = e^x$$

Section 5.1

1. Characteristic polynomial $(\lambda - 3)(\lambda - 2)^2$ eigenvalues $\lambda = 2, 3$ eigenvectors: $\lambda = 2$,

$\hat{v} = \begin{pmatrix} 0 \\ 1 \\ 0 \end{pmatrix}; \lambda = 3, \hat{v} = \begin{pmatrix} -1 \\ 1 \\ 1 \end{pmatrix}$

3. Characteristic polynomial $(\lambda - 2)^3$ eigenvalue $\lambda = 2$ eigenvectors: $\lambda = 2, \hat{v} = \begin{pmatrix} 1 \\ 0 \\ 0 \end{pmatrix}$

5. Characteristic polynomial $(\lambda - 1)(\lambda - 5)(\lambda + 5)$ eigenvalues $\lambda = 1, 5, -5$ eigenvectors:

$\lambda = 1, \hat{v} = \begin{pmatrix} \frac{1}{7} \\ \frac{4}{7} \\ 1 \end{pmatrix}; \lambda = 5, \hat{v} = \begin{pmatrix} -1 \\ 0 \\ 1 \end{pmatrix}; \lambda = -5, \hat{v} = \begin{pmatrix} 1 \\ -2 \\ 1 \end{pmatrix}$

7. Characteristic polynomial $(\lambda - 1)(\lambda - 2)^2$ eigenvalues $\lambda = 2, 3$ eigenvectors: $\lambda = 1$,

$\hat{v} = \begin{pmatrix} -2 \\ 1 \\ 1 \end{pmatrix}; \lambda = 2, \hat{v} = \begin{pmatrix} -1 \\ 0 \\ 1 \end{pmatrix} \hat{v} = \begin{pmatrix} 0 \\ 1 \\ 0 \end{pmatrix}$

9. Characteristic polynomial $(\lambda + 3)(\lambda - 4)(\lambda - 3)^2$ eigenvalues $\lambda = 2, 3$ eigenvectors:

$\lambda = 3, \hat{v} = \begin{pmatrix} 0 \\ 0 \\ 1 \\ 0 \end{pmatrix}; \lambda = 4, \hat{v} = \begin{pmatrix} 6 \\ 1 \\ 0 \\ 0 \end{pmatrix}; \lambda = -3, \hat{v} = \begin{pmatrix} -1 \\ 1 \\ 0 \\ 0 \end{pmatrix}$

13. (a) No
    (b) Yes

Section 5.2

1. Cannot be diagonalized

3. Cannot be diagonalized

5. $P = \begin{pmatrix} \frac{1}{7} & -1 & 1 \\ \frac{4}{7} & 0 & -5 \\ 1 & 11 & 1 \end{pmatrix}$ $D = \begin{pmatrix} 1 & 0 & 0 \\ 0 & 5 & 0 \\ 0 & 0 & -5 \end{pmatrix}$ is one correct answer

7. $P = \begin{pmatrix} -2 & -1 & 0 \\ 1 & 0 & 1 \\ 1 & 1 & 0 \end{pmatrix}$ $D = \begin{pmatrix} 1 & 0 & 0 \\ 0 & 2 & 0 \\ 0 & 0 & 2 \end{pmatrix}$ is one correct answer

9. Cannot be diagonalized

Section 5.3

1. $P_\mathcal{B} = \begin{pmatrix} 2 & 1 \\ 1 & 1 \end{pmatrix}$, $T = \begin{pmatrix} 2 & -1 \\ 5 & 3 \end{pmatrix}$, $[T]_\mathcal{B} = P_\mathcal{B}^{-1} T P_\mathcal{B} = \begin{pmatrix} -10 & -7 \\ 23 & 15 \end{pmatrix}$

3. $P_\mathcal{B} = \begin{pmatrix} 2 & 0 & 3 \\ 0 & 4 & 3 \\ 1 & -1 & 0 \end{pmatrix}$, $T = \begin{pmatrix} 1 & 0 & 1 \\ 1 & 2 & 1 \\ 2 & 3 & 0 \end{pmatrix}$, $[T]_\mathcal{B} = \begin{pmatrix} 8 & 28 & 33 \\ 4 & 16 & 18 \\ -13/3 & -19 & -21 \end{pmatrix}$

Section 5.4

1. Eigenvalues are 1, 5, −5 with eigenvectors $\begin{pmatrix} 1 \\ 4 \\ 7 \end{pmatrix}$, $\begin{pmatrix} -1 \\ 0 \\ 1 \end{pmatrix}$, $\begin{pmatrix} 1 \\ -2 \\ 1 \end{pmatrix}$, respectively.

$$x_1(t) = c_1 e^t - c_2 e^{5t} + c_3 e^{-5t}$$
$$x_2(t) = 4c_1 e^t - 2c_3 e^{-5t}$$
$$x_3(t) = 7c_1 e^t + c_2 e^{5t} + c_3 e^{-5t} \quad c_1 = \frac{1}{4}, \quad c_2 = -\frac{5}{4}, \quad c_3 = \frac{1}{2}$$

3. $a = -2$

Section 6.2

1. (a) $\|f\|^2 = \int_0^1 \left(x^2 + x\right)^2 dx = \dfrac{31}{30}$

(c) $\|f - g\|^2 = \int_0^1 \left[\left(x^2 + x\right) - \left(x - 1\right)\right]^2 dx = \dfrac{28}{15}$

3. $\left\langle \begin{pmatrix} x_1 \\ y_1 \end{pmatrix} + \begin{pmatrix} x_2 \\ y_2 \end{pmatrix}, \begin{pmatrix} x_3 \\ y_3 \end{pmatrix} \right\rangle = \left\langle \begin{pmatrix} x_1 + x_2 \\ y_1 + y_2 \end{pmatrix}, \begin{pmatrix} x_3 \\ y_3 \end{pmatrix} \right\rangle = \left(x_1 + x_2\right)^2 x_3 + \left(y_1 + y_2\right) y_3$

$\left\langle \begin{pmatrix} x_1 \\ y_1 \end{pmatrix}, \begin{pmatrix} x_3 \\ y_3 \end{pmatrix} \right\rangle + \left\langle \begin{pmatrix} x_2 \\ y_2 \end{pmatrix}, \begin{pmatrix} x_3 \\ y_3 \end{pmatrix} \right\rangle = \left(x_1\right)^2 x_3 + \left(y_1 + y_2\right) y_3 + \left(x_2\right)^2 x_3 + \left(y_1 + y_2\right) y_3$

5. (a) $\langle \hat{u}, \hat{v} \rangle = 7 + 29i$; $\left|\langle \hat{u}, \hat{v} \rangle\right| \approx 29.83$; $\|\hat{u}\| = \sqrt{15}$, $\|\hat{v}\| = \sqrt{62}$

(b) $\langle \hat{u}, \hat{v} \rangle = 23 + 3i$; $\left|\langle \hat{u}, \hat{v} \rangle\right| \approx 23.19$; $\|\hat{u}\| = \sqrt{30}$, $\|\hat{v}\| = \sqrt{50}$

Section 6.3

1. (a) Orthogonal $\begin{pmatrix} 7 \\ -3 \end{pmatrix} = \dfrac{19}{17} \begin{pmatrix} 1 \\ -4 \end{pmatrix} + \dfrac{25}{17} \begin{pmatrix} 4 \\ 1 \end{pmatrix}$

(c) Orthonormal $\begin{pmatrix} 0 \\ 3 \end{pmatrix} = \dfrac{-6}{\sqrt{5}} \begin{pmatrix} \dfrac{1}{\sqrt{5}} \\ -\dfrac{2}{\sqrt{5}} \end{pmatrix} + \dfrac{-6}{\sqrt{5}} \begin{pmatrix} -\dfrac{2}{\sqrt{5}} \\ -\dfrac{1}{\sqrt{5}} \end{pmatrix}$

(e) Neither orthogonal nor orthonormal

Section 6.4

1. The sets of vectors in these answers are orthogonal but not orthonormal.

(a) $\hat{v}_1 = \begin{pmatrix} 1 \\ 0 \end{pmatrix}$, $\hat{v}_2 = \begin{pmatrix} 0 \\ 7 \end{pmatrix}$

(c) $\hat{v}_1 = \begin{pmatrix} 2 \\ 7 \end{pmatrix}$, $\hat{v}_2 = \begin{pmatrix} -119/53 \\ 34/53 \end{pmatrix}$

(e) $\hat{v}_1 = \begin{pmatrix} -2 \\ 5 \\ -5 \end{pmatrix}$, $\hat{v}_2 = \begin{pmatrix} 35/27 \\ 61/27 \\ 47/27 \end{pmatrix}$, $\hat{v}_3 = \begin{pmatrix} -90/53 \\ 27/106 \\ 99/106 \end{pmatrix}$

(g) $\hat{v}_1 = \begin{pmatrix} 1 \\ 1 \\ 1 \end{pmatrix}$, $\hat{v}_2 = \begin{pmatrix} 7/3 \\ 13/3 \\ -20/3 \end{pmatrix}$, $\hat{v}_3 = \begin{pmatrix} 33/103 \\ -27/103 \\ -6/103 \end{pmatrix}$

(i) $\hat{v}_1 = \begin{pmatrix} 2 \\ 3 \\ 0 \\ 0 \end{pmatrix}$, $\hat{v}_2 = \begin{pmatrix} 33/13 \\ -22/13 \\ 6 \\ 3 \end{pmatrix}$, $\hat{v}_3 = \begin{pmatrix} 1359/706 \\ -453/353 \\ -544/353 \\ 515/707 \end{pmatrix}$, $\hat{v}_4 = \begin{pmatrix} 729/5831 \\ -486/5831 \\ 27/343 \\ -1809/5831 \end{pmatrix}$

3. An orthogonal set of vectors is

$$\hat{v}_1 = 2 + x, \quad \hat{v}_2 = \frac{13x^2 - 8x - 3}{13}, \quad \hat{v}_3 = \frac{5x^2 + 2x - 2}{13}$$

The norms are

$$\|\hat{v}_1\|^2 = \frac{26}{3}, \quad \|\hat{v}_2\|^2 = \frac{88}{195}, \quad \|\hat{v}_3\|^2 = \frac{2}{33}$$

so an orthonormal set is

$$\sqrt{\frac{3}{26}}(2+x), \quad \sqrt{\frac{195}{88}}\left(\frac{13x^2 - 8x - 3}{13}\right), \quad \sqrt{\frac{33}{2}}\left(\frac{5x^2 + 2x - 2}{13}\right)$$

5. An orthogonal set is $\hat{v}_1 = \begin{pmatrix} -1 \\ 3 \end{pmatrix}$, $\hat{v}_2 = \begin{pmatrix} 9/2 \\ 1/2 \end{pmatrix}$

$$\|\hat{v}_1\|^2 = 28 \, \|\hat{v}_2\|^2 = 90/4$$

Section 6.5

1. $\begin{pmatrix} -23\sqrt{2}/8 & 7\sqrt{2}/8 \\ -3\sqrt{2}/8 & 3\sqrt{2}/8 \end{pmatrix}$

Section 6.6

1. (a) $W^\perp = span\left\{ \begin{pmatrix} 4 \\ -\dfrac{9}{2} \\ 1 \end{pmatrix} \right\}$

(b) $W^\perp = span\left\{ \begin{pmatrix} -\dfrac{1}{2} \\ 0 \\ 1 \end{pmatrix} \right\}$

(c) $W^\perp = span\left\{ \begin{pmatrix} -2/3 \\ 0 \\ 0 \\ 1 \end{pmatrix}, \begin{pmatrix} 1 \\ -4 \\ 1 \\ 0 \end{pmatrix} \right\}$

Section 6.7

1. $\begin{pmatrix} \dfrac{8}{41} \\ \dfrac{-10}{41} \end{pmatrix}$

3. Orthonormal basis of $W = \left\{ \begin{pmatrix} \dfrac{1}{\sqrt{3}} \\ -\dfrac{1}{\sqrt{3}} \\ \dfrac{1}{\sqrt{3}} \end{pmatrix}, \begin{pmatrix} \dfrac{1}{\sqrt{6}} \\ \dfrac{2}{\sqrt{6}} \\ \dfrac{1}{\sqrt{6}} \end{pmatrix} \right\}$ $proj_W \hat{v} = \begin{pmatrix} \dfrac{7}{2} \\ 2 \\ \dfrac{7}{2} \end{pmatrix}$

5. Orthonormal basis of $W = \left\{ \begin{pmatrix} \dfrac{1}{2} \\ \dfrac{1}{2} \\ \dfrac{1}{2} \\ \dfrac{1}{2} \end{pmatrix}, \begin{pmatrix} \dfrac{2}{\sqrt{18}} \\ -\dfrac{3}{\sqrt{18}} \\ \dfrac{2}{\sqrt{18}} \\ -\dfrac{1}{\sqrt{18}} \end{pmatrix} \right\}$ $proj_W \hat{v} = \begin{pmatrix} \dfrac{43}{36} \\ -\dfrac{29}{12} \\ \dfrac{43}{36} \\ -\dfrac{35}{36} \end{pmatrix}$

7. Orthonormal basis of $W = \left\{ \begin{pmatrix} \dfrac{-1}{2} \\ \dfrac{1}{2} \\ \dfrac{-1}{2} \\ \dfrac{1}{2} \end{pmatrix} , \begin{pmatrix} \dfrac{1}{\sqrt{12}} \\ \dfrac{3}{\sqrt{12}} \\ \dfrac{1}{\sqrt{12}} \\ -\dfrac{1}{\sqrt{12}} \end{pmatrix} \right\}$ $proj_W \, \hat{v} = \begin{pmatrix} 0 \\ 4 \\ 0 \\ 0 \end{pmatrix}$

Section 7.1

1. (a) $\hat{w} = \begin{pmatrix} 4 \\ -2 \end{pmatrix}$

   (c) $\hat{w} = \begin{pmatrix} 1 \\ 0 \\ 0 \end{pmatrix}$

   (e) $\hat{w} = \begin{pmatrix} i \\ 3+2i \\ 5-4i \end{pmatrix}$

3. (a) $\hat{w} = \begin{pmatrix} 5 \\ 3 \\ -4 \end{pmatrix}$

   (b) $P_B = \begin{pmatrix} 1 & 1 & 2 \\ 1 & 1 & 0 \\ 1 & 0 & 0 \end{pmatrix}$ $P_B\left[\hat{w}\right]_B = \left[\hat{w}\right]; \; \left[\hat{w}\right]_B = P_B^{-1}\left[\hat{w}\right] = \begin{pmatrix} -4 \\ 7 \\ 1 \end{pmatrix}$

Section 7.2

1. (a) Matrix of $T = \begin{pmatrix} 1 & 2 & -3 \\ 3 & 0 & 5 \end{pmatrix}$; $T^*\left(\hat{w}\right) = \begin{pmatrix} a+3b \\ 2a \\ -3-5b \end{pmatrix}$; matrix of $T^* = \begin{pmatrix} 1 & 3 \\ 2 & 0 \\ -3 & 5 \end{pmatrix}$

   (c) Matrix of $T = \begin{pmatrix} 4 \\ -3 \end{pmatrix}$; $T^*\left(\hat{w}\right) = \left(4a - 3b\right)$; matrix of $T^* = \left(4 \; -3\right)$

   (e) Matrix of $T = \begin{pmatrix} 3 & 4+5i \\ 6i & -2i \\ 6-2i & 1-i \end{pmatrix}$

$$T^*(\hat{w}) = \begin{pmatrix} 3a - 6ib + (6+2i)c \\ (4-5i)a + 2ib + (1+i)c \end{pmatrix}$$

$$\text{matrix of } T^* = \begin{pmatrix} 3 & -6i & 6+2i \\ 4-5i & 2i & 1+i \end{pmatrix}$$

Section 7.3

1. $\hat{v}_1 = \begin{pmatrix} \dfrac{1}{\sqrt{2}} \\ -\dfrac{1}{\sqrt{2}} \end{pmatrix}$ $\hat{v}_2 = \begin{pmatrix} \dfrac{1}{\sqrt{2}} \\ \dfrac{1}{\sqrt{2}} \end{pmatrix}$ $D = \begin{pmatrix} 1 & 0 \\ 0 & 3 \end{pmatrix}$

3. $\hat{v}_1 = \begin{pmatrix} 1/\sqrt{2} \\ -1/\sqrt{2} \end{pmatrix}$ $\hat{v}_2 = \begin{pmatrix} 1/\sqrt{2} \\ 1/\sqrt{2} \end{pmatrix}$ $D = \begin{pmatrix} 1 & 0 \\ 0 & 5 \end{pmatrix}$

5. $\hat{v}_1 = \begin{pmatrix} 1 \\ -1 \\ 0 \end{pmatrix}$ $\hat{v}_2 = \begin{pmatrix} 1 \\ 1 \\ 1 \end{pmatrix}$ $\hat{v}_3 = \begin{pmatrix} 1 \\ 1 \\ -2 \end{pmatrix}$ $D = \begin{pmatrix} 2 & 0 & 0 \\ 0 & 3 & 0 \\ 0 & 0 & 6 \end{pmatrix}$

Index

Printed and bound by CPI Group (UK) Ltd, Croydon, CR0 4YY

17/10/2024

01775672-0004